JIG AND FIXTURE DESIGN

JIG AND FIXTURE DESIGN

DESIGN

Fifth Edition

EDWARD G. HOFFMAN

DELMAR
CENGAGE Learning

Australia • Brazil • Japan • Korea • Mexico • Singapore • Spain • United Kingdom • United States

Jig and Fixture Design, fifth edition
Edward G. Hoffman

Vice President, Technology and Trades SBU:
  Alar Elken

Editorial Director: Sandy Clark

Senior Acquisitions Editor: Jim DeVoe

Senior Development Editor: John Fisher

Marketing Director: Cyndi Eichelman

Channel Manager: Fair Huntoon

Marketing Coordinator: Sarena Douglass

Production Director: Mary Ellen Black

Production Manager: Andrew Crouth

Production Editor: Stacy Masucci

Senior Art/Design Coordinator:
  Mary Beth Vought

Editorial Assistant: Mary Ellen Martino

For product information and technology assistance, contact us at
Cengage Learning Customer & Sales Support, 1-800-354-9706
For permission to use material from this text or product,
submit all requests online **www.cengage.com/permissions**
Further permissions questions can be emailed to
permissionrequest@cengage.com

ISBN-13: 978-1-4018-1107-5

ISBN-10: 1-4018-1107-8

Delmar
Executive Woods
5 Maxwell Drive
Clifton Park, NY 12065
USA

Cengage Learning is a leading provider of customized learning solutions with office locations around the globe, including Singapore, the United Kingdom, Australia, Mexico, Brazil, and Japan. Locate your local office at **www.cengage.com/global**

Cengage Learning products are represented in Canada by Nelson Education, Ltd.

To learn more about Delmar, visit **www.cengage.com/delmar**

Purchase any of our products at your local college store or at our preferred online store **www.cengagebrain.com**

Notice to the Reader

Printed in the United States of America
11 12 13 14 15   18 17 16 15 14

Contents

Preface

INTRODUCTION

The world's demand for manufactured goods is growing at a staggering rate. Industry has responded to this demand with many new and sometimes radical ways of producing products. From the evolution of Computer Numerical Control and Computer Aided Manufacturing to today's modern manufacturing concepts, such as Flexible Manufacturing Systems (FMS), Just In Time (JIT), and Statistical Process Control (SPC), the art of manufacturing has undergone many dramatic changes and advances. These changes have created a dire need for more cost-effective and efficient workholding methods and devices. As more manufacturing companies shift their emphasis toward a zero-parts inventory system to keep costs down and profits up, the need for efficient and cost-effective workholders is becoming increasingly important.

Despite the many advancements and changes in cutting tools, machine tools, and production methods, the basic requirement of holding the workpiece has remained constant. Every part produced must be held while it is machined, joined, or inspected or has any number of other operations performed on it. So, whether the operation requires a simple drill press, a multiple-axes computer, or numerically controlled machining center, the workpiece must be accurately located and securely held throughout the operation.

The part, not the process, is the primary consideration in workholding.

Just as no single machine tool will perform every required operation, no individual jig or fixture can possibly hold every part. However, each workholder variation has basic similarities to other types and styles of jigs and fixtures. The subject of this text is these similarities rather than the differences. This text helps the reader develop a thorough understanding and working knowledge of how and why jigs and fixtures are designed and built as they are. To do this, the discussion starts with the fundamentals of jigs and fixtures and works through the various elements and considerations of design.

Throughout the text, two fundamental tool design principles are constantly stressed: simplicity and economy. To be effective, a workholder must save money in production. To this end, the construction of the tool must be as cost-effective as possible while ensuring that the tool has the capacity to perform all of the intended functions.

ORGANIZATION OF CONTENT

The three-part approach of each unit is directed toward making the material understandable and logical: (1) the *introduction* phase presents the basic concepts, ideas, and fundamentals; (2) the *explanation* phase describes the particular techniques of design and fabrication as they apply to each type of workholder; and (3) the *applications* at the end of many units require the reader to apply the lessons learned by solving problems in tool design. The reader becomes familiar with working with part drawings and production plans showing the sequencing of operations in the shop.

The 20 units of this text are divided into four major sections. The first section (Units 1–5) gives the reader an overview of the basic types and functions of jigs and fixtures, as well as a detailed description of the way these workholders are designed and built. The reader learns the basic elements of supporting, locating, and clamping the part and then is introduced to the basic principles of workholder construction. This section provides the background information for the more advanced study later in the text.

The second section (Units 6–8) introduces the reader to the primary considerations of design economics and the basic methods used to initiate and prepare the design drawings. In keeping with the worldwide standard of measurement and modern drafting practices, the SI system (International System of Units) is introduced and explained. The process of geometric dimensioning and tolerancing is also presented; many manufacturing drawings in industry use this system of dimensioning, and the reader should be comfortable with its basic principles. New material covering the applications of Computer Aided Design (CAD) and how it is applied to jig and fixture design has also been added to this section to inform the reader of the developments in this important area of tool manufacturing.

The third section (Units 9–14) introduces and explains the processes involved in designing and constructing the basic types and forms of jigs and fixtures. From simple template and plate-type workholders to more detailed and complex channel and box-type tooling, each basic style is thoroughly explained and illustrated.

The final section (Units 15–21) covers the specialized workholding topics in manufacturing as they relate to jig and fixture design. Covering power workholding methods and equipment, modular workholding systems and low-cost tooling practices, and

designing jigs and fixtures for inspection and welding and for numerically controlled machine tools, this section has been expanded to include those areas of tooling technology that will service the needs and requirements of industry for years to come. Finally, a unit on tooling materials covers the properties of commonly used materials and the effects of these properties and heat treatment on workholder design.

A glossary is provided for ready reference and as an aid to the reader in mastering the terminology of workholder design.

FEATURES

This edition includes a more integrated approach to the global nature that challenges the tool designer. Concepts include the use of design teams in accomplishing the task of competitively delivering good design in a time efficient manner. This requires the use of CAD as a universal language that crosses boundaries and languages, and yet provides an ergonomically designed tool capable of passing the test of OSHA standards. Vendor supplied libraries of tooling components have allowed the design process to proceed at a much faster pace and is an excellent companion to the Machinery's handbook, which is now on CD as well.

At the heart of the team is the concept of concurrent engineering. Communicating as part of the design team may require good computer skills and video conferencing capabilities, since the end function of the tooling may be half way around the world. The design team may be comprised of individuals from more than one country. This is a departure from the traditional methods that were used when global partnerships were not as predominant. Many areas still work through the tool design process in very familiar fashion; however, the trend toward concurrent engineering is used in many industry areas where outsourcing is very common.

The following are the features of this new edition:

• Easy-to-read presentation with numerous illustrations and many new photographs showing the variety of tools and workholders available as well as typical applications.

- Drawings updated to the requirements of the ANSI Y14.5M-1994 (R1999) dimensioning standard.
- A discussion of geometric dimensioning and tolerancing, added to introduce the basic principles. Numerous examples are included, as well as applications in the review portion of the unit.
- New information on the use of the computer for Computer Aided Design (CAD) and Computer Aided Manufacturing (CAM), as well as the methods and systems used for designing jigs and fixtures using fixturing component libraries.
- Expanded information on tooling for numerically controlled machines.
- Expanded information about low-cost jigs and fixtures describing many new and innovative products useful for reducing quantity production runs and one-of-a-kind machining.
- New information on setup reduction for workholding to minimize fixturing costs while increasing production.
- Expanded information on modular tooling where commercially available tooling can be adapted and modified to meet a number of requirements.
- Metric dimensioning on approximately 20 percent of the part drawings.
- A new glossary for quick reference to new terminology.
- An *Instructor's Guide* to accompany the text. The guide contains the answers to the reviews at the end of each unit of the text.

ABOUT THE AUTHOR

Dr. Edward G. Hoffman, president of Hoffman & Associates, an engineering consulting firm based in Colorado Springs, Colorado, is a tool engineering consultant, technical writer, editor, and lecturer. He has written 17 books on tool and manufacturing engineering subjects and currently writes 40 magazine columns per year for several trade journals.

Dr. Hoffman's 25 years of industrial experience include positions as both a journeyman toolmaker and a tool engineer. His experience also includes 10 years of teaching and lecturing for colleges, technical societies, and trade schools. He has helped thousands of engineers and managers in problem-solving sessions at seminars conducted by the Society of Manufacturing Engineers and at numerous other universities and colleges, trade associations, and in-plant technical seminars.

Dr. Hoffman holds a B.S. degree in industrial management and an M.S. and Ph.D. in manufacturing engineering. He is past chairman of the Society of Manufacturing Engineers, Pikes Peak Chapter #213, in Colorado Springs. Additionally, he is a member of SME's Tool Engineering Council, and he is a certified manufacturing engineer (CMfgE) and a certified advanced metrication specialist (CAMS).

Contributing to this edition are Mike Turner and LaVonne Vichlach from Hawkeye Community College in Waterloo, Iowa. Mike has taught at the college since 1983 and is a Tool and Die Maker, having completed his apprenticeship in 1976. He has developed and conducted courses for local industries, including John Deere, Traer Manufacturing, Northeast Machine Tool, Viking Pump, General Machine Tool and Bertch Cabinets. Mike is a member of the Administrative Circle at Hawkeye Community College and serves as the program chair of skilled trades. He is a co-recipient of the President's Award for his work on the innovative Exploring Manufacturing Careers Consortium (EMC^2 Program), a school to work program.

LaVonne has taught at HCC for 10 years and is a graduate of both the University of Northern Iowa and Hawkeye Community College. She has 15 years of experience at John Deere Engine Works and has worked as a machinist, a CNC programmer, an Engineering and Senior Engineering Analyst, and as a Process Engineer. She assisted students in chartering a student chapter of the Society of Manufacturing Engineers at HCC. LaVonne teaches across several disciplines and has developed e-resource materials for several courses.

ACKNOWLEDGMENTS

The following instructors reviewed the revised manuscript:

Clyde Avery, Carritos College, Norwalk, CA

John Campbell, Cayuga Community College, Auburn, NY

Terry Foster, Pima Community College, Tucson, AZ

Jeff Szymanski, Milwaukee Area Technical College, Milwaukee, WI

Lavonne Vichlach, Hawkeye Community College, Waterloo, IA

Basic Types and Functions of Jigs and Fixtures

UNIT I

Purpose of Tool Design

OBJECTIVES

After completing this unit, the student should be able to:

• List the objectives of tool design.
• Identify the source of specified design data.

TOOL DESIGN

Tool design is the process of designing and developing the tools, methods, and techniques necessary to improve manufacturing efficiency and productivity. It gives industry the machines and special tooling needed for today's high-speed, high-volume production. It does this at a level of quality and economy that will ensure that the cost of the product is competitive. Since no single tool or process can serve all forms of manufacturing, tool design is an ever-changing, growing process of creative problem solving.

TOOL DESIGN OBJECTIVES

The main objective of tool design is to lower manufacturing costs while maintaining quality and increased production. To accomplish this, the tool designer must satisfy the following objectives:

• Provide simple, easy-to-operate tools for maximum efficiency.
• Reduce manufacturing expenses by producing parts at the lowest possible cost.
• Design tools that consistently produce parts of high quality.
• Increase the rate of production with existing machine tools.
• Design the tool to make it foolproof and to prevent improper use.
• Select materials that will give adequate tool life.
• Provide protection in the design of the tools for maximum safety of the operator.

TOOL DESIGN IN MANUFACTURING

Manufacturing for global competitiveness clearly requires the success of concurrent engineering. Concurrent engineering is a process that allows the design team to be involved in a comprehensive plan for product design and production. Concurrent engineering allows the tool design team member to be involved in product design and production where their knowledge of fixtures and manufacturing processes will result in fewer design errors. Concurrent engineering teams consist of product designers, process planning engineers, tool designers, quality control engineers, production management, and machining technicians. Companies may vary job titles and team compositions to suit their internal company structure.

Team members contribute based on their area of expertise. The product, a method for manufacturing, tooling concepts, and a quality plan are developed that suits the selected manufacturing facility. In this way, problems are not discovered on the production floor, but are corrected early in the concurrent process. This ultimately saves time and money while speeding up the process of getting product to market earlier. Concurrent engineering allows a company to have a distinct economic advantage in a global market.

The tool designer develops a plan for maintaining the concepts developed by the team with respect to economic guidelines. Expert computer systems are now part of the design environment, and they support an integrated approach for tracking time and money allocated for the project and provide immediate information at any point in the concurrent process.

PLANNING THE DESIGN

The designer is responsible for managing information resources that impact the tool design. Product design changes are continuously reviewed to determine tooling changes that might be necessary. Last-minute costly changes are eliminated or minimized. The team meets regularly to provide any necessary updates or changes in the production plan. This is time wisely spent and results in an efficient and cost-effective tool design. The design process is not as linear as it used to be. Communication models between team members include e-mail and electronic transfer of materials and may make use of sophisticated technology such as teleconferencing. Team members may consist of customers, designers, and builders in different locations that may take them halfway around the world.

Part Drawings

The tool designer receives a duplicate of the part geometry that will be used to make the part (Figure 1–1). Many part prints are transmitted electronically and may include a solid model. The solid model allows the designer to view the three-dimensional part geometry. The task of tool design begins with a more complete understanding of the part. A prototype, or a single manufactured part used for evaluation purposes, can be made available. A prototype goes one more step beyond the solid computer model. The prototype, a single physical part provided prior to formal production, is a valuable tool for understanding more complex part geometries. Prototypes are manufactured using conventional Computer Numerical Control (CNC) machine tools or some of the newer technologies such as stereolithography or a layered object manufacture, more commonly referred to as a LOM. Both the stereolithography and LOM develop the part geometry using a system of layering the medium and solidifying or cutting out that layer with a laser. The result is a solid object made one layer at a time where the layers may be no more than .003 thick. Whether analyzing the prototype and the part drawing or just the part drawing, the designer must consider the following factors that directly influence the design choices. These factors are:

- Overall size and shape of the part
- Type and condition of the material used for the part
- Type of machining operation to be performed
- Degree of accuracy
- Number of pieces to be made
- Locating and clamping surfaces

Production Plan

The production plan (Figure 1–2) is an itemized list of the manufacturing operations and the sequence of the operations chosen by the process planning engineer. The production plan can take many forms, depending on the needs of each company. At the least, it should include a brief description of each machining operation and the machine tool designated for these operations.

The tool designer also uses this plan to assist in the design. The production plan can include the following:

- Type and size of machine tool specified for each operation
- Type and size of cutters specified for each operation

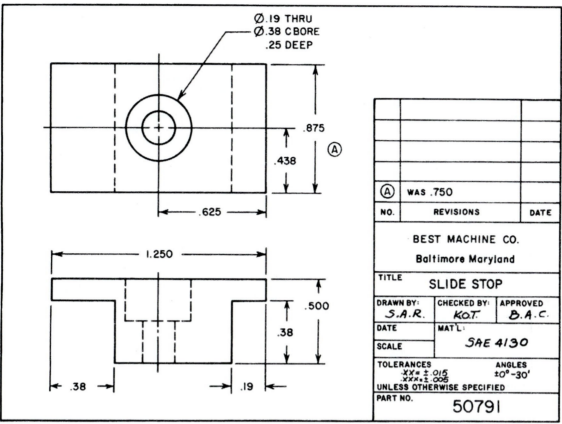

Ø.19 THRU
Ø.38 C BORE
.25 DEEP

.875

.438

.625

1.250

.500

.38

.38

.19

Ⓐ	WAS .750	
NO.	REVISIONS	DATE

BEST MACHINE CO.

Baltimore Maryland

TITLE	SLIDE STOP	

DRAWN BY: S.A.R.	CHECKED BY: K.O.T.	APPROVED B.A.C.

DATE	MAT'L:	
SCALE	SAE 4130	

TOLERANCES .XX= ±.015 .XXX=±.005 UNLESS OTHERWISE SPECIFIED	ANGLES ±0°-30'

PART NO.   50791

Figure 1-1 Part drawing.

- Sequence of operations
- Previous machining operations performed on the part

In addition to the part drawing and production plan, the tool designer is informed of the amount of time and money that is available to spend on the design. Using this information and a little creativity and experience, the tool designer begins to study the design alternatives.

Alternatives

One of the first steps in problem solving is determining the alternative solutions. The same process is used in tool design to ensure that the best method is chosen. During this phase of the design, the tool designer must analyze all important information in order to answer the following questions:

- Should special tooling be used or existing equipment modified?

- Should multiple-spindle or single-spindle machines be used?
- Should the tool be single-purpose or multipurpose?
- Will the savings justify the cost of the tool?
- What type of gauge, if any, should be used to check each operation?

Answering these questions and others related to the specific task, the tool designer develops alternative solutions. From these alternative solutions, the most efficient, dependable, and cost-effective design is chosen.

CHALLENGES TO THE TOOL DESIGNER

The tool designer has many manufacturing responsibilities. In addition to technical design duties, the tool designer may be responsible for obtaining materials, toolroom supervision, and tool inspection. The tool designer should understand the extent of these additional duties.

	BEST MACHINE COMPANY		
	Baltimore, Maryland		

PART # 50791	PART NAME Slide Stop	QUANTITY 7500	ORDER # 13762
DWG # D-50791	PROCESS PLANNER R.E. Tucker	REVISION # DATE	Page 1 of 1

OPR #	DESCRIPTION	DEPT.	MACHINE TOOL
1.	Cutoff – .875 X .500 stock to 1.250 length.	#68 Cutoff Rm.	Abrasive Cutoff Saw #68-19
2.	Drill – ∅.19 hole thru	#66 Drilling	Drill Press #66-141
3.	Counterbore – ∅.38 .25 Deep	#66 Drilling	Drill Press #66-141
4.	Mill – .38 X .38 and .19 X .38 shoulders	#37 Milling	Horiz. Mill #37-804
5.	Deburr	#7 Finishing	Tumbler #7-1053
6.	Inspect – Visual and dimensional	#7 Finishing	None
6.	Receiving gauge (1) Pin gauges (2)		#I-50791-3 #I-50791-1/2
4.	Side milling cutters (2)	4 x .500 x 1	Fixture #S-50791-1
3.	Counterbore with pilot	.375 x .19	Jig #J-50791-1
2.	Drill	.187 (3/16)	Jig #J-50791-1
1.	Cutoff wheel	10 x .062	None
OPR #	TOOL DESCRIPTION	SIZE	SPEC. TOOL

Figure 1–2 Production plan.

Design

In this phase, the tool designer is responsible for developing the drawings and sketches of the tool design ideas. Design drawings are usually subject to approval by a chief designer. However, in smaller companies, the tool designer often makes the tooling decisions.

Supervision

The extent of a tool designer's supervision is normally determined by the size of the company (Figure 1–3). Supervision for a single section, such as design or toolmaking, or for the entire tooling department, may become the tool designer's responsibility. In either case, the ability to lead others is helpful.

One resource a tool designer may often use to help resolve design problems is the group of skilled people in the toolroom. The toolroom is the area in a shop where the machine tools and the skilled workforce are found. These skilled trades employees are capable of taking the prints for the individual components of a tool and manufacturing them, assembling the parts, and verifying their accuracy. A variety of machine tools including manual mills, lathes, grinders, jig mills, machining centers and in some cases their CNC counterparts might be found in a typical toolroom. Regardless of the level of skill a designer possesses, these skilled toolmakers can often see solutions that may not be obvious to the designer. For this reason, it is always a good idea to build a good working relationship with your toolmakers. In tool design, a cooperative relationship between the designer and the toolmakers is essential. Not only does working together make the task at hand easier, but also using the available expertise makes more sense than trying to do the job alone.

Procurement

Often a tool designer is responsible for obtaining the materials to make the tool. In these situations, the tool designer normally relies on vendors or salespeople to supply materials and parts that meet the design specifications. When selecting a vendor, a good practice is to choose the company that offers the most service to its customers. Services such as design assistance and problem solving, where their product is involved, are important factors to consider before making a final selection. Another point to consider is whether the vendor can supply special parts or components when necessary. Generally, the specialty vendors can furnish special items for much less than those items cost to make in-house. Since most specialty vendors offer these services, the decision should be made on a basis of which vendor can meet the designer's needs in the most timely, efficient, and dependable manner.

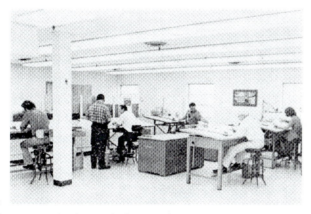

A   B

Figure 1-3 The tool design departments in most manufacturing organizations use a combination of drawing boards (A) and computer-aided design (CAD) workstations (B) to create the necessary tool design drawings (*Photo courtesy of Advanced Technologies Center*).

Inspection

Many times the tool designer is required to inspect the finished tool to ensure that it meets specifications. This inspection, or functional tryout, is normally conducted in two phases. First, the tool itself is inspected for compliance with the tool drawing. Second, several test parts are produced with the tool and are carefully checked to ensure that they conform to the specifications shown on the part print. After the tool has been turned over to the production department, the tool designer should make periodic checks during production to ensure that the specified tolerances are maintained (Figure 1–4).

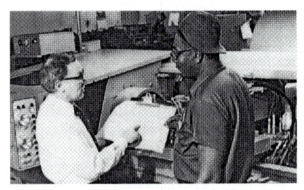

Figure 1–4 The tool designer consults the machinist to determine how well the jig or fixture performs.

REQUIREMENTS TO BECOME A TOOL DESIGNER

To perform the functions of a tool designer, an individual must have the following skills:

- The ability to make mechanical drawings and sketches
- An understanding of modern manufacturing methods, tools, and techniques
- A creative mechanical ability
- An understanding of basic toolmaking methods
- A knowledge of technical mathematics through practical trigonometry
- CAD drafting skills
- File management
- Electronic communication skills
- Geometric dimensioning and tolerancing

SUMMARY

The following important concepts were presented in this unit:

- Tool design is the process of designing and developing tooling devices, methods, and procedures to aid in improving overall manufacturing efficiency and productivity.
- The primary objective of tool design is lowering manufacturing costs while maintaining consistent quality and increased production.
- The tool design function is a well-integrated position within the concurrent engineering team, requiring skills in computer technology and multiple communication mediums.
- Tool designers use part drawings and production plans in developing alternative design solutions for efficient, dependable, and cost-effective tool designs.
- Tool designers, in addition to designing tooling, may also be responsible for toolroom supervision, procurement, and tool inspection.
- To become a tool designer, an individual must be able to make mechanical drawings and sketches, understand manufacturing techniques and toolmaking methods and equipment, have a creative mechanical ability, and have a working knowledge of shop mathematics through practical trigonometry.

REVIEW

1. List the seven objectives of tool design.
2. Determine the source of the following data by indicating *1* for the part drawing, *2* for the production plan, and *3* for additional instructions.
   a. Time allocation
   b. Overall size and shape of the part
   c. Required accuracy
   d. Sequence of operations
   e. Type and size of machines used
   f. Money available

g. Number of pieces
h. Previous machining
i. Locating surfaces
j. Material specifications
k. Type of cutters needed
l. Type of machining required

3. What does the term *concurrent* mean and how is it applied to the design of tooling?
4. Describe a toolroom.
5. List the skills of a tool designer.

UNIT 2

Types and Functions of Jigs and Fixtures

OBJECTIVES

After completing this unit, the student should be able to:

- Identify the classes of jigs and fixtures.
- Identify the types of jigs and fixtures.
- Choose a class and type of jig and fixture for selected operations on sample parts.

JIGS AND FIXTURES

Jigs and *fixtures* are production-workholding devices used to manufacture duplicate parts accurately. The correct relationship and alignment between the cutter, or other tool, and the workpiece must be maintained. To do this, a jig or fixture is designed and built to hold, support, and locate every part to ensure that each is drilled or machined within the specified limits.

Jigs and fixtures are so closely related that the terms are sometimes confused or used interchangeably. The difference is in the way the tool is guided to the workpiece.

A *jig* is a special device that holds, supports, or is placed on a part to be machined. It is a production tool made so that it not only locates and holds the workpiece but also guides the cutting tool as the oper-

ation is performed. Jigs are usually fitted with hardened steel bushings for guiding drills or other cutting tools (Figure 2–1A).

As a rule, small jigs are not fastened to the drill press table. If, however, holes above .25 inch in diameter are to be drilled, it is usually necessary to fasten the jig to the table securely.

A *fixture* is a production tool that locates, holds, and supports the work securely so the required machining operations can be performed. Set blocks and feeler or thickness gauges are used with fixtures to reference the cutter to the workpiece (Figure 2–1B). A fixture should be securely fastened to the table of the machine upon which the work is done. Though largely used on milling machines, fixtures are also designed to hold work for various operations on most of the standard machine tools.

Fixtures vary in design from relatively simple tools to expensive, complicated devices. Fixtures also help to simplify metalworking operations performed on special equipment.

CLASSES OF JIGS

Jigs may be divided into two general classes: boring jigs and drill jigs. *Boring jigs* are used to bore holes that either are too large to drill or must be made an odd size (Figure 2–2). *Drill jigs* are used to drill,

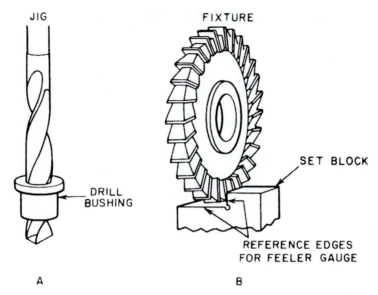

Figure 2-1 Referencing the tool to the work.

ream, tap, chamfer, counterbore, countersink, reverse spotface, or reverse countersink (Figure 2–3). The basic jig is almost the same for either machining operation. The only difference is in the size of the bushings used.

TYPES OF JIGS

Drill jigs may be divided into two general types, open and closed. *Open jigs* are for simple operations where work is done on only one side of the part. *Closed,* or

box, jigs are used for parts that must be machined on more than one side. The names used to identify these jigs refer to how the tool is built.

Template jigs are normally used for accuracy rather than speed. This type of jig fits over, on, or into the work and is not usually clamped (Figure 2–4). Templates are the least expensive and simplest type of jig to use. They may or may not have bushings. When bushings are not used, the whole jig plate is normally hardened.

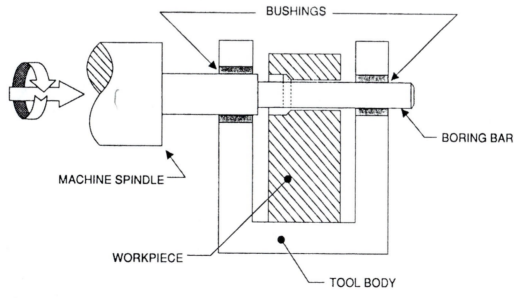

Figure 2-2 Boring jig.

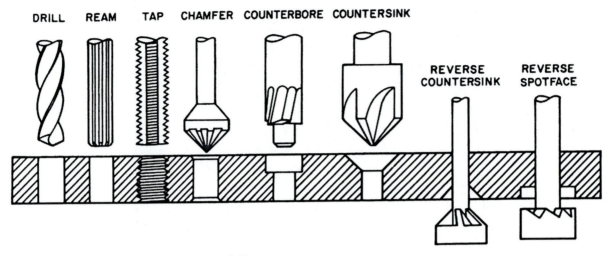

Figure 2-3 Operations common to a drill jig.

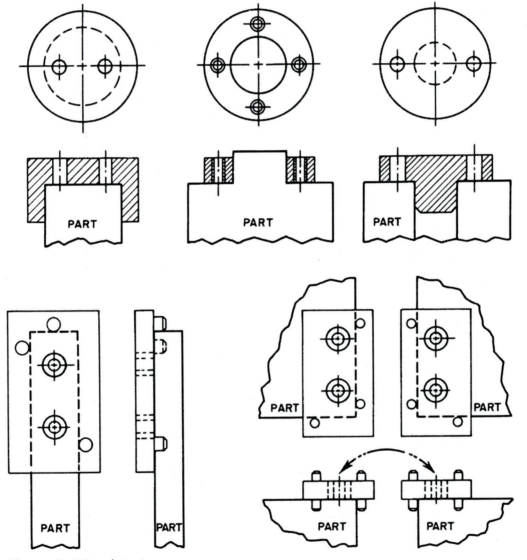

Figure 2-4 Template jigs.

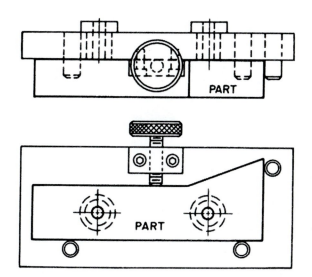

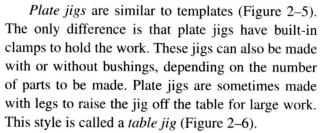

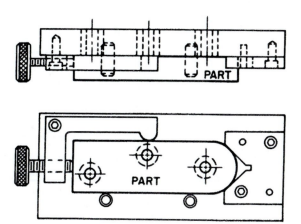

Figure 2-5 Plate jig.

Plate jigs are similar to templates (Figure 2–5). The only difference is that plate jigs have built-in clamps to hold the work. These jigs can also be made with or without bushings, depending on the number of parts to be made. Plate jigs are sometimes made with legs to raise the jig off the table for large work. This style is called a *table jig* (Figure 2–6).

Sandwich jigs are a form of plate jig with a back plate (Figure 2–7). This type of jig is ideal for thin or soft parts that could bend or warp in another style of jig. Here again, the use of bushings is determined by the number of parts to be made.

Angle-plate jigs are used to hold parts that are machined at right angles to their mounting locators (Figure 2–8). Pulleys, collars, and gears are some of the parts that use this type of jig. A variation is the *modified angle-plate jig*, which is used for machining angles other than 90 degrees (Figure 2–9). Both of these examples have clearance problems with the cutting tool. As the drill exits the product being drilled, it has little or no room for the drill point to clear the product completely, produce a round hole all the way through the part wall, and avoid drilling the part locator. This is most noticeable in Figure 2–9, where an angled hole requires additional clearance to the relieved portion of the part locator. Additional clearance here would allow the drill to complete the hole and avoid drilling the relieved portion of the locator. The part locator will most likely be hardened and the

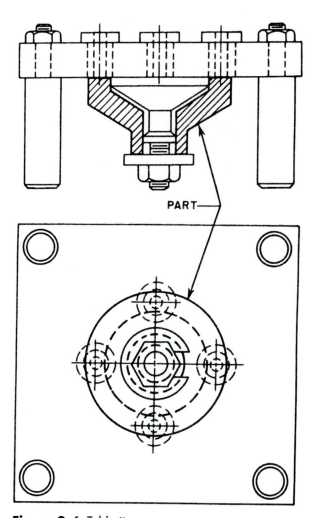

Figure 2-6 Table jig.

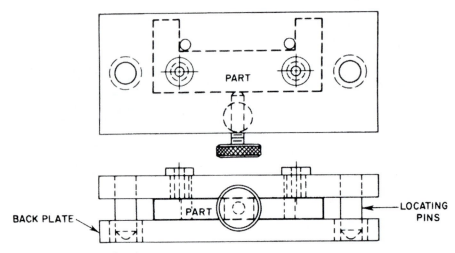

Figure 2-7 Sandwich jig.

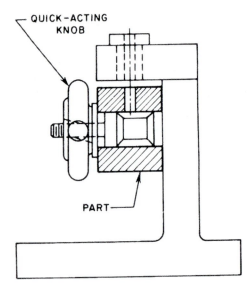

Figure 2-8 Angle-plate jig.

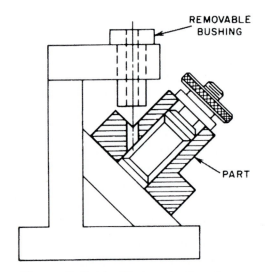

Figure 2-9 Modified angle-plate jig.

drill will be lost as a result of any attempted drilling. Additional clearance on the relieved diameter of the part locator may be possible. A larger clearance hole in the locator could also be added if the relieved diameter cannot be reduced. The additional design consideration added to the locator would include the feature to provide the correct orientation of this clearance hole or machined relief to line up with the bushing location.

Box jigs, or *tumble jigs*, usually totally surround the part (Figure 2–10). This style of jig allows the part to be completely machined on every surface without the need to reposition the work in the jig.

Channel jigs are the simplest form of box jig (Figure 2–11). The work is held between two sides and machined from the third side. In some cases, where jig feet are used, the work can be machined on three sides.

Leaf jigs are small box jigs with a hinged leaf to allow for easier loading and unloading (Figure 2–12). The main differences between leaf jigs and box jigs are size and part location. Leaf jigs are normally smaller than box jigs and are sometimes made so that they do not completely surround the part. They are usually equipped with a handle for easier movement.

Indexing jigs are used to accurately space holes or other machined areas around a part. To do this, the

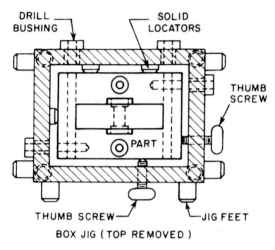

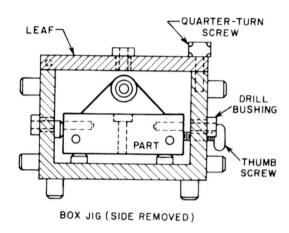

Figure 2-10 Box or tumble jig.

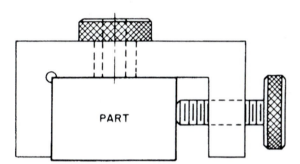

Figure 2-11 Channel jig.

jig uses either the part itself or a reference plate and a plunger (Figure 2–13). Larger indexing jigs are called *rotary jigs.*

Trunnion jigs are a form of rotary jig for very large or odd-shaped parts (Figure 2–14). The part is first put into a box-type carrier and then loaded on the trunnion. This jig is well suited for large, heavy parts that must be machined with several separate plate-type jigs.

Pump jigs are commercially made jigs that must be adapted by the user (Figure 2–15). The lever-activated plate makes this tool very fast to load and unload. Since the tool is already made and only needs to be modified, a great deal of time is saved by using this jig.

Multistation jigs are made in any of the forms already discussed (Figure 2–16). The main feature of this jig is how it locates the work. While one part is

drilled, another can be reamed and a third counter-bored. The final station is used for unloading the finished parts and loading fresh parts. This jig is commonly used on multiple-spindle machines. It could also work on single-spindle models.

There are several other jigs that are combinations of the types described. These complex jigs are often so specialized that they cannot be classified. Regardless of the jig selected, it must suit the part, perform the operation accurately, and be simple and safe to operate.

TYPES OF FIXTURES

The names used to describe the various types of fixtures are determined mainly by how the tool is built. Jigs and fixtures are made basically the same way as far as locators and positioners are concerned. The main construction difference is mass. Because of the increased tool forces, fixtures are built stronger and heavier than a jig would be for the same part.

Plate fixtures are the simplest form of fixture (Figure 2–17). The basic fixture is made from a flat plate that has a variety of clamps and locators to hold and locate the part. The simplicity of this fixture makes it useful for most machining operations. Its adaptability makes it popular.

The *angle-plate fixture* is a variation of the plate fixture (Figure 2–18). With this tool, the part is normally machined at a right angle to its locator. While

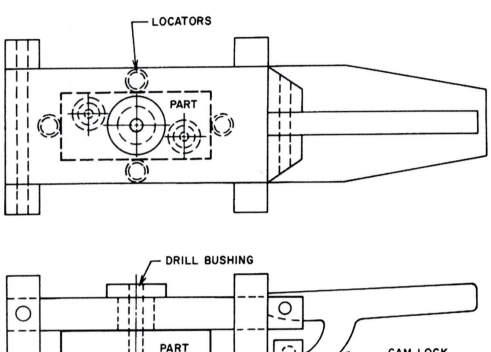

Figure 2–12 Leaf jig.

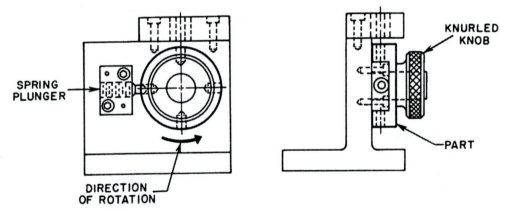

Figure 2–13 Indexing jig.

most angle-plate fixtures are made at 90 degrees, there are times when other angles are needed. In these cases, a modified angle-plate fixture can be used (Figure 2–19).

Vise-jaw fixtures are used for machining small parts (Figure 2–20). With this type of tool, the standard vise jaws are replaced with jaws that are formed to fit the part. Vise-jaw fixtures are the least expen-

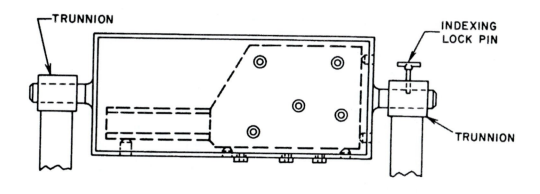

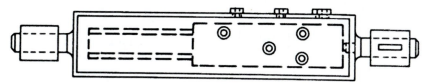

Figure 2-14 Trunnion jig.

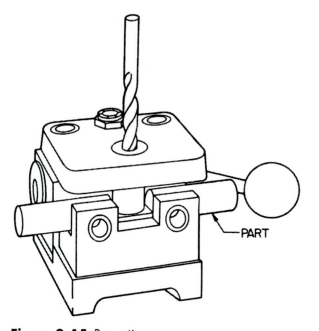

Figure 2-15 Pump jig.

The parts shown in Figure 2–22 are examples of the uses of an indexing fixture.

Multistation fixtures are used primarily for high-speed, high-volume production runs, where the machining cycle must be continuous. *Duplex fixtures* are the simplest form of multistation fixture, using only two stations (Figure 2–23). This form allows the loading and unloading operations to be performed while the machining operation is in progress. For example, once the machining operation is complete at station 1, the tool is revolved and the cycle is repeated at station 2. At the same time, the part is unloaded at station 1 and a fresh part is loaded.

Profiling fixtures are used to guide tools for machining contours that the machine cannot normally follow. These contours can be either internal or external. Since the fixture continuously contacts the tool, an incorrectly cut shape is almost impossible. The operation in Figure 2–24 shows how the cam is accurately cut by maintaining contact between the fixture and the bearing on the milling cutter. This bearing is an important part of the tool and must always be used.

sive type of fixture to make. Their use is limited only by the sizes of the vises available.

Indexing fixtures are very similar to indexing jigs (Figure 2–21). These fixtures are used for machining parts that must have machined details evenly spaced.

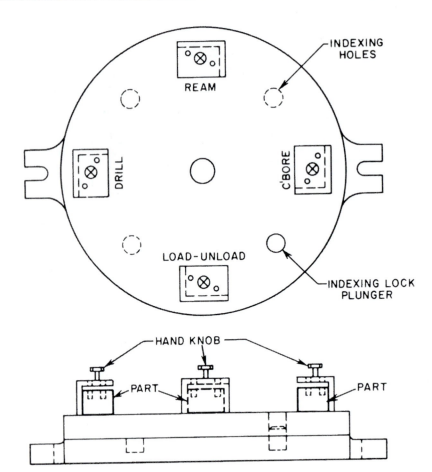

Figure 2-16 Multistation jig.

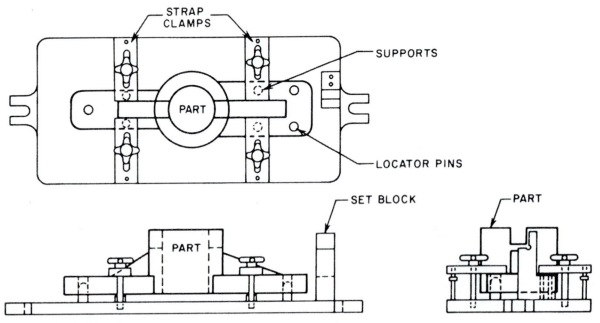

Figure 2-17 Plate fixture.

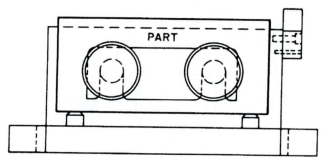

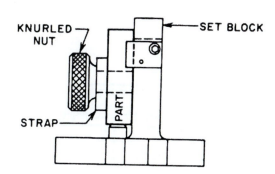

Figure 2-18 Angle-plate fixture.

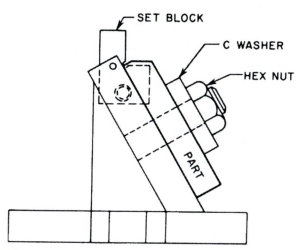

Figure 2-19 Modified angle-plate fixture.

CLASSIFICATION OF FIXTURES

Fixtures are normally classified by the type of machine on which they are used. Fixtures can also be identified by a subclassification. For example, if a fixture is designed to be used on a milling machine, it is called a *milling fixture*. If the task it is intended to perform is straddle milling, it is called a *straddle-milling fixture*. The same principle applies to a lathe fixture that is designed to machine radii. It is called a *lathe-radius fixture*.

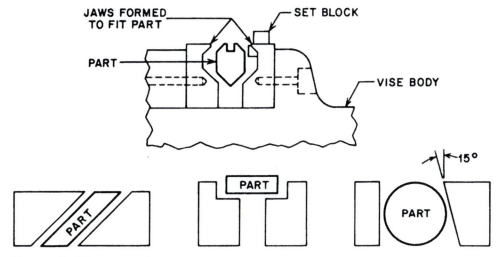

Figure 2-20 Vise-jaw fixture.

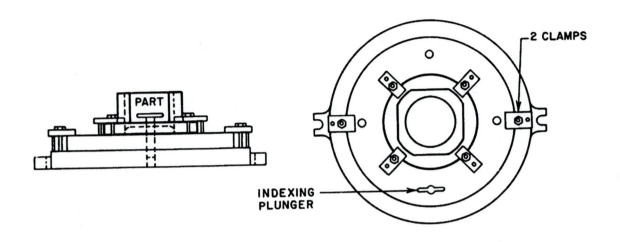

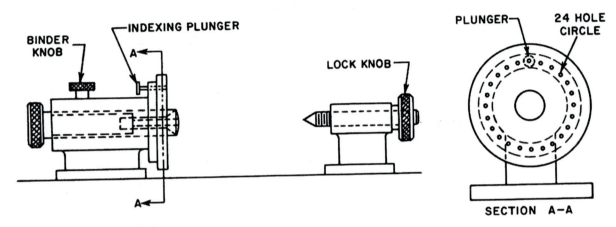

Figure 2–21 Indexing fixture.

Figure 2–22 Parts machined with an indexing fixture.

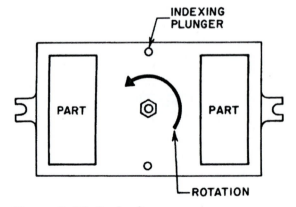

Figure 2–23 Duplex fixture.

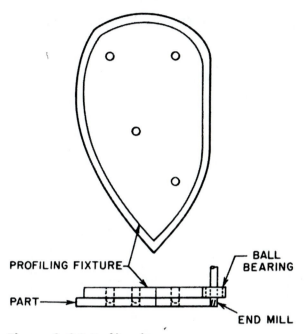

Figure 2-24 Profiling fixture.

The following is a partial list of production operations that use fixtures:

Assembling	Lapping
Boring	Milling
Broaching	Planing
Drilling	Sawing
Forming	Shaping
Gauging	Stamping
Grinding	Tapping
Heat treating	Testing
Honing	Turning
Inspecting	Welding

SUMMARY

The following important concepts were presented in this unit:

- Jigs and fixtures are production workholding devices designed to hold, support, and locate a workpiece.
  - A jig guides the cutting tool with a drill bushing.
  - A fixture references the cutting tool with a set block and feeler, or thickness gauges.
- Jigs are divided into two general classes: drill jigs and boring jigs.

- The type of jig is determined by how it is built. The two types of jigs are open and closed.
  - Template, plate, table, sandwich, and angle-plate jigs are all open jigs.
  - Box, channel, and leaf jigs are all closed jigs.
- Other variations, such as indexing, rotary, trunnion, pump, and multistation jigs, are made as either open or closed jigs.
- Fixture types are determined by the way they are built. The most common types are plate, angle-plate, vise-jaw, indexing, and multistation fixtures.
- Fixture classes are determined by the machine tools on which they are used and sometimes by the operations performed. A fixture used for a straddle-milling operation is classed as a mill fixture, but it may also be classed as a straddle-milling fixture.

REVIEW

1. What is the difference between a jig and a fixture?
2. How are jigs and fixtures normally identified?
3. What are set blocks used for?
4. What class of jig would normally be used to tap holes?
5. A gang-milling fixture is actually what class of tool?
6. Analyze the following part drawings and operations to be performed and select the best jig or fixture for each.
   A. Figure 2–25. Operation: Mill a slot .250 inch by .250 inch.
      1. Box fixture
      2. Duplex fixture
      3. Vise-jaw fixture

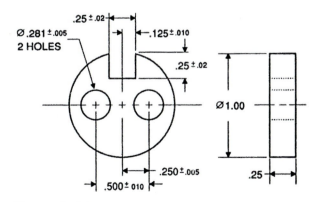

Figure 2-25

B. Figure 2–26. Operation: Drill four .500-inch-diameter holes.
   1. Plate jig
   2. Angle-plate jig
   3. Channel jig

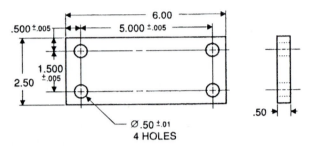

Figure 2–26

C. Figure 2–27. Operation: Drill four holes (two .62-inch and two .25-inch).
   1. Channel jig
   2. Plate jig
   3. Box jig

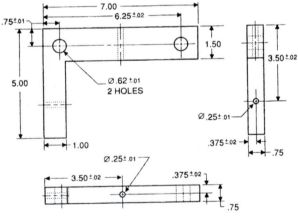

Figure 2–27

D. Figure 2–28. Operation: Drill four holes, .50 inch in diameter.
   1. Box jig
   2. Angle-plate jig
   3. Template jig

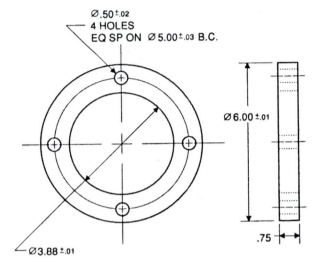

Figure 2–28

E. Figure 2–29. Operation: Mill a shoulder .75 inch by .75 inch by .38 inch.
   1. Plate fixture
   2. Angle-plate fixture
   3. Indexing fixture

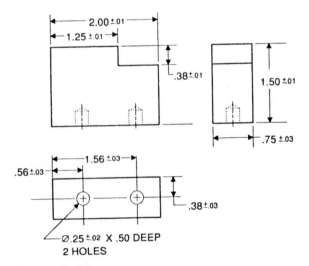

Figure 2–29

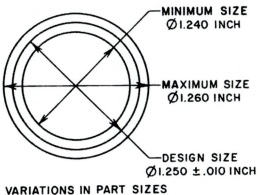

VARIATIONS IN PART SIZES

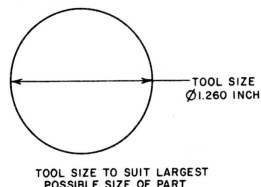

TOOL SIZE TO SUIT LARGEST
POSSIBLE SIZE OF PART

Figure 3-3 Part and tool size relationship.

largest size, the diameter would be 1.260 inches. Any parts made within these sizes would be correct. If the tool is made to fit the part at its design size of 1.250 inches, the parts between 1.250 inches and 1.260 inches, while correct, will not fit into the tool. To prevent this, the tool must be made to fit the parts at their largest or smallest limits of size, depending on how the part is located.

Foolproofing

Foolproofing is a means by which the tool designer ensures that the part will fit into the tool only in its correct position. The part in Figure 3–4A must be machined on the tapered end, so the tool designer includes a pin to prevent the part from being loaded incorrectly. This pin foolproofs the tool. The part in Figure 3–4B shows a hole that must be drilled with

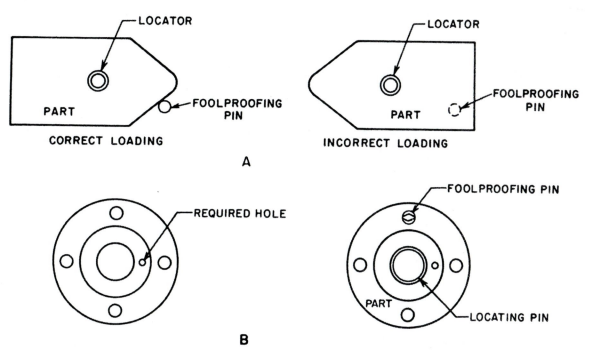

Figure 3-4 Foolproofing.

reference to the holes in the flange. A simple pin placed in one of these holes makes it impossible to load the tool incorrectly.

Other foolproofing devices are just as simple. If foolproofing devices are not simple, they tend to complicate an otherwise easy task.

Duplicate Locators

The use of duplicate locators should always be avoided. Figure 3–5 shows examples of duplicate locators. Locator duplication not only costs more but also could cause inaccuracies.

For example, the flange in Figure 3–5A is located on both the underside of the flange and the bottom of the hub. Since these are parallel surfaces, only one is needed and the other should be eliminated. If the reference surface is the flange, as in Figure 3–5B, the hub locator is not necessary. If the hub is the reference surface, as in Figure 3–5C, the flange locator is unnecessary. To correct this, the tool designer must first determine which surface is to be referenced. Only then should the locators for that surface be specified.

Locational inaccuracies develop because of the difference in position and location tolerances between the tool and the work (Figure 3–6). Locating the part from both its outside edge and the holes can create problems. First, the location of the pins in the tool is fixed and cannot be changed to suit each part. Second, the location of the holes in the part is variable within limits. When a part is placed in the tool that is at either extreme of the part tolerance, it may not fit. To eliminate this possibility, the hole locator can be made

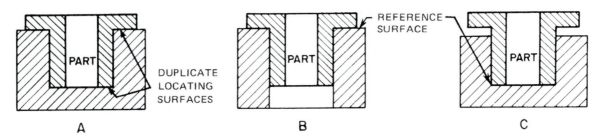

Figure 3-5 Duplicate locators.

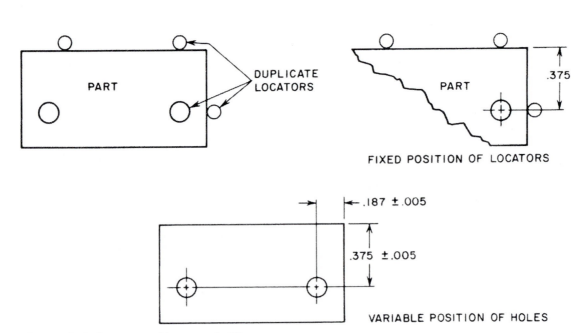

Figure 3-6 Position and locational differences.

smaller to accommodate the variation, but if this is done, the effectiveness of the hold locator is minimized and the locator becomes useless. To avoid this problem, the tool designer must specify whether the part is to be located from its holes or its edges, never both.

PLANES OF MOVEMENT

An unrestricted object is free to move in any of twelve possible directions. Figure 3–7 shows an object with three axes, or planes, along which movement may occur. An object is free to revolve around or move parallel to any axis in either direction. To illustrate this, the planes have been marked X-X, Y-Y, and Z-Z. The directions of movement are numbered from one to twelve.

Restricting Movement

To accurately locate a part in a jig or fixture, movements must be restricted. This is done with locators and clamps.

The fixture for the part in Figure 3–8 illustrates the principle of restricting movement. By placing the part on a three-pin base, five directions of move-

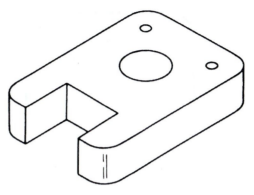

Figure 3–8 Adjusting block.

ment (2, 5, 1, 4, and 12) are restricted (Figure 3–9). Using pin- or button-type locators minimizes the chance of error by limiting the area of contact and raising the part above the chips. Flat bases may also be used, but these should be installed rather than machined into the base. Installed locators are less expensive to use because they take less time to install and are replaceable. If button or flat locators are used, the most important consideration is keeping the part above the chips and in constant contact with all three locators.

To restrict the movement of the part around the Z-Z-axis and in direction eight, two more pin-type locators are positioned (Figure 3–10). To restrict direction seven, a single-pin locator is used (Figure 3–11). The remaining directions, nine, ten, and eleven, are restricted by a clamping device. This three-two-one, or six-point, locating method is the most common external locator for square or rectangular parts.

When a workpiece having holes is located, the holes provide an excellent method of locating the complete part. As shown in Figure 3–12, the center hole is used as a primary locator, and one of the other holes is used as a secondary locator. Here the primary locator is a round pin, and the secondary locator is a diamond pin. As shown, the base plate with the round pin positioned in the center hole will restrict nine degrees of movement (1, 2, 4, 5, 7, 8, 10, 11, and 12). The diamond pin, located as shown, further restricts another two degrees of movement (6 and 3). Together, these locators restrict eleven degrees of movement. The only direction the workpiece can move in

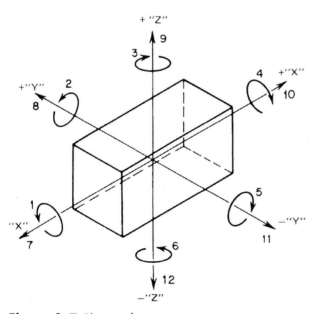

Figure 3–7 Planes of movement.

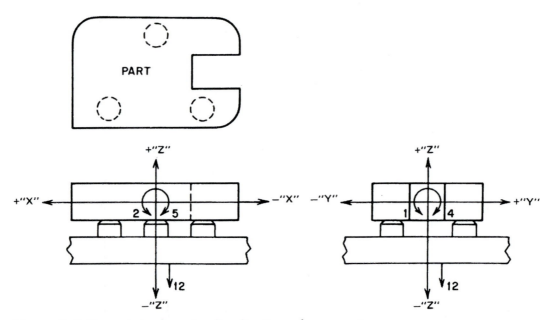

Figure 3–9 Three-pin base restricts five directions of movement.

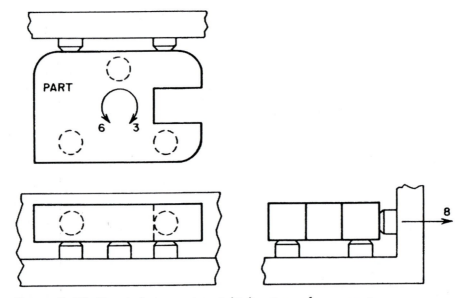

Figure 3–10 Five-pin base restricts eight directions of movement.

is straight up, so the clamping device is actually holding only one direction of movement.

LOCATING THE WORK

Parts are made in almost every possible shape and size. The tool designer must be able to accurately locate each part regardless of how it is made. To do this, the tool designer must know the various types of locators and how each should be used to get the best part placement with the least number of locators.

Locating from a Flat Surface

There are three primary methods of locating work from a flat surface: solid supports, adjustable supports, and equalizing supports. These locators set the vertical position of the part, support the part, and prevent distortion during the machining operation.

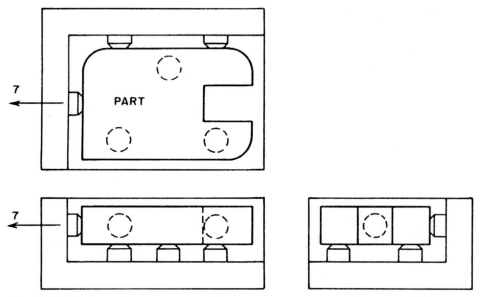

Figure 3–11 Six-pin base restricts nine directions of movement.

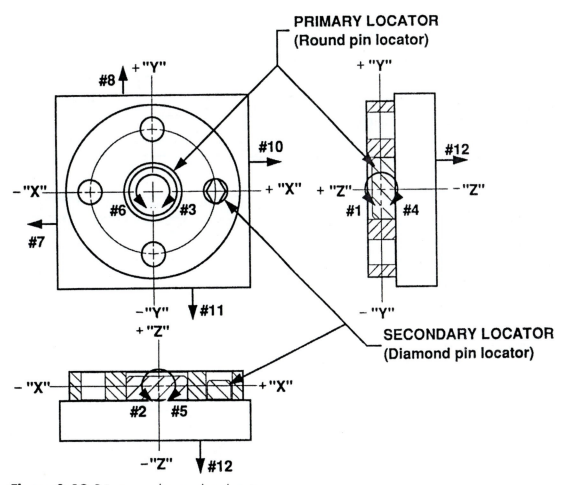

Figure 3–12 Primary and secondary locators.

Solid supports are the easiest to use. They can be either machined into the tool base or installed (Figure 3–13). This type of support is normally used when a machined surface acts as a locating point.

Adjustable supports are used when the surface is rough or uneven, such as in cast parts. There are many styles of adjustable supports. A few of the more common are the threaded (Figure 3–14A), spring (Figure 3–14B), and push types (Figure 3–14C). The threaded style is the easiest and most economical, and it has a larger adjustment range than the others. Adjustable locators are normally used with one or more solid locators to allow any adjustment needed to level the work.

Equalizing supports are also a form of adjustable support (Figure 3–15). They provide equal support through two connected contact points. As one point is depressed, the other raises and maintains contact with the part. This feature is especially necessary on uneven cast surfaces.

The terms *locator* and *support* are used interchangeably when the devices used under a workpiece are discussed. The locating devices used to reference the edges of a part are called *locators* or *stops*.

Before choosing a support, the tool designer must consider the shape and surface of the part and the type of clamping device to be used. The support selected must be strong enough to resist both the clamping pressure and the cutting forces. The clamps should be positioned directly over the supports to avoid distorting or bending the part.

Locating from an Internal Diameter

Locating a part from a hole or pattern is the most effective way to accurately position work. Nine of the twelve directions of movement are restricted by using a single pin, and eleven directions of movement are restricted with two pins. When possible, it is logical to use holes as primary part locators.

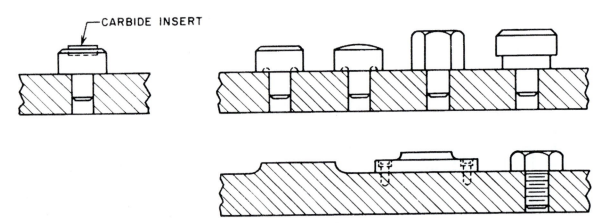

Figure 3–13 Solid supports.

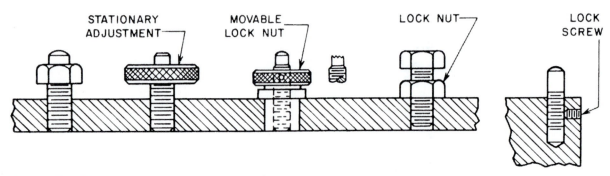

Figure 3–14A Adjustable supports, threaded type.

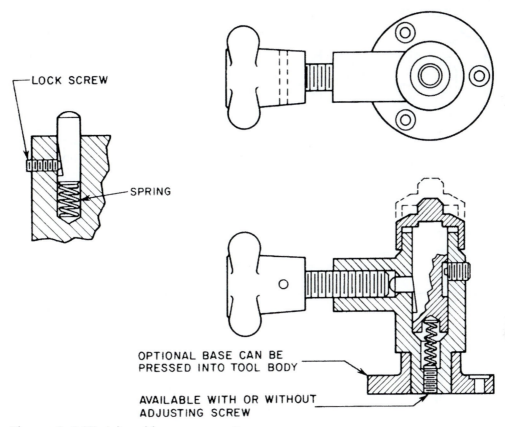

Figure 3-14B Adjustable supports, spring type.

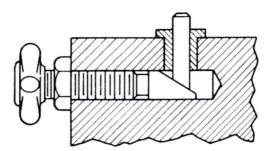

Figure 3-14C Adjustable support, push type.

Several types of locators are used for locating work from holes. Figure 3–16 shows a few locators used for large holes. When large holes locate the work, fasten the internal locator with both screws and dowels. Under normal conditions, two dowels and two screws are needed to hold the locator. With more force, it is better to use larger dowels and screws rather than to increase their number.

With *shank-type locators*, it is a good practice to use the press-fit locator rather than the threaded locator for accuracy. Threaded locators are useful in areas

where the construction of the tool will not permit the other type to be pressed out. Another type has the advantages of the press-fit and the locking properties of a thread.

Pin-type locators are used for smaller holes and for aligning members of the tool (Figure 3–17). When the pins are used for alignment, special bushings should also be used so that they can be replaced when they wear. Pins used for part location are made with either tapered ends or rounded ends, allowing the parts to be installed and removed easily (Figure 3–18).

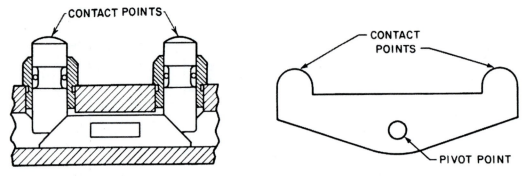

Figure 3-15 Equalizing supports.

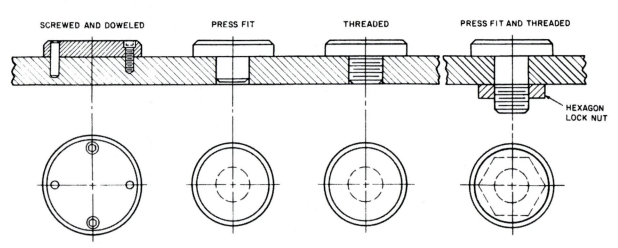

Figure 3-16 Internal locators.

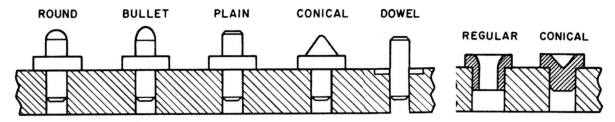

Figure 3-17 Pin locators and bushing.

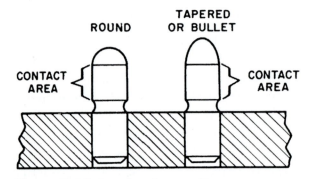

Figure 3-18 Round and tapered locators.

The main difference between the pins used for location and the pins used for alignment is the amount of bearing surface. Alignment pins usually have a longer area of contact. Locating pins usually have a contact area of one-eighth to one-half of the part thickness. More than this makes placement and removal operations difficult.

Another style of pin common to jigs and fixtures is the *diamond* or *relieved pin*, which is normally used along with the round pin to reduce the time it takes to load and unload the tool. It is easier to locate a part on one round pin and one diamond pin than to locate it on two round pins. In use, the round pin locates the part and the diamond pin prevents the movement around the pin (Figure 3–19). Notice the direction of movement the part has around the round pin. By installing the diamond pin as shown, this movement is restricted.

To be effective, the diamond pin must always be placed to resist this movement. Figure 3–20 shows how two diamond pins could be used to locate a part. Notice how each restricts the direction of movement of the other. Two diamond pins should be used to locate a part when the part has adequate locational tolerance.

In addition to the diamond pin relieved locator, other types are used for some workholders. A few examples of relieved locators are shown in Figure 3–21. The specific design of any relieved locator is determined by the workpiece and the type of location required. Relieved locators reduce the area of contact between the workpiece and the locator. Decreasing the contact area has little or no effect on the overall locational accuracy; however, reducing the contact area helps make the jig or fixture easier to load and unload and lessens the problems caused by dirt, chips, and burrs.

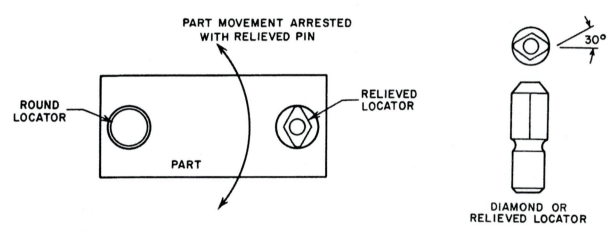

Figure 3–19 Locating with one relieved locator.

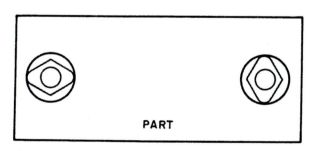

Figure 3–20 Locating with two relieved locators.

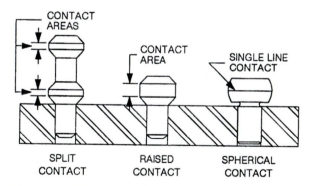

Figure 3–21 Relieved locators.

The *split contact locator*, shown in Figure 3–22, is a type of relieved locator used for thick workpieces. Here, rather than using the complete thickness of the part for location, the locator is relieved in the middle, and only the top and bottom areas of the locator contact the workpiece. This design provides full location and makes the locator less likely to bind in the workpiece.

The *raised contact locator*, shown in Figure 3–23, is an example of relieving a locator for better function. Here the top and bottom contact areas of the locator have been removed. This design reduces the contact area and raises the point where the locator and workpiece touch. Moving this contact point off the base plate, to the middle of the workpiece, helps reduce the effects of dirt, chips, or burrs. The raised contact design supplies a complete locating surface and reduces the chance of the locator binding in the hole.

To increase the overall locational efficiency, a great number of locator designs have been examined. However, there is only one style that will not bind in any locating hole. This is the *spherical locator*. A spherical locator greatly reduces the contact area by removing all the material not directly in contact with the workpiece.

Spherical locators are impossible to bind because, unlike with cylindrical locators, the distance between the opposite sides of the contact area is always the same. As shown in Figure 3–24A, if a part is loaded on a cylindrical pin, it can and will bind unless the centerline of the hole is precisely aligned with the centerline of the locator. This phenomenon is caused by the effect of the hypotenuse of the triangle formed by the centerlines if they are not perfectly aligned. The diameter of the pin shown at d is smaller than the

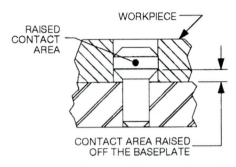

Figure 3-23 Raised contact relieved locator.

elliptical form of the pin shown at D. This difference in diameter is what causes the cylindrical pin to bind.

However, as shown in Figure 3–24B, a spherical locator always has the same diameter regardless of how the centerlines of the part and the locator vary. This can be seen by the three positions of the diameter lines. Regardless of where the diameter, d, is mea-

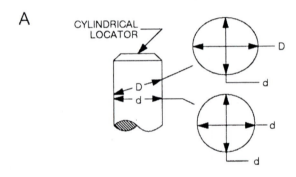

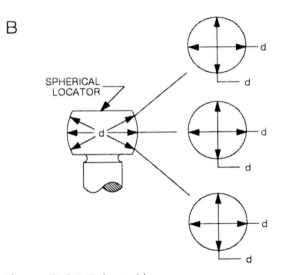

Figure 3-24 Spherical locators.

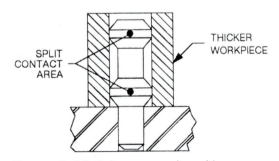

Figure 3-22 Split contact relieved locator.

sured, it is always the same size. This design results in a locator that contacts the workpiece only with a thin, single line of contact around the locator. While such locators cannot bind in the hole, they also have their own problems. Since a spherical locator contacts the workpiece only with a single-line contact, locator wear can become a real problem. Should any wear occur on this locator, the locational accuracy is immediately affected. To minimize this problem, these locators should be used only where sufficient tolerance for wear is permitted. To reduce the effects of wear, the spherical surface may also be hardened.

To achieve both the nonjamming characteristics of a spherical locator and the extended service life of a relieved locator, a modified form of locator can be used. Two examples of this type are shown in Figure 3–25. The locator shown at the left is a form of raised contact relieved locator that has a very thin contact band. This design offers more resistance to wear, while the 45-degree relief angles make the locator less likely to bind or jam during loading and unloading operations. The locator shown at the right has a similar design, but it uses a relief groove in place of the lower relief angle.

Locating from an External Profile

Locating work from an external profile, or outside edge, is the most common method of locating work in the early stages of machining. Profile locators position the work in relation to an outside edge or the outside of a detail, such as a hub or a boss. The following are examples of the most common ways a part can be located from its profile.

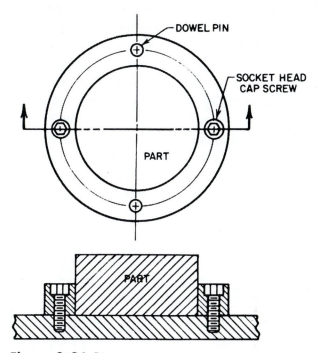

Figure 3–26 Ring nest.

Nesting locators position a part by enclosing it in a depression, or recess, of the same shape as the part. Nesting is the most accurate locating device for profile location. Since the nest must conform to the shape of the part, nests are very expensive to design for complicated shapes. The most common type is the ring nest, which is normally used for cylindrical profiles (Figure 3–26). The *full nest* completely encloses shapes other than cylindrical (Figure 3–27).

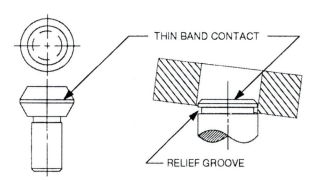

Figure 3–25 Alternative relieved locator designs.

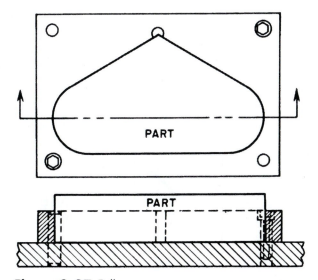

Figure 3–27 Full nest.

The *partial nest* is a variation of the full nest and encloses only a part of the workpiece (Figure 3–28).

Vee locators are used mainly for round work. They can locate flat work with rounded or angular ends and flat discs (Figure 3–29). The vee-block locator is normally used to locate round shafts or other workpieces with cylindrical sections (Figure 3–30). One advantage vee locators have over other locators is their centralizing feature. When using a vee locator, be sure it is positioned to allow for the differences in part sizes (Figure 3–31).

Fixed-stop locators are used for parts that cannot be placed in either a nest or a vee locator. They are either machined into the tool body (Figure 3–32), or installed (Figure 3–33).

Installed locators are normally more economical to use because of the time it takes to make the

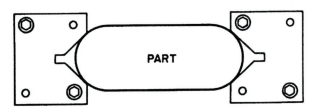

Figure 3–29 Vee-block locators.

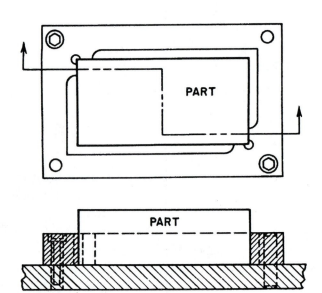

Figure 3–28 Partial nest.

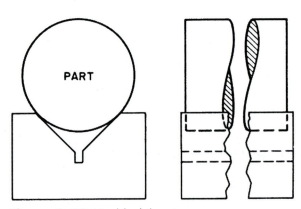

Figure 3–30 Vee-block locators.

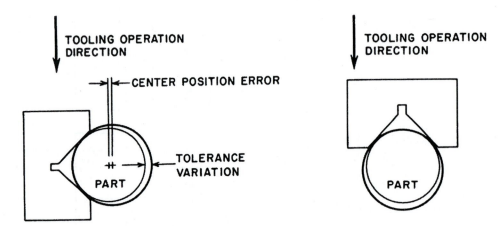

Figure 3–31 Positioning vee locator to allow for differences in part size.

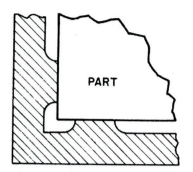

Figure 3–32 Machined fixed-stop locators.

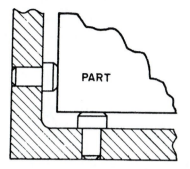

Figure 3–33 Installed fixed-stop locators.

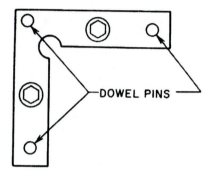

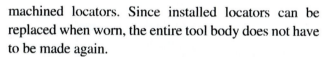

Figure 3–34 Doweled locators.

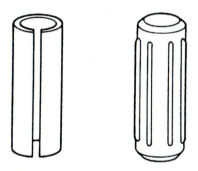

Figure 3–35 Split and grooved dowels.

machined locators. Since installed locators can be replaced when worn, the entire tool body does not have to be made again.

The most common type of fixed locator is the *dowel pin*. Dowel pins also attach other locational devices, such as the block in Figure 3–34.

When possible, *split pins* or *groove pins* should be used in place of dowel pins to reduce the cost. Split pins and groove pins hold as well as dowel pins (Figure 3–35). However, they do not require a reamed hole, so they are not as accurate. When the tolerance permits, these pins should be used to reduce the time and cost.

Adjustable-stop locators can also be used to keep the cost of a tool to a minimum (Figure 3–36). Since these stops are adjustable, their position on the tool body does not have to be as closely controlled.

One common way to locate parts is to use both fixed stops and adjustable stops. The tool in Figure 3–37 shows how the fixed locator is used to reference the end of the part while the adjustable locators are used on both sides. Using adjustable locators for this jig allows the part to be positioned exactly. If adjustment is necessary because of wear or misalignment, it can be easily corrected.

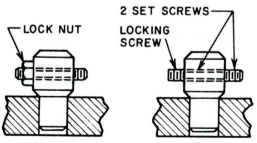
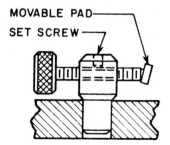
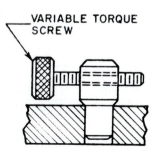

Figure 3–36 Adjustable-stop locators.

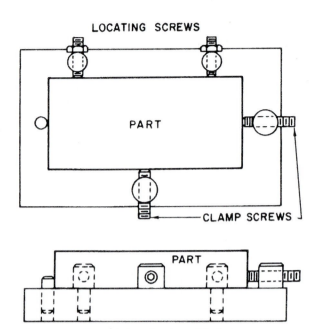

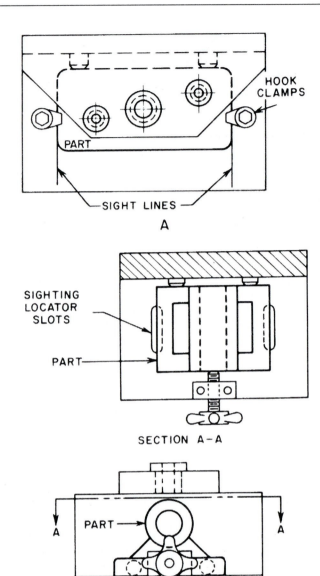

Figure 3-37 Solid and adjustable locators.

Another advantage of the adjustable-type locator is its ability to double as a clamp. This is done by replacing the adjustment screw with a knurled-head screw (Figure 3–38).

Sight locators align rough parts in a tool for approximate machining (Figure 3–39). There are two methods of referencing a part by sight location: by lines etched on the tool, as in Figure 3–39A, or by slots, also shown in Figure 3–39B. In both cases, the part is aligned with the marks until it is in the approximate center. It is then clamped and machined.

Ejectors

Ejectors are used to remove work from close-fitting locators, such as full nests or ring nests. These devices speed up the unloading of the part from the tool, which reduces the in-tool time and increases the

Figure 3-39 Sight locators.

production rate. Figure 3–40 shows two styles of ejectors common to both jigs and fixtures.

Spring-Stop Buttons and Spring-Locating Pins

Spring-stop buttons and *spring-locating pins*, while not locating devices, do aid in properly locating a part (Figure 3–41). These devices are used to push the part against the locators to ensure proper contact during the clamping operation. The three styles of spring-stop

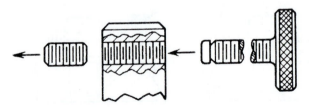

Figure 3-38 Adjustable locator used as a clamp.

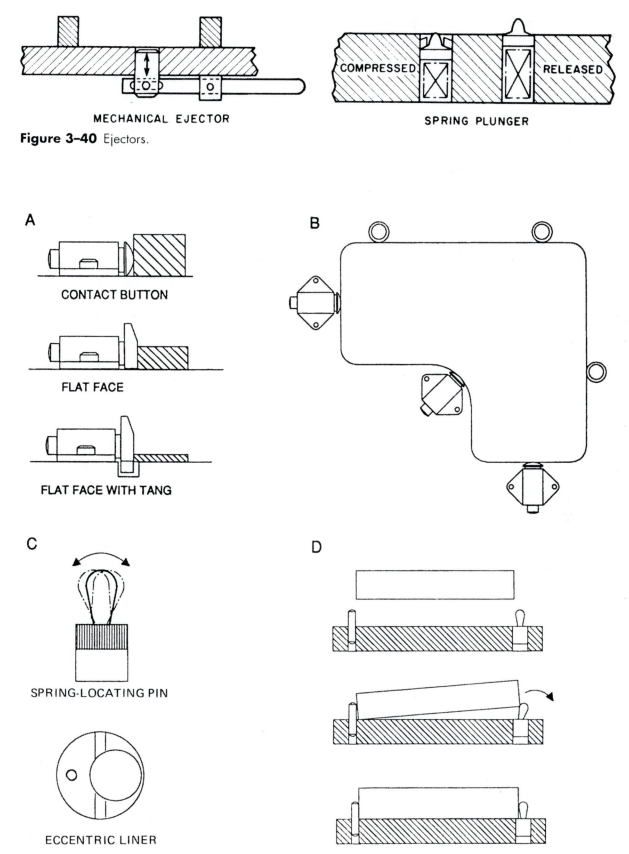

MECHANICAL EJECTOR

COMPRESSED   RELEASED

SPRING PLUNGER

Figure 3–40 Ejectors.

A

CONTACT BUTTON

FLAT FACE

FLAT FACE WITH TANG

B

C

SPRING-LOCATING PIN

ECCENTRIC LINER

D

Figure 3–41 Spring-stop buttons and spring-locating pins.

buttons most commonly available are the contact button, flat face, and flat face with tang, shown at view A. The application of these units is shown at view B. Here the part is held against the locating pins with three spring-stop buttons. Using these buttons not only helps position the part but also eliminates the need for a third hand when clamping some parts.

The spring-locating pins, shown at view C, are another variation of this idea that is very useful for smaller parts or in a confined space. These pins may be installed in a hole or mounted in an eccentric liner. This liner permits the position of the spring pin to be adjusted to hold parts with looser tolerances. The application of these spring pins is shown at view D. The first step is to position the part over the workholder. The part is then butted against the solid locator and pushed down against the spring pin. When fully seated, the spring-locating pin pushes the workpiece against the solid locator, as shown.

SUMMARY

The following important concepts were presented in this unit:

- To achieve proper location, the locators must properly *reference* the part and ensure the *repeatability* of location throughout the production cycle.
  - *Referencing* is the process of properly positioning the part with respect to the cutter or other tool.
  - *Repeatability* is the feature of location that permits the parts to be made within their stated tolerances, part after part, throughout the production run.
- The major concerns in locating a part that the designer must keep in mind are the position of the locators, locational tolerances, foolproofing the location, and avoiding duplicate location.
- There are twelve planes of movement in any workpiece. When locating a part, the designer must select the locational method that will restrict a maximum number of these planes of movement.
  - The three-two-one, or six-point, location method will restrict a total of 9 planes of movement.
  - Locating a part by a single center hole will restrict nine planes of movement with a single locator. By positioning a second locator in

another hole in the part, a total of eleven planes of movement can be restricted with two locators.
  - In either case, the remaining unrestricted planes of movement are restricted with the clamping device.
- Locators positioned under a part are referred to as supports. Locators that position a part by its edges are called *locators* or *stops*.
  - The three primary supports are solid, adjustable, and equalizing.
  - The major types of locator used to locate parts on an internal diameter are shank-mounted locators and pin locators. Pin locators may also be used to locate a part from its external edges.
- Other devices used to locate parts include nesting locators, vee locators, fixed-stop locators, adjustable-stop locators, and sight locators.
- Ejectors are devices that are used to help remove a part from close-fitting locators. Ejectors are made in many variations. The mechanical-level ejector and spring-type ejector are two examples of these devices.

REVIEW

1. Where should locators contact the part? Why?
2. What is repeatability?
3. What percentage of the part tolerance must be applied to the tool?
4. Why should the tool be foolproof?
5. What is duplicate locating?
6. Select the proper locator from the choices listed to best locate the following sample parts.
   A. Figure 3–42. Operation: Drill two .19-inch holes as shown.
      1. Groove pad and dowels
      2. Vee pad and dowels
      3. Ring nest and vee block

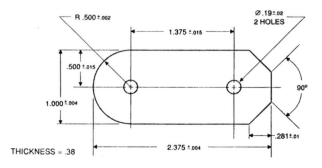

Figure 3–42

B. Figure 3–43. Operation: Mill two .22 × .22-inch slots as shown.
1. Pin locators
2. Nest locators
3. Adjustable locators

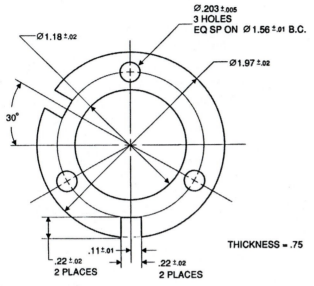

Figure 3–43

C. Figure 3–44. Operation: Mill a vee groove as shown.
1. Vee pad and dowel pins
2. Adjustable locator and nest
3. Diamond and dowel pins

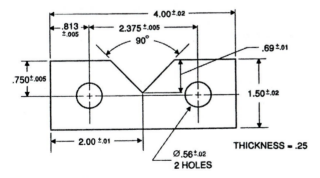

Figure 3–44

D. Figure 3–45. Operation: Drill three .25-inch holes as shown.
1. Diamond pin and equalizer
2. Solid and adjustable locators
3. Adjustable supports and diamond pins

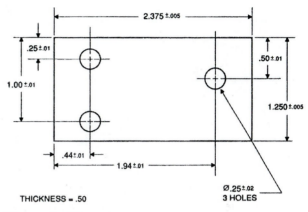

Figure 3–45

E. Figure 3–46. Operation: Drill two .25-inch holes as shown.
1. Dowel pin and adjustable support
2. Diamond pin and round pin locators
3. Round pin locator and fixed support

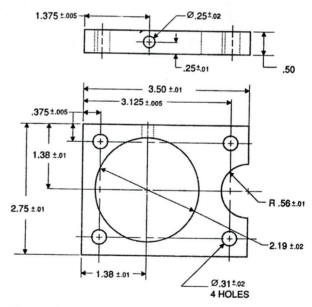

Figure 3–46

F. Figure 3–47. Operation: Drill two .44-inch holes as shown.

1. Ring nest
2. Adjustable locators
3. Sight locators

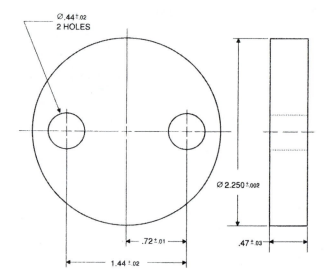

Figure 3–47

G. Identify the locators shown in Figure 3–48.

a. Conical locator
b. Relieved locator
c. Bar locator
d. Round locator
e. Adjustable locator
f. Solid support
g. Plain locator   # 4
h. Nesting locator
i. Diamond pin locator
j. Adjustable support
k. Bullet locator

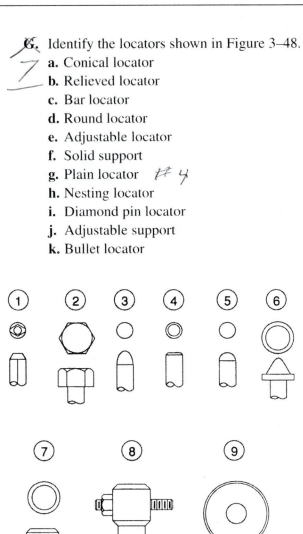

Figure 3–48

UNIT 4

Clamping and Workholding Principles

OBJECTIVES

After completing this unit, the student should be able to:

• Explain the basic principles of workholders.

• Identify the types of workholding devices.

• Match the characteristics and applications to a particular type of clamping device.

WORKHOLDERS

The term *workholder*, as applied to jig and fixture design, has two similar but different meanings. In general terms, the complete jig and fixture family is described as workholders. Here the term is used to identify the complete group of production tools. When clamping is discussed, the term *workholder* identifies the parts of a jig or fixture that clamp, chuck, hold, or grip the part. The use of the word normally determines which meaning is intended.

The main purpose of a workholder, or clamping device, is to securely hold the position of the part against the locators throughout the machining cycle. To do this, the clamp used must meet the following conditions:

• The clamp must be strong enough to hold the part and to resist movement.

• The clamp must not damage or deform the part.

• The clamp should be fast-acting and allow rapid loading and unloading of parts.

To use the proper clamp for each job, the tool designer must know and understand the basic principles of clamping, as well as the devices commonly used to hold the work.

BASIC RULES OF CLAMPING

The function of a clamp is to hold a part against the locators during the machining cycle. To be effective and efficient, clamps must be planned into the tool design.

Positioning the Clamps

Clamps should always contact the work at its most rigid point. This prevents the clamping force from bending or damaging the part. The part must be supported if the work is clamped at a point where the force could bend the part. The flange in Figure 4–1 shows this point. The ideal place to clamp the part is from its center hole. If it is held by the outer edge, the part must be supported (Figure 4–2).

Clamps are also positioned so they do not interfere with the operation of the tool or machine. It is important that the clamps be placed so that the operator can work easily yet safely.

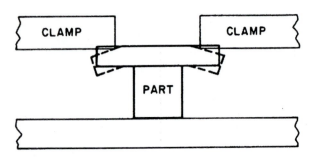

Figure 4–1 Flange ring.

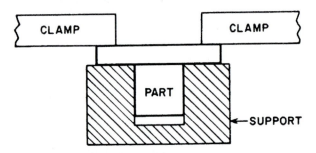

Figure 4–2 Supporting the flange ring.

Tool Forces

Tool forces are forces generated by the cutting action. They are caused by resistance of the workpiece being cut or sheared by the tool. To clamp a part correctly, the tool designer must know how tool forces, or cutting forces act in reference to the tool. A properly designed tool can use the cutting forces to its own advantage. The drill jig in Figure 4–3 is an example of how the cutting force is used to hold the work.

Most of the force is in a downward direction against the base of the tool. The forces that must be resisted cause the part to revolve around the drill axis, which in turn causes the part to climb the drill when the drill breaks through the opposite side of the part. In this drill jig, the forces that cause the part to revolve are restricted and held by the locators. This leaves the climbing action to be restrained by the clamp. The climbing force is a small fraction of the drilling force and is treated as such when clamped.

This same principle is used for all operations. The necessary clamping pressure is reduced a great deal when the bulk of the tool forces are directed at a solid part of the tool body.

Clamping Forces

Clamping force is the force required to hold a part against the locators. Clamping prevents the part from shifting or being pulled from the jig or fixture during the machining operation. The type and amount of clamping force needed to hold a part is usually determined by the tool forces working on the part and how the part is positioned in the tool. Sometimes the operation being performed is a factor. In the case of the bored ring in Figure 4–4, if the ring is clamped as shown at view A, the part can bend. If it is held as at view B, this possibility is reduced.

Clamping pressure, as a general rule, should only be enough to hold the part against the locators. The locators should resist the bulk of the thrust. If a part

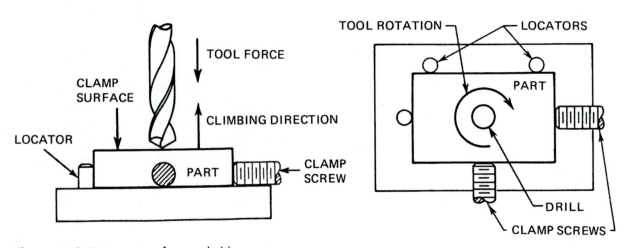

Figure 4–3 Using cutting force to hold a part.

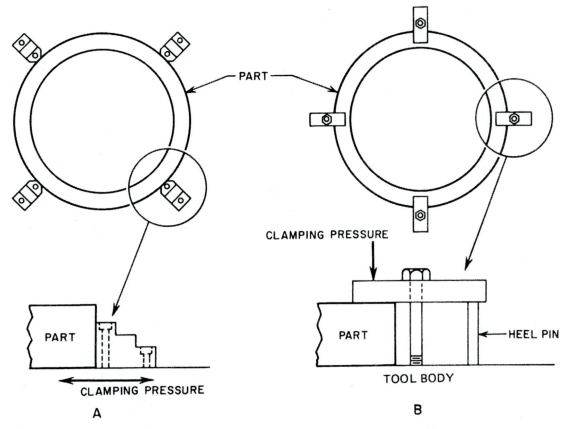

Figure 4–4 Clamping forces.

must be clamped with a great deal of pressure, the tool should be redesigned so that the tool thrust is directed at the locators and the tool body. Clamps must never be expected to hold all of the tool thrust.

TYPES OF CLAMPS

Various methods of clamping are common to both jigs and fixtures. The type of clamp the tool designer chooses is determined by the shape and size of the part, the type of jig or fixture being used, and the work to be done. The tool designer should choose the clamp that is the simplest, easiest to use, and most efficient.

Strap Clamps

Strap clamps are the simplest clamps used for jigs and fixtures (Figure 4–5). Their basic operation is the same as that of a lever. Strap clamps can be grouped into three classes, each representing a form of lever (Figure 4–6).

Figure 4–6A shows the first clamp, which has the fulcrum between the work and the effort. This is the principle of a first-class lever. The second clamp, shown in Figure 4–6B, places the work between the fulcrum and the effort, as with a second-class lever. The third clamp, shown in Figure 4–6C, uses the principle of a third-class lever by placing the effort between the work and the fulcrum.

Most strap clamps use the third-class lever arrangement. When these strap clamps are used, the spacing of the three elements is also important. The distance between the fastener (effort) and the workpiece should always be less than that between the fastener and the heel pin (fulcrum). This increases the mechanical advantage of the clamp and increases the holding force on the workpiece.

Thus, when a strap clamp is used, the force on the workpiece is always proportional to the position of the fastener with respect to the workpiece and the heel support. If the stud were positioned exactly in

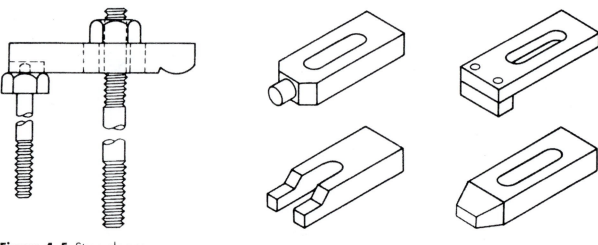

Figure 4-5 Strap clamps.

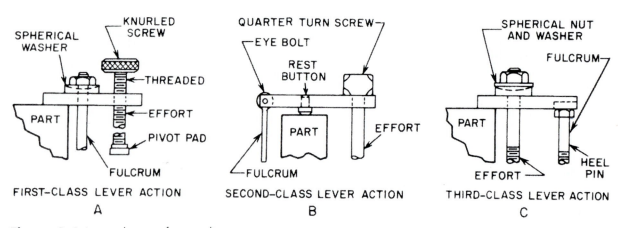

Figure 4-6 Lever classes of strap clamps.

the center of the strap, the pressure generated by the fastener would be distributed equally between the workpiece and the heel support (Figure 4–7). This arrangement may be desired for some clamping operations, such as clamping two parts side by side, as in Figure 4–8. In most cases, when a single part is clamped, this is not the most efficient method.

The arrangement shown in Figure 4–9 is a better way to clamp a single part. Here the fastener is positioned so that one-third of the strap length is between the fastener and the workpiece and two-thirds of the strap is between the fastener and the heel support. The clamping pressure on the workpiece with this setup is twice as great as that on the heel support.

Strap clamps are used in almost every area of jig and fixture design and construction. Some more

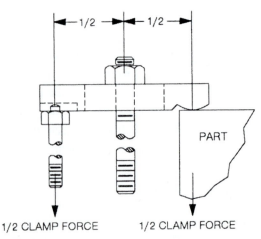

Figure 4-7 Positioning the fastener in the center of the strap applies equal pressure at both ends of the clamp.

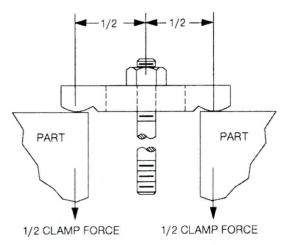

Figure 4-8 The fastener placed in the center of the strap is often used for clamping two parts side by side.

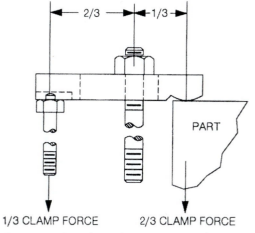

Figure 4-9 When clamping a single part, a better placement is with one-third of the strap length between the fastener and the workpiece and two-thirds between the fastener and the heel support.

common types are the hinge clamp, the sliding clamp, and the latch clamp (Figure 4–10).

The fulcrum is positioned so that the clamp bar is parallel to the base of the tool at all times. Because of the slight differences in part thickness, this is not always possible. To make up for these differences, spherical washers or nuts are used. Spherical washers and nuts provide a positive base for clamping elements, and reduce unnecessary stresses to the threaded members (Figure 4–11).

Strap clamps can be operated by either manual devices or power-driven devices. Manual devices include hex nuts, hand knobs, and cams (Figure 4–12). Power devices include hydraulic systems and pneumatic systems (Figure 4–13).

The holding power of a strap clamp is determined by the size of the threaded member binding the clamp. The chart in Figure 4–14 lists the recommended clamping pressures for the six most common UNC and ISO metric clamp screws. The values shown are based on standard commercial bolts of 50,000 psi (pounds per square inch) minimum tensile strength. Commercial bolts and studs made particularly for jig and fixture clamps normally have a minimum tensile strength value of 75,000 to 100,000 psi. The values given for these clamp bolts should be adjusted to reflect the difference in tensile strength.

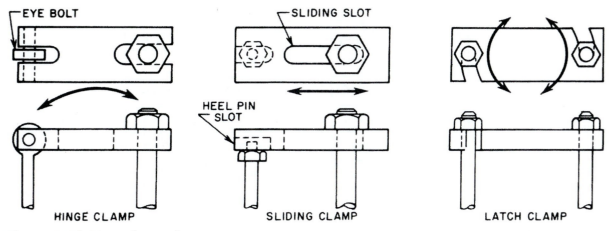

Figure 4-10 Types of strap clamp.

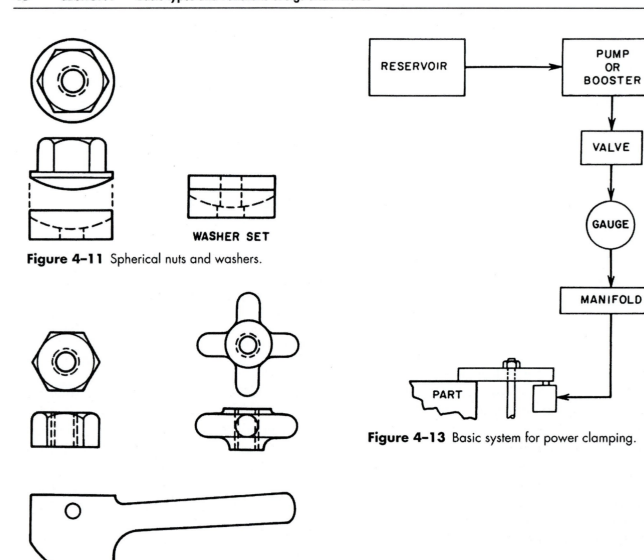

Figure 4-11 Spherical nuts and washers.

WASHER SET

Figure 4-12 Mechanical holding devices.

Figure 4-13 Basic system for power clamping.

SCREW SIZE		HOLDING FORCE	
METRIC	U.S. CUSTOMARY	METRIC	U.S. CUSTOMARY
M 6	$\frac{1}{4}$ - 20 UNC	TO 710 NEWTONS	TO 160 LB
M 8	$\frac{5}{16}$ - 18 UNC	TO 1110 NEWTONS	TO 250 LB
M 10	$\frac{3}{8}$ - 16 UNC	TO 1670 NEWTONS	TO 375 LB
M 12	$\frac{1}{2}$ - 13 UNC	TO 3110 NEWTONS	TO 700 LB
M 16	$\frac{5}{8}$ - 11 UNC	TO 4890 NEWTONS	TO 1100 LB
M 20	$\frac{3}{4}$ - 10 UNC	TO 7120 NEWTONS	TO 1600 LB

Figure 4-14 Recommended holding pressures.

Although standard high-strength fastening devices may be used for many workholding tasks, the commercially available jig and fixture studs, bolts, nuts, washers, and other hardware should be used whenever possible. At no time should the long threaded rods available from a hardware store be used for workholding. These rods do not have sufficient strength to be used for jig and fixture applications.

Screw Clamps

Screw clamps are widely used for jigs and fixtures. They offer the tool designer almost unlimited application potential, lower costs, and, in many cases, less complex designs. The only disadvantage in using screw clamps is their relatively slow operating speeds. The basic screw clamp uses the torque developed by a screw thread to hold a part in place—either by direct pressure or by its action on another clamp (Figure 4–15).

There are variations of the screw-type clamp. Numerous commercial clamps have been developed that include the advantages of the screw clamp and reduce its disadvantages. The following are the commercial components that improve the efficiency of the screw clamp.

Swing Clamps

Swing clamps combine the screw clamp with a swinging arm that pivots on its mounting stud. The holding power with this clamp is generated by the screw. The rapid action needed is achieved by the swinging arm (Figure 4–16).

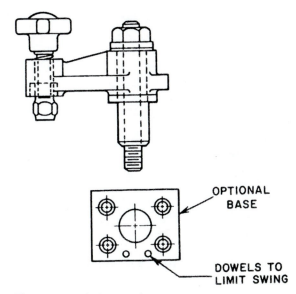

Figure 4–16 Swing clamp.

Hook Clamps

Hook clamps are similar to swing clamps but they are much smaller (Figure 4–17). They are useful in tight places or where several small clamps rather than one large clamp must be used. A variation of the hook clamp is shown in Figure 4–18. The modified hook clamp is made to be operated from the opposite side of the tool. It is useful in tight places or where a safety hazard may be present if the tool is operated from the front side.

Quick-Acting Knobs

Quick-acting knobs are useful for increasing the output of low-cost tools. These knobs are made so that when pressure is released, they can be tilted and

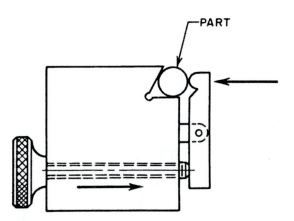

Figure 4–15 Indirect clamping with a screw clamp.

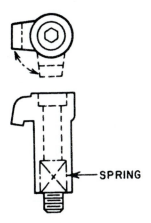

Figure 4–17 Hook clamp.

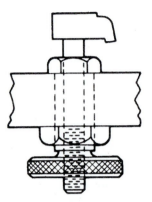

Figure 4–18 Modified hook clamp.

slid off a stud (Figure 4–19). The knob is slid over the stud until it contacts the part. It is then tilted to engage the threads and is turned until tight.

Several other accessories are commercially available to increase the efficiency and effectiveness of the screw clamp. The common accessories and attachments are shown in the design section of this text.

Cam-Action Clamps

Cam-action clamps, when properly selected and used, provide a fast, efficient, and simple way to hold work (Figure 4–20). Because of their construction and basic operating principles, the use of cam-action clamps is limited in some types of tools.

Cam clamps, which apply pressure directly to the work, are not used when a strong vibration is present. This might cause the clamp to loosen, creating a dangerous condition. Direct-pressure cam clamps must be positioned to resist the natural tendencies of the clamp to shift or move the work when the clamp is engaged. To prevent this movement, the clamp is

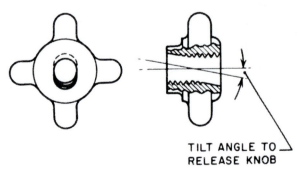

Figure 4–19 Quick-acting knob.

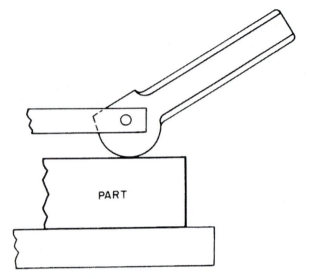

Figure 4–20 Direct-pressure cam clamp.

always positioned so that the work is pushed into the locators when pressure is applied.

Commercial cam clamp assemblies use cam action rather than screw threads to bind strap clamps. This indirect clamping has all the advantages of cam action (Figure 4–21). It also decreases the possibility of loosening or shifting the work when clamping.

Three basic cam types are used for clamping mechanisms: flat eccentric, flat spiral, and cylindrical.

Flat eccentric cams are the easiest to make and can operate in either direction from their center position. The basic eccentric cam locks when the cam reaches its high center position (Figure 4–22). This limits the full lock range to a rather small area. Movement beyond

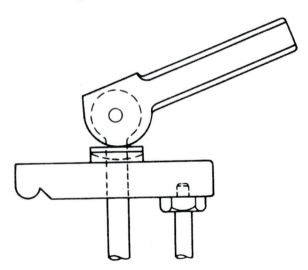

Figure 4–21 Indirect-pressure cam clamp.

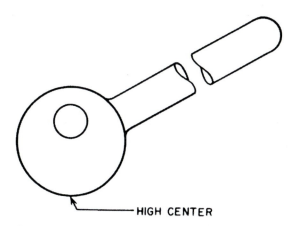

Figure 4-22 Flat eccentric cam.

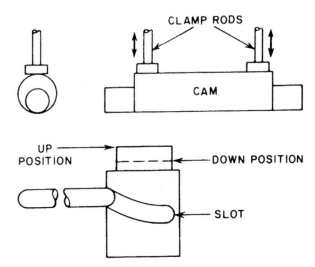

Figure 4-24 Cylindrical cams.

this high center position automatically loosens the clamp. For this reason, eccentric cams do not hold as well as spiral-type cams do.

Flat spiral cams are the most common style of cam clamp used for jigs and fixtures. Commercial cam clamps use the spiral design rather than the eccentric design because of its superior holding properties and wider locking range (Figure 4-23).

Cylindrical cams are also used in many jig and fixture applications. They actuate the clamp by a lobe or through a groove cut into the surface of the cylinder (Figure 4-24). The quick-acting cam clamp is one commercial variation that uses the cylindrical cam principle to combine fast action with positive holding (Figure 4-25).

Wedge Clamps

Wedge clamps apply the basic principle of the inclined plane to hold work in a manner similar to a cam. These clamps are normally found in two general forms, flat wedges and conical wedges.

Flat wedges, or *flat cams*, hold the part by using a binding action between the clamp and a solid portion

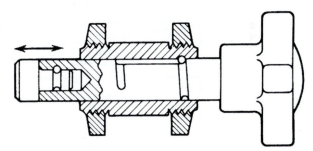

Figure 4-25 Quick-acting cam clamp.

of the tool body (Figure 4-26). Wedges having a slight angle, from 1 to 4 degrees, normally hold the work without additional attachments. This type of wedge is considered to be self-holding. Large-angle, or self-releasing, wedges are used where more movement must be made (Figure 4-27). Since they will not hold by themselves, another device such as a cam or screw must be used to hold them in place.

Conical wedges, or *mandrels*, are used for holding work through a hole (Figure 4-28). Mandrels are available in solid form and expansion form. Solid mandrels are limited in use to one size of hole. Expansion mandrels are made to fit a range of sizes.

Toggle-Action Clamps

Toggle-action clamps, shown in Figure 4-29, are made with four basic clamping actions: hold down, squeeze, pull, and straight-line. Toggle clamps are fast-acting. Because of how they are made, they have the natural ability to move completely free of the work, thus

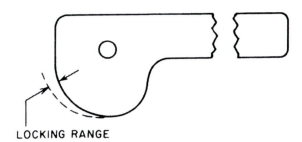

Figure 4-23 Flat spiral cam.

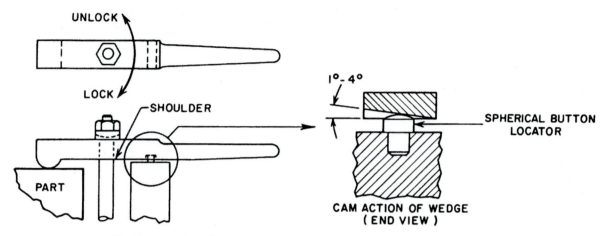

Figure 4–26 Self-holding wedge clamp.

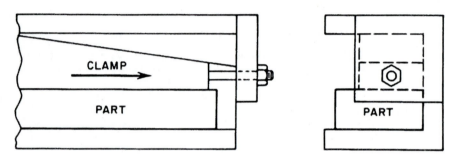

Figure 4–27 Self-releasing wedge clamp.

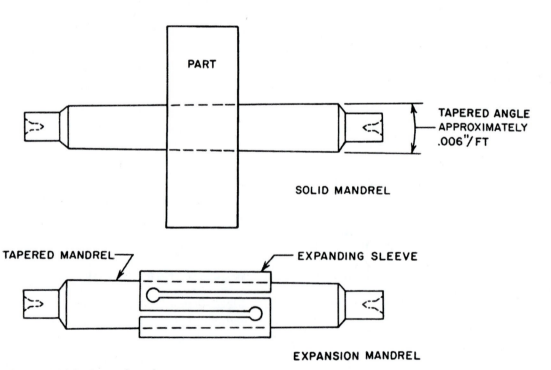

Figure 4–28 Conical wedges.

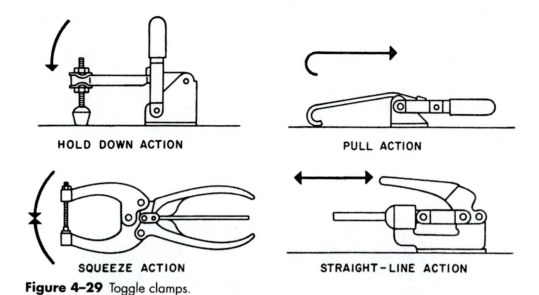

HOLD DOWN ACTION

PULL ACTION

SQUEEZE ACTION

STRAIGHT-LINE ACTION

Figure 4–29 Toggle clamps.

allowing for faster part changes. Another advantage is their high ratio of holding force to application force.

Toggle clamps operate on a system of levers and three pivot points. When the clamp is locked, the pivots are in line, as in Figure 4–30A. When retracted, the pivots and levers are positioned, as in Figure 4–30B. These toggle clamps can be used in a variety of ways with the special mounts, spindles, and handles that are available.

Power Clamping

Power-activated clamps are an alternative to manually operated clamping devices. Power clamping systems normally operate under hydraulic power or pneumatic power, or with an air-to-hydraulic booster. The system used is determined by the type of power supply available. The air-to-hydraulic system is preferred because it can operate from the regular shop line pressure and no pumps or high-pressure valves are needed.

The advantages of power clamps are better control of clamping pressures, less wear on moving parts of the clamp, and faster operating cycles. The main disadvantage is cost, but this is easily offset by increased production speeds and higher efficiency. Typical applications of power clamps are shown in Figure 4–31.

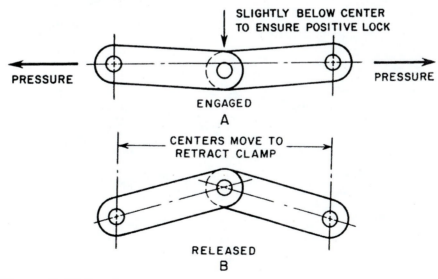

SLIGHTLY BELOW CENTER
TO ENSURE POSITIVE LOCK

PRESSURE ← → PRESSURE

ENGAGED
A

CENTERS MOVE TO
RETRACT CLAMP

RELEASED
B

Figure 4–30 Toggle action.

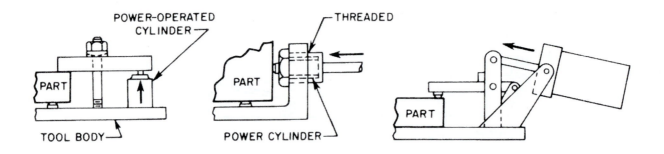

Figure 4-31 Power clamps.

Chucks and Vises

Commercially available chucks and vises offer the tool designer devices that, when modified, greatly reduce tooling costs. Quite often, because of tight budgets, one chuck must be used for several tools. To allow this, the tool designer simply modifies the jaws to suit the job at hand, Figure 4–32. Blank chuck jaws are easily modified to suit practically every clamping need, Figure 4–33. Blank vise jaws can also be modified to suit each job, Figure 4–34.

Using standard chucks and vises for special tools can save the tool designer a great deal of time and money while increasing the efficiency of the job.

NONMECHANICAL CLAMPING

Nonmechanical clamping is a term that is typically applied to the group of workholding devices used to hold parts by means other than direct mechanical

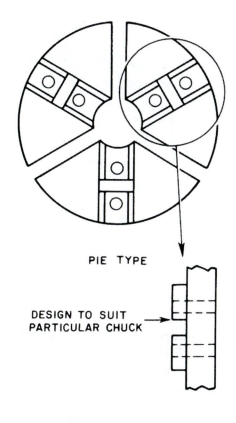

PIE TYPE

DESIGN TO SUIT PARTICULAR CHUCK

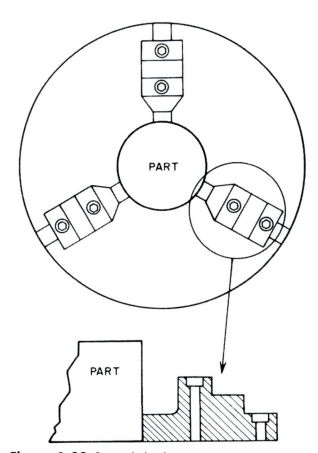

Figure 4–32 Special chuck jaws.

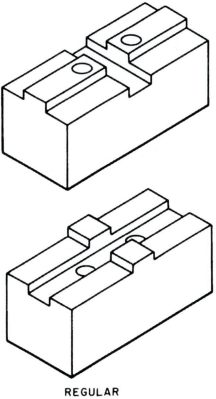

REGULAR

Figure 4–33 Blank chuck jaws.

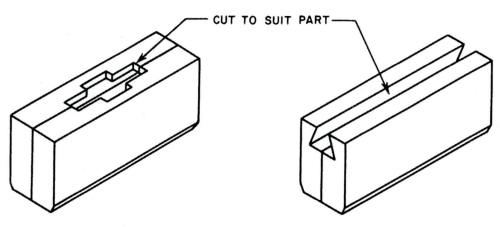

Figure 4–34 Modified vise jaws.

contact. This form of clamping is most often employed for workpieces that cannot be held with other devices due to the size, shape, or configuration of the fixtured parts. Other situations in which non-mechanical clamps are frequently used are where the clamping forces must be applied evenly across the entire part to minimize any possible workpiece distortion. The two principal forms of nonmechanical clamping used for production manufacturing are magnetic clamping and vacuum clamping. *Magnetic chucks* are most often used to hold ferrous metals or workpieces made from other magnetic materials. However, there are also some magnetic chuck setups that can be used for holding nonmagnetic workpieces. When fixturing nonmagnetic materials, such as most nonferrous or nonmetallic materials, mechanical devices like that shown in Figure 4–35, can be used to assist in securely holding the parts.

As shown in Figure 4–36, the two basic styles of magnetic chucks are the electromagnetic chuck and the permanent magnetic chuck. Electromagnetic chucks are electrically operated, while permanent magnetic chucks use a mechanical lever, handle, or other device to activate the magnet. Both types of chucks are available in square, rectangular, or round shapes. While most square and rectangular chucks typically use a parallel pole pattern, round chucks are available with parallel pole, concentric pole, or radial pole patterns (Figure 4–37). Other styles and shapes

of magnetic chucks are also available for specialized applications. Figure 4–38 shows a keyway-milling fixture where a magnetic "V"-block is used to hold a shaft.

A variety of magnetic chuck accessories are also used for fixturing some parts. Some of the more common elements include parallels and "V"-blocks (Figure 4–39). These accessories are not normally magnetic themselves, but are designed to permit the magnetic force from the chuck to be transferred through the accessory into the workpiece.

Another type of auxiliary device frequently used for magnetic chuck setups is packing. While many parts are held in place on the chuck by basic magnetic energy, others may require packing to prevent lateral movement. Packing is generally used when heavy tool forces are expected or where there is a limited contact area between the workpiece and the chuck. This packing can take a variety of forms and, depending on the operation or the workpiece, can range from simple blocks to custom-made elements. As shown in Figure 4–40, the simplest packing uses either standard parallels or pieces of scrap stock positioned around the part. For more complex parts or detailed setups, machined packing elements like those shown in Figure 4–41 can also be made to conform to the complete outer shape of the workpiece.

Vacuum chucks are another style of chuck used to clamp difficult parts. While these chucks can hold

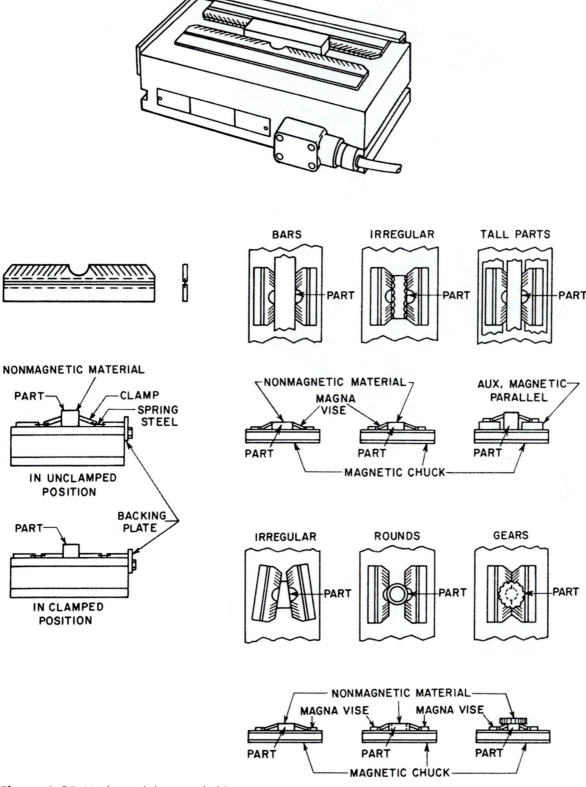

Figure 4–35 Mechanical device to hold nonmagnetic parts.

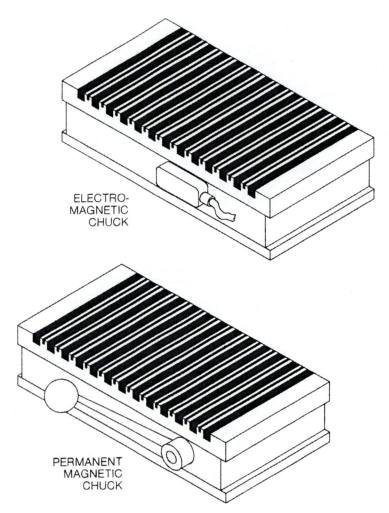

Figure 4–36 Magnetic chucks.

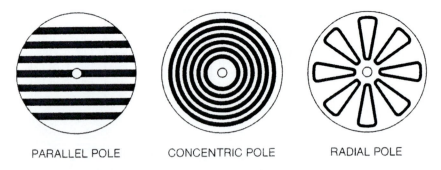

PARALLEL POLE   CONCENTRIC POLE   RADIAL POLE

Figure 4–37 Magnetic pole patterns commonly found in round chucks.

almost any type of nonporous material, they are typically used for nonmagnetic materials or for parts that must be clamped uniformly. Vacuum chucks, like magnetic chucks, equalize the clamping pressure over the entire clamping surface and are suitable for almost every machining operation. In operation the holding forces are generated by a vacuum pump that draws out the air between the chuck face and the workpiece.

Figure 4-38 Magnetic keyway-milling fixture.

Figure 4-39 Parallels and "V"-blocks.

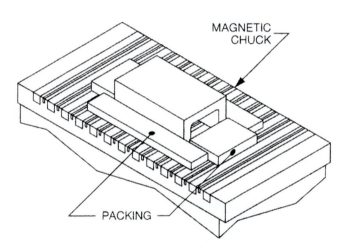

Figure 4-40 Simple packing.

Figure 4-41 Machined packing elements made to conform to the complete outer shape of the workpiece.

A variety of standard top plate designs are available for vacuum chucks. The two common patterns used with most of these chucks are the groove and vein-type and the hole-type. The groove and vein-type plates have a series of grooves cut into the face of the top plate to carry the vacuum forces across the face of the plate. This design is typically used for holding large regularly shaped parts. The hole-type plates are made with a series of holes, rather than grooves to apply the vacuum force to the workpiece.

These holes may have a standard grid pattern or a specialty shaped pattern as required by the shape of the parts.

Most standard vacuum chucks have these patterns across the entire mounting face of the chuck. This usually requires a mask to cover over the grooves or holes that are not directly covered by the mounted workpiece. This mask prevents the vacuum from being lost through any uncovered ports during the clamping cycle. While a variety of different materials may be used to construct these masks, metals or other non-porous materials are the most commonly used.

One style of vacuum chuck that helps eliminate this extra masking step is the Dunham vacuum chuck (Figure 4–42). These chucks operate like other vacuum chucks but rather than having a groove or hole pattern on the chuck face, the chucks are furnished with a series of drilled and tapped holes across the mounting surface of the chuck. The holes act as individual vacuum valves that can be opened or closed by loosening or tightening the valve screws. This permits the vacuum to be applied to only those areas of the chuck that are under the loaded parts.

This design not only simplifies using these chucks, but also makes setups much faster and eliminates the need to fabricate a mask. In addition to the rectangular chuck shape, these are also available in round face chucks. The arrangement shown in Figure 4–43 shows a typical setup where this type vacuum chuck is used for a milling operation.

Another workholding unit, well suited for specialty setups is the Vacumag Plate (Figure 4–44). This

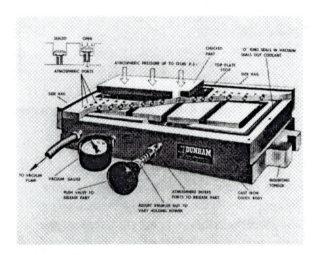

Figure 4–42 Vacuum chuck.

Figure 4–43 Vacuum chuck used for milling operation.

Figure 4–44 A Vacumag Plate.

vacuum chuck accessory is designed to be mounted on a magnetic chuck to hold either ferrous or nonferrous workpieces with vacuum forces. The Vacumag arrangement reduces setup time by allowing magnetic only or vacuum only workholding as needed, without changing or moving either chuck. An additional benefit of this plate design is in addition to the vacuum, the magnetic forces can also be used to securely hold the mask in position.

Most workpieces can be fixtured using one of the standard vacuum chuck designs. However, when vacuum clamping is required and none of the normal chucks is suitable for a particular part, it may be necessary to construct a custom chuck plate. Here, the plates are normally made from either steel or aluminum and simply machined to suit both the chuck and the workpiece. These plates may be made with custom details or special hole patterns as necessary for the parts to be fixtured.

In cases where the mounting surface of the workpiece cannot be easily machined, such as with some castings, a cast chuck plate may be fabricated (Figure 4–45). While these cast plates can be made from a variety of materials, cast epoxy resins are some of the most popular materials for these applications. With these materials the basic shape of the part is cast into the plate and once dried and cured, the epoxy is machined to add the holes, grooves, or other details needed to accommodate the vacuum system. To ensure the proper seal between the workpiece and the chuck, it may also be necessary to add an "O"-ring to act as a seal. One style of available seal that can be easily added to the custom plate is the Flap Seal (Figure 4–46). Flap seals are available as neoprene extrusions and can be easily installed in almost any custom chuck plate.

When neither magnetic chucks nor vacuum chucks will meet the fixturing requirements, another option is the ElectraLock clamp (Figure 4–47). This clamp relies on the actions of a shape memory alloy rather than a conventional nonmechanical power source. The metal actuator used to operate these clamps expands and contracts due to a crystalline phase transformation.

The shape memory actuator technology behind these clamps is based on the phase transformation of a nickel titanium alloy as it moves from martensitic to austenitic phase. This transformation has been refined to produce a trainable, repeatable, and predictable result. The actuator, within the clamp, works when the alloy is electrically heated, expanding the material and releasing clamping pressure.

Figure 4-45 A cast chuck plate.

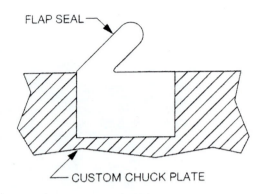

Figure 4-46 A Flap Seal.

Figure 4-47 An ElectraLock clamp.

This new technology makes these clamps ideal for palletized fixturing applications. Using electric power to open the clamp provides reliable mechanical clamping pressure when the clamp is disconnected. In use the clamping power supply is simply plugged into a standard electrical socket. When the fixture must be moved, the power supply is unplugged and the clamps remain locked until reconnected.

SPECIAL CLAMPING OPERATIONS

The clamps discussed so far in this unit are generally used to hold symmetrical shapes or parts that lend themselves to conventional clamping. Some operations present the tool designer with real problems in creative clamping. Clamping odd shapes and multiple clamping are two of these problems.

Clamping Odd Shapes

Several methods can be used to hold odd-shaped work. The best method is to make the clamps and locators conform to the shape of the part. Machining intricate details into the tool body is one way of doing this, but because of high cost, it is rare. The alternative to machining is casting the special shapes. The most popular compounds for casting are epoxy resins and low-melt alloys.

Epoxy resins are useful for casting special vise or chuck jaws. They can be used alone or mixed with a filler material such as metal filings, sand, or ground glass. Epoxy resins are easily shaped by placing the part in a shell filled with the compound (Figure 4–48). A releasing agent is applied to the part so that it can be removed easily once the epoxy has hardened.

Low-melt alloys of bismuth, lead, tin, and antimony are used to pour-case special shapes (Figure 4–49). In this method, the part is suspended in the shell, and the low-melt alloy is poured around the part. Wooden blocks are used as spacers, which are removed before the cast jaws are used.

Multiple Clamping Devices

Many times production operations call for making more than one part at a time. The tool designer must know how to design clamps that are capable of holding several parts.

Designing a clamp to hold more than one workpiece requires imagination. Using the basic ideas and rules for single-part clamping, the tool designer can easily design clamps to hold any number of parts. The main points to remember in multiple clamping are that the clamping pressure must be equal on all parts and that the clamp must have only one operating, or locking, point. Clamps that do not apply equal pressure can damage the parts being machined or create a dangerous situation if a part pops out of the tool during the machining cycle. Using more than one operating point reduces the gains made by using a multiple-type tool. Examples of multiple-type tools are shown in Figure 4–50.

CLAMPING ACCESSORIES

Several commercially made accessories increase both the application and the effectiveness of clamps. Clamp

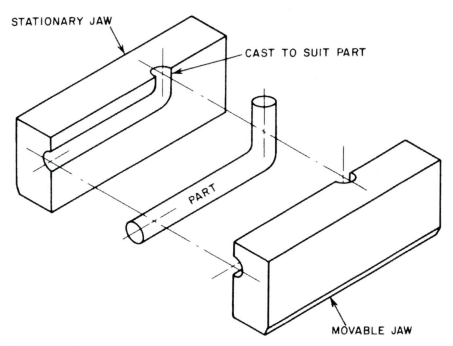

Figure 4–48 Cast-vise jaws of epoxy resin.

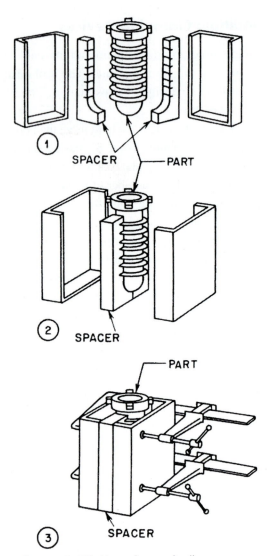

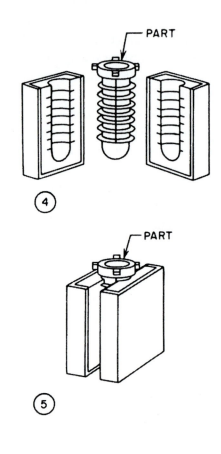

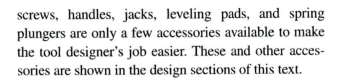

Figure 4-49 Using low-melt alloys.

screws, handles, jacks, leveling pads, and spring plungers are only a few accessories available to make the tool designer's job easier. These and other accessories are shown in the design sections of this text.

SUMMARY

The following important concepts were presented in this unit:

- The term *workholder* is used to identify the complete jig or fixture as well as the devices used to actually hold the part.
- The main purpose of a clamp is to hold the part against the locators throughout the production cycle.

  - Clamps must be strong enough to hold the part and resist movement.
  - Clamps must not damage or deform the part.
  - Clamps must be fast-acting to permit rapid loading and unloading of parts.
- When clamping a part, the designer must keep these major concerns in mind: the position of the clamps, the tool forces acting on the part, and controlling and directing the clamping forces.
- The primary manual clamping devices are strap clamps, screw clamps, cam-action clamps, wedge clamps, and toggle-action clamps.
- Other forms of clamping used for jigs and fixtures include power clamping, chucks and vises,

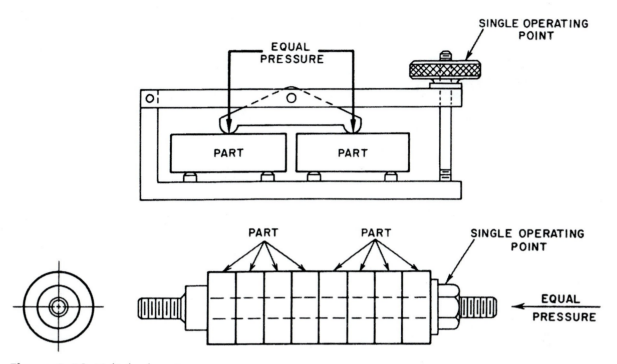

Figure 4–50 Multiple clamping.

and magnetic and vacuum chucks used for nonmechanical clamping.

- Special arrangements may be used for clamping multiple parts or for holding odd-shaped parts. Here cast low-melt alloys and cast epoxy resin tools are very useful.
- Commercial clamping components increase both the application and the effectiveness of the clamping operation.

REVIEW

1. Where should the clamp contact the part?
2. Why must clamped areas have support?
3. What causes tool forces?
4. How can tool forces be used to advantage?
5. How are the type and amount of necessary clamping forces determined?
6. Where should the bulk of the tool thrust be directed?
7. How much clamping force should be used?
8. What should be done if the clamps cannot hold the part?
9. What is the purpose of spherical washers?
10. What determines the allowable force applied to a bolt?

11. Match the letter values to the numbers indicating the characteristic or application of that clamp type.

a. Strap clamp

b. Screw clamp

c. Toggle clamp

d. Cam clamp

e. Power clamp

f. Vacuum chuck

g. Molded clamp

h. Magnetic chuck

i. Wedge clamp

j. Multiple clamp

1. Generally for ferrous metals
2. Uses epoxy or low-melt alloys
3. Either flat or conical
4. Lever action
5. Must have single operating point
6. Pivot and lever action
7. Used only for plastics
8. Best pressure control
9. Could loosen when vibrated
10. Operates on spring pressure
11. Equalizes clamping pressure
12. Uses thread-generated torque

12. Identify the workholding devices shown in Figure 4–51.
- **a.** Strap clamp
- **b.** Spiral-wedge clamp
- **c.** Cam clamp
- **d.** Multiple-part clamp
- **e.** Flat-wedge clamp
- **f.** Screw clamp
- **g.** Toggle clamp
- **h.** Nest clamp
- **i.** Conical-wedge clamp

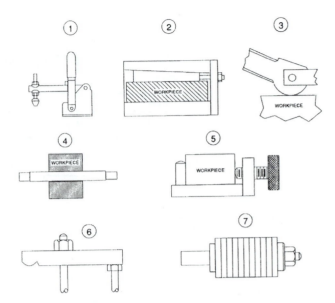

Figure 4–51

UNIT 5

Basic Construction Principles

OBJECTIVES

After completing this unit, the student should be able to:

- Identify the characteristics of tool bodies.
- Identify various drill bushings.
- Describe the proper placement and clearance for drill bushings.
- Identify common jig and fixture hardware.

TOOL BODIES

The tool body provides a rigid base for mounting the locators, supports, clamps, and other details needed to reference, locate, and hold the part while work is performed. The basic size, shape, material, and method used to construct the tool body are determined by the part to be machined.

As a rule, the size and shape of the tool body are determined by the size of the part and the operation to be performed. The choice of material and method depends on economy, required rigidity, accuracy, and projected tool life.

Tool bodies are made in three general forms: cast, welded, and built-up. The materials used for tool bod-ies are steel, cast iron, aluminum, magnesium, epoxy resins, and wood.

Cast Tool Bodies

Cast tool bodies are made of cast iron, cast aluminum, or cast resins. The main advantages of cast bodies include stability, savings in machine time, and good distribution of material. Cast tool bodies are also best for part nesting and offer vibration dampening. The main disadvantage of cast tool bodies is higher cost due to the required pattern and longer *lead time*. Lead time is the time spent between design and fabrication of a part or assembly. The time needed to obtain stock, tool up, and begin actual production is part of the lead time.

Welded Tool Bodies

Welded tool bodies are usually made from steel, aluminum, or magnesium. Their main advantages include high strength and rigidity, design versatility and ease of modification, and short lead time. Their main disad-vantage is the added cost of the secondary machining.

Built-Up Tool Bodies

Built-up tool bodies are the most common form of tool body and can be made from almost any material, such as steel, precast sections, aluminum, magnesium, and wood. The main advantages of built-up tool bodies are adaptability, design versatility, ease of modification, and short lead time. Another important advantage is the use of standard parts—a built-up tool body requires a slightly longer lead time than that used by the welded tool body because of the added time needed to drill and tap the mounting holes and check the fit of the parts before using the tool.

PREFORMED MATERIALS

Preformed materials can greatly reduce the cost of any tool body. Since preformed materials are available in a variety of sizes and shapes, the time required to machine a tool body is also greatly reduced. The most common preformed materials used to construct tool bodies are precision-ground flat stocks, cast-bracket materials, precision-ground drill rods, structural steel sections, and precast tool bodies.

Precision-Ground Flat Stock

Precision-ground flat stock is available in a variety of sizes from .016 inch by .500 inch through 2.000 inch by 4.000 inch, and in lengths up to 36 inches. It is made in three general classes: low-carbon, oil-hardened, and air-hardened tool steel. All precision-ground

flat stock is normally manufactured to a tolerance of ±.001 inch in thickness and width.

Cast-Bracket Materials

Cast-bracket materials are available in various shapes and sizes. They include cast iron, cast aluminum, and cast steel, (Figure 5–1). These sections are usually made in 25-inch lengths; the toolmaker can then cut off the required amount for each job.

Precision-Ground Drill Rod

Precision-ground drill rod is available in different diameters and in 36-inch lengths. The rods are ideal for locators, stops, or other details that require an accurate, round contour. Since drill rod can be hardened, parts can be heat-treated if necessary after fabrication.

Structural Steel Sections

Structural steel sections are normally rolled in lengths of 12 to 20 feet. This gives the tool designer an alternative to higher priced materials. Structural steel sections are not as accurately finished as ground sections. When the tolerance permits, these sections are useful in tool construction (Figure 5–2).

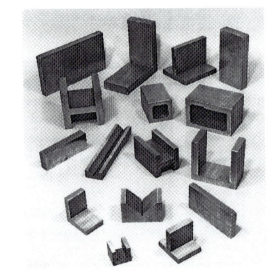

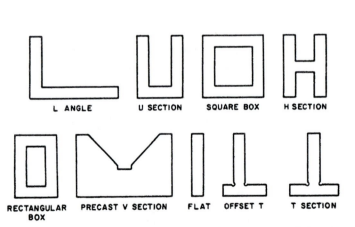

Figure 5–1 Cast-bracket materials (*Photo courtesy of Carr Lane Manufacturing Co.*).

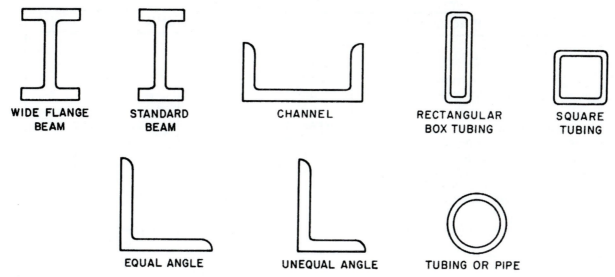

Figure 5–2 Structural steel sections.

Precast Tool Bodies

Precast tool bodies meet many tooling needs (Figure 5–3). They save hours of construction time. The body is already made, so the only requirement is installing the supports, locators, clamps, and bushings.

DRILL BUSHINGS

Drill bushings are used to locate and guide drills, reamers, taps, counterbores, countersinks, spotfacing tools, and any other rotating tools commonly used to make or modify a hole. Figure 5–4 shows the standard

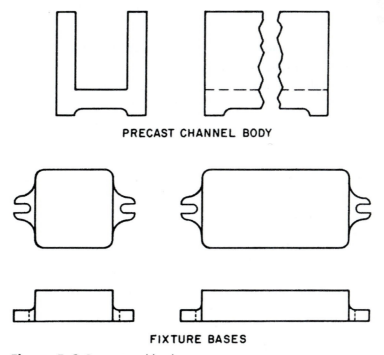

Figure 5–3 Precast tool bodies.

COMMERCIAL SIZE DESIGNATIONS OF DRILL BUSHINGS		
P BUSHING TYPE DESIGNATOR S – SLIP RENEWABLE F – FIXED RENEWABLE L – HEADLESS LINER HL – HEAD TYPE LINER P – HEADLESS PRESS FIT H – HEAD TYPE PRESS FIT U – UNGROUND OUTSIDE DIA (USED WITH TYPE DESIGNATER)	$\frac{1}{4}$ INSIDE DIA. $\frac{3}{4}$ OUTSIDE DIA. EXPRESSED AS DECIMAL- FRACTION-NUMBER OR LETTER SIZE OF NOMINAL DIAMETER	**1"** LENGTH EXPRESSED IN STANDARD INCREMENTS OF $\frac{1}{4}$, $\frac{5}{16}$, $\frac{3}{8}$, $\frac{1}{2}$, $\frac{5}{8}$, $\frac{3}{4}$, 1, $1\frac{1}{4}$, $1\frac{3}{8}$, $1\frac{1}{2}$ ETC. LENGTH IS PROPORTIONAL TO DIAMETERS

ANSI STANDARD DESIGNATIONS OF DRILL BUSHINGS				
2500 INSIDE DIA EXPRESSED AS DECIMAL-FRACTION NUMBER OR LETTER SIZE OF NOMINAL DIAMETER	**P** BUSHING TYPE DESIGNATOR SAME AS COMMERCIAL EXCEPT FOR LOCATION OF "U"	**48** OUTSIDE DIA EXPRESSED IN $\frac{1}{64}$ INCREMENTS $\frac{3}{4}$ = $\frac{48}{64}$ OR **48**	**16** LENGTH EXPRESSED IN $\frac{1}{16}$ INCREMENTS 1" = $\frac{16}{16}$ OR **16**	**U** ADD HERE FOR UNFINISHED OUTSIDE DIA OMIT FOR FINISHED OUTSIDE DIA

Figure 5–4 Designing sizes for drill bushings.

method for designating the size of drill bushings. Drill bushings are usually hardened and ground to exact sizes to ensure the needed repeatability in the jig.

Types of Bushings

The most common drill bushings are renewable bushings, press-fit bushings, and liner bushings. There are also bushings for special operations.

Renewable bushings are commonly divided into two groups, slip and fixed. They are used where bushings are changed many times during the jig life.

Slip-renewable bushings are used when more than one operation is performed in the same location, such as drilling and reaming (Figure 5–5). They are used with a liner bushing and are held in place by the radial lock and the bushing head. When another size bushing is required for a hole, the first bushing is

removed by turning it counterclockwise and lifting. The new bushing is installed by placing it in the hole with the radial lock aligned and turning it clockwise, Figure 5–6.

Fixed-renewable bushings are used where only one operation is performed in each hole but where several bushings must be used during the life of the tool (Figure 5–7). Fixed-renewable bushings are also fit into a liner. They are held by a mechanical clamp and take considerably longer to remove than slip-

Figure 5–5 Slip-renewable bushing.

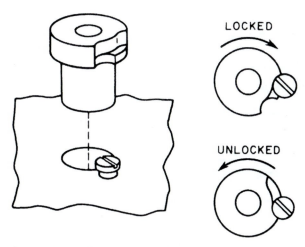

Figure 5-9 Press-fit bushings.

Figure 5-6 Slip-renewable bushing.

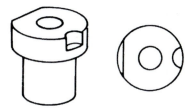

Figure 5-10 Liner bushings.

Figure 5-7 Fixed-renewable bushing.

They are used to provide a hardened hole where renewable bushings can be located. Since the liner bushing is hardened, there is little chance of affecting the accuracy of the tool by changing the bushings. Several variations of these bushings are available for special applications.

Special-purpose bushings allow for special jobs. The great diversity of drilling operations common to manufacturing demands a variety of special bushings (Figure 5–11).

Template bushings are used for installations in thin jig plates. This type of bushing is pressed into the hole, and a lock ring is installed on the opposite side (Figure 5–12).

Oil-groove bushings permit positive and complete lubrication of the bushing for continuous high-speed drilling operations (Figure 5–13).

Knurled and *serrated bushings,* shown in Figure 5–14, are made with a knurled outer surface. Although both styles are knurled, the terms *serrated* and *knurled* have different meanings. Knurled bushings are made with a diamond pattern knurl, while serrated bushings have a straight knurl pattern.

Knurled bushings, shown at view A, are used for applications where the bushing is cast, potted, or embedded into a jig plate. The diamond pattern of this knurl, combined with the grooves around the bushing, afford the maximum gripping area for the bushing. Depending on the application, these bushings are normally cast into a jig plate with epoxy resins, low-melt alloys, or a similar castable material.

renewable bushings take. The most common types of fixed-renewable clamps are shown in Figure 5–8, along with the matching bushing head type.

Press-fit bushings are made in two general forms, head or headless. They are intended for use in limited production tooling, where no bushing change is required (Figure 5–9). Since press-fit bushings are pressed directly into the jig plate, repeated changes can affect their accuracy.

Liner bushings are available in head or headless types and are pressed into the jig plate (Figure 5–10).

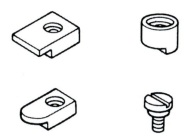

Figure 5-8 Common types of fixed-renewable bushing clamps.

Figure 5–11 Special-purpose bushings.

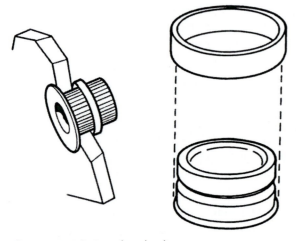

Figure 5–12 Template bushing.

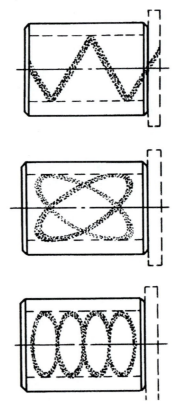

Figure 5–13 Oil-groove bushings.

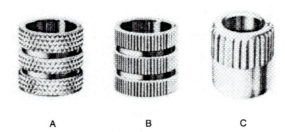

A   B   C

Figure 5–14 Serrated and knurled bushings.

Serrated bushings, shown at views B and C, are also used for applications where the bushings are cast into the jig plate. This is especially true with the bushing design shown at view B. Here the grooved body helps to maintain the grip of the bushings. However, in addition to cast-in applications, the bushing style shown at view C may also be used for press-fit installations in jig plates made of soft materials, such as aluminum or magnesium alloys. For press-fit installations, the straight knurl design of the serrated bushing is usually preferred over the knurled bushing's diamond pattern. This is due mainly to the broaching, or cutting, action of the diamond pattern as it is pressed into the mounting hole. The serrated bushing displaces the material in the hole and yields a tighter fit and a more secure mount for the bushing.

Extended-range bushings are used where regular bushings are too short to properly support or guide the tool (Figure 5–15). Standard bushings are available in sizes up to 5.000 inches (175 mm) long.

Carbide bushings are also available in most sizes for extended service in high-speed production. The additional cost of carbide bushings is justified for shorter runs where the material being machined is very abrasive.

Installing Drill Bushings

Correctly fitting drill bushings are important in jig work. If the drill bushings are not properly installed, they could fall out during use or could bind and break the tool. The correct method of installation is shown in Figure 5–16.

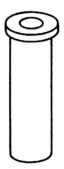

Figure 5–15 Extended-range bushing.

Proper sizing of mounting holes also includes roundness. Holes designed to receive drill bushings must be made under size and must be perfectly round to allow the bushing to fit correctly.

Jig Plates

A *jig plate* is the part of a drill jig that holds and positions the drill bushings. The thickness of the jig plate is normally dependent on the size of the bushings used. As a rule, the bushing should be long enough to support and guide the tool properly. A length between one and two times the tool diameter is usually sufficient to prevent inaccuracy (Figure 5–17). The wall thickness of the bushing should be able to easily withstand all the cutting forces and maintain tool accuracy.

Bushing Clearance

For most applications, the end of the bushing should not touch the work. A clearance of one to one and a half times the tool diameter is sufficient for the required chip clearance (Figure 5–18). Exceptions to this rule occur when extreme accuracy is called for, when accuracy in secondary operations is necessary, and when one is drilling into irregular surfaces. In these cases, the bushing should be as close to the work as possible to permit the required precision (Figure 5–19). Occasionally a bushing must be altered, as shown in the example with the inclined and curved surfaces. Here it is important to remember that if the end of the bushing is modified, sufficient bearing area for the drill must be maintained inside the bushing. For this

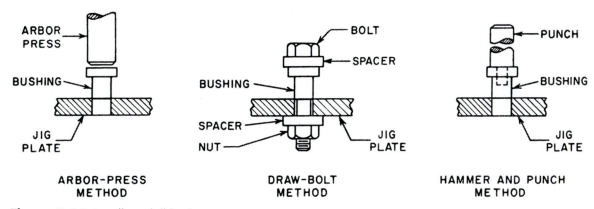

Figure 5–16 Installing drill bushings.

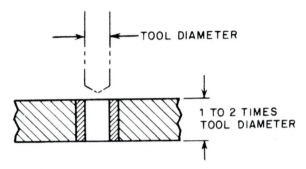

Figure 5-17 Jig plate.

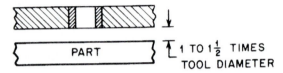

Figure 5-18 Bushing clearance.

reason, it is always best to use a straight bushing having the same inside diameter all the way through. Do not use counterbored bushings for this type of application.

Proper bushing clearance is important to the overall function of any jig. The chips cause the bushing to wear rapidly if the bushings are unnecessarily close. If they are placed too far away, precision is lost.

Burr Clearance

When installing bushings, another important factor to remember is burr clearance. In any drilling operation, two burrs are produced, primary and secondary (Figure 5–20). The *primary burr* is made on the side opposite the drill bushing; the *secondary burr* is produced at the point where the drill enters the work. These burrs must be considered and sufficient clearance must be provided.

Another problem facing the tool designer is placing bushings for holes that are close together (Figure 5–21). In these cases, thin-wall bushings can sometimes be used. Figure 5–22 shows another method of grinding flats on adjacent bushings to allow for clearance. When necessary, holes can be drilled and reamed and the bushing can be alternated from one hole to the other (Figure 5–23).

SET BLOCKS

Setting the cutters for fixtures requires a different method from that for jigs. Set blocks and feeler gauges are used to set the relationship between the

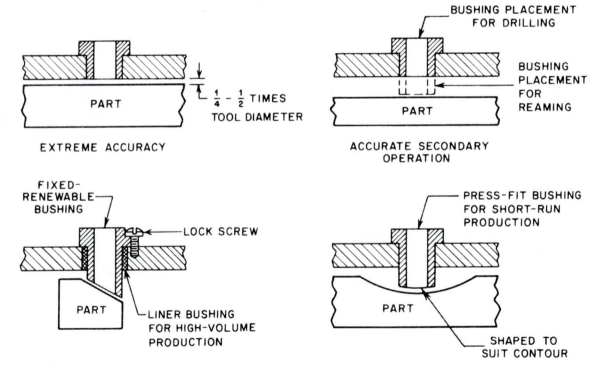

Figure 5-19 Bushing clearance in special cases.

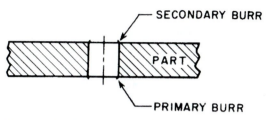

Figure 5-20 Primary and secondary burrs.

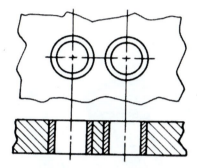

Figure 5-21 Placing bushings for holes.

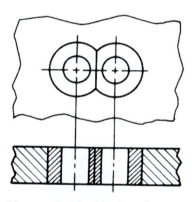

Figure 5-22 Grinding flats on adjacent bushings.

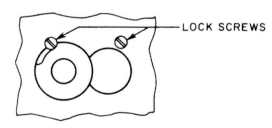

Figure 5-23 Drilling and reaming holes.

work and the tool for machining operations such as milling, turning, and grinding.

Set blocks, also known as setup gauges, are normally located directly on the fixture. The surface used to reference the cutter is controlled by the type of operation being performed.

Using feeler gauges to ensure the proper clearance prevents damage and wear to the set block when setting the cutter. Typical uses of set blocks are illustrated in Figure 5–24. One point to keep in mind when designing set blocks is the allowance for the feeler gauge. They should be made thick enough to resist bending or warping during use. Stock thicknesses between .030 and .100 inch, or larger, prevent bending and are easily mounted on the tool. Another convenience is to etch the size of the feeler and the tool part number directly on the feeler gauge. Then the operator knows the right gauge is being used to set the cutters.

FASTENING DEVICES

Many types of fasteners are used in building jigs and fixtures. Screws, nuts, bolts, washers, lock rings, keys, and pins are all used to make tools. A point to remember when designing tools is to use standard hardware. Special bolts and fasteners cost more money and add little to the tool value.

Cap Screws

The *socket-head cap screw* is the most common type of screw used in jig and fixture work. This screw provides superior holding power with easy installation and minimal space requirements (Figure 5–25).

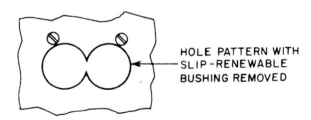

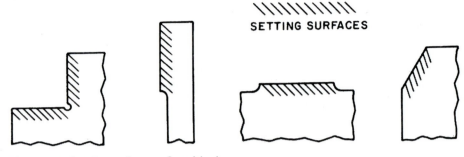

SETTING SURFACES

Figure 5–24 Typical uses of set blocks.

Figure 5–25 Cap screw.

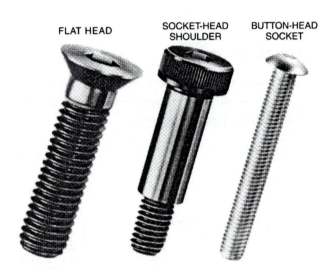

FLAT HEAD   SOCKET-HEAD   BUTTON-HEAD
                 SHOULDER   SOCKET

Figure 5–26 Various cap screws.

Variations of this screw form are shown in Figure 5–26. These screws are available in plain or self-locking styles.

Set Screws

The set screw is another type of screw widely used in jig and fixture work. Standard set screws are available in many sizes and point styles (Figure 5–27 and Figure 5–28).

Thread Inserts

Thread inserts provide renewable threaded holes in materials that cannot normally hold threads, such as epoxy resins or soft aluminum (Figure 5–29). Thread inserts also allow replacement of threaded holes in other materials that could wear because of heavy or prolonged use.

Figure 5–27 Standard set screws.

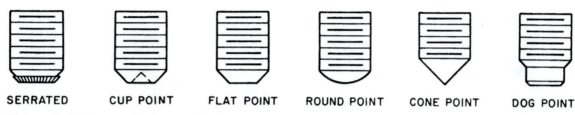

SERRATED   CUP POINT   FLAT POINT   ROUND POINT   CONE POINT   DOG POINT

Figure 5–28 Point styles for standard set screws.

Figure 5–29 Threaded insert.

Nuts and Washers

A variety of nuts and washers are also commercially available to assist the tool designer. The most common types are shown in Figure 5–30.

Special-Purpose Bolts and Nuts

Jigs and fixtures, while normally designed around standard hardware items, do occasionally require special T-bolts, slot nuts, and studs, which are commercially available. T-bolts, slot nuts, and studs are used primarily to hold special tools to machine tables. They are available in a variety of sizes to fit most machine tables (Figure 5–31).

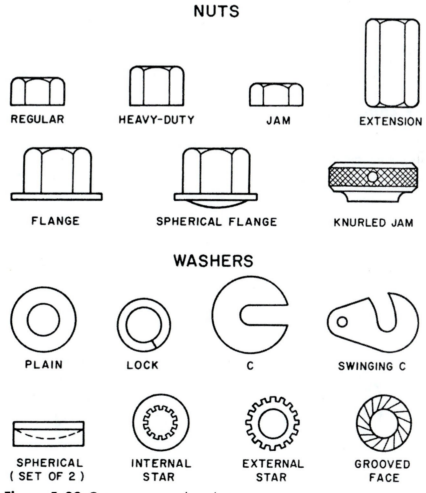

NUTS

REGULAR   HEAVY-DUTY   JAM   EXTENSION

FLANGE   SPHERICAL FLANGE   KNURLED JAM

WASHERS

PLAIN   LOCK   C   SWINGING C

SPHERICAL (SET OF 2)   INTERNAL STAR   EXTERNAL STAR   GROOVED FACE

Figure 5–30 Common nuts and washers.

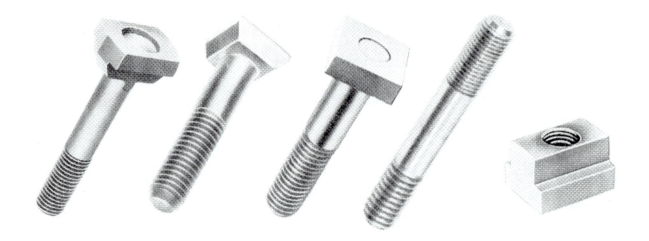

Figure 5-31 Special-purpose bolts and nuts.

Retaining Rings

Another type of fastener that can save many hours of work is the retaining ring, which is available in internal and external styles. When properly installed, these rings provide sufficient strength for most tooling applications (Figure 5-32).

Interchangeable Fixture Keys

The *interchangeable fixture key* is a special device that can save countless hours of machining time. This key is used to precisely locate the base of the tool in the table "T"-slot.

The conventional method of installing standard fixture keys requires milling a groove in the tool base and fitting rectangular keys to the slot. The keys are then held in place with cap screws (Figure 5-33). Using the interchangeable key shown in Figure 5-34 requires only two holes to be drilled and reamed. Then the key is placed in the hole and locked into position with a hex wrench.

Dowel Pins

Dowel pins are normally used with screws to keep mated parts aligned. The five most common dowel pins are plain, tapered, pull, grooved, and spring (Figure 5-35).

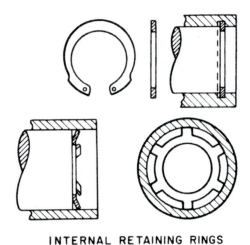

INTERNAL RETAINING RINGS

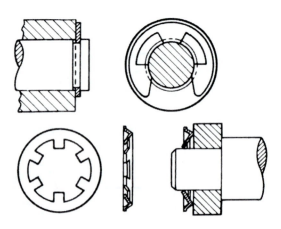

EXTERNAL RETAINING RINGS

Figure 5-32 Retaining rings.

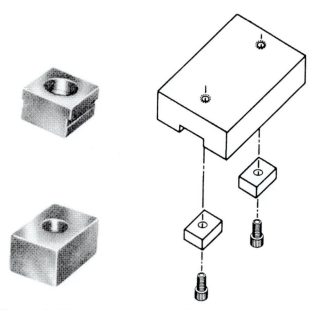

Figure 5–33 Conventional installation of standard fixture key.

The tapered dowel is self-holding. Some tapered dowels have threaded portions that aid in installing or removing the pins (Figure 5–36).

Pull dowels are used in blind holes where frequent disassembly is needed. Two types of pull dowel are shown in Figure 5–37 and Figure 5–38.

Plain, grooved, and spring dowels are used basically in the same manner. The difference among the three is the degree of precision they require. Plain dowels require an accurately drilled and reamed hole for installation. When possible, the use of groove and spring dowels should be specified where extreme precision is not necessary. This saves machining time and money.

Jig Pins

Jig pin is a term used to describe the general family of locating, alignment, and clamping pins widely used for

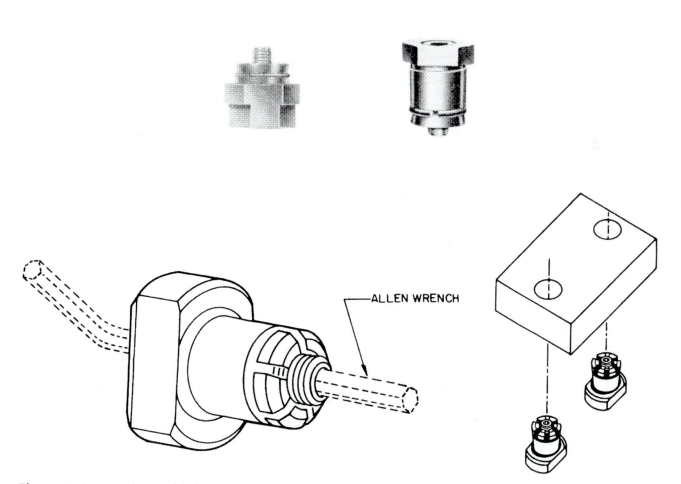

Figure 5–34 Interchangeable fixture key.

Figure 5-35 Common dowel pins.

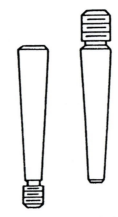

Figure 5-36 Tapered dowels with threaded portions.

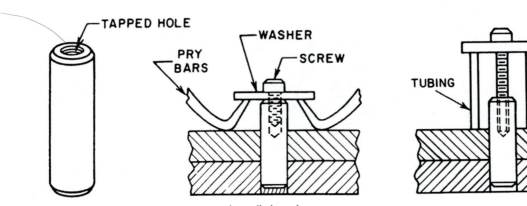

Figure 5-37 Removing a straight pull dowel.

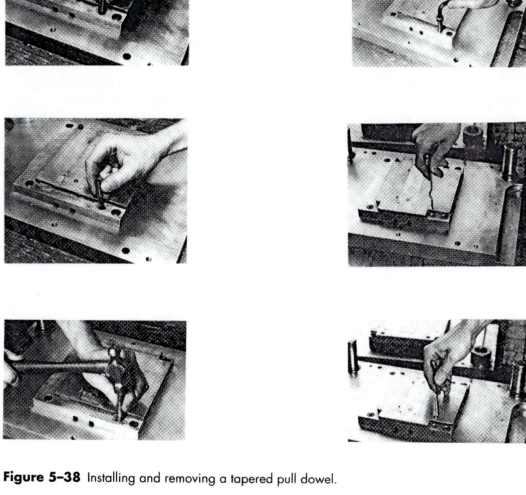

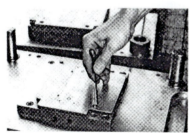

Figure 5–38 Installing and removing a tapered pull dowel.

tooling applications. When building any jig or fixture, properly locating, positioning, and aligning the workpiece or the various elements of the tool are primary considerations. Likewise, any tool with removable elements can present a variety of locating requirements. In fact, ensuring that the removable elements are correctly installed each time the tool is assembled can be just as important as locating the workpiece.

To simplify jig and fixture construction, there are a wide range of commercially available locating devices made for these operations. Most of these locators are intended for very specific types of workpieces or locating processes. However, jig pins are one general category of locator that are well suited for a range of general locating tasks. Although jig pins are commercially available in a wide array of styles and types, most can be classified as simple alignment pins or locking/clamping pins.

Plain alignment pins, as shown in Figure 4–39, are the simplest form of jig pins. The major styles of plain alignment pins are "L" pins, "T" pins, jig pins, and shoulder pins. These pins are made with a precise locating diameter to ensure proper alignment. Typically the locating diameter on these pins is held to within –.005" of the normal size. Each pin size is made with either a bullet or a tapered nose for easy referencing and insertion in the mounting holes.

The "L" pins and "T" pins are almost identical, the only difference being the shape of the handle. Jig pins are made with a shoulder to keep the handle raised above the mounting surface. Depending on the manufacturer, these may be made with an "L," a "T," or a sliding handle. Shoulder pins, like jig pins are made with a shoulder. However, the shoulder pin design typically has a longer shoulder and is often made without a handle. Here the hole in the end of the pin is designed to attach the pin to the workholder with a cable.

The specific mount used for any alignment pin is determined by the application. In some cases, alignment pins are installed in drilled and reamed holes. For more precision or for longer production runs, they may also be installed in hardened bushings. Slotted locator bushings, as shown in Figure 5–40, are often used when the workpiece holes are slightly misaligned. These bushings are generally used in combination with standard round bushings. The design of these slotted bushings permit up to .12" of movement in only one direction. Slotted locator bushings are installed and aligned in the mounting plate with a dowel pin (Figure 5–41). These bushings are available for either press-fit installations or with a knurled outside diameter for cast-in-place applications. Here the slotted bushings are cast into the mounting plate with either a plastic epoxy compound or a low-melt alloy.

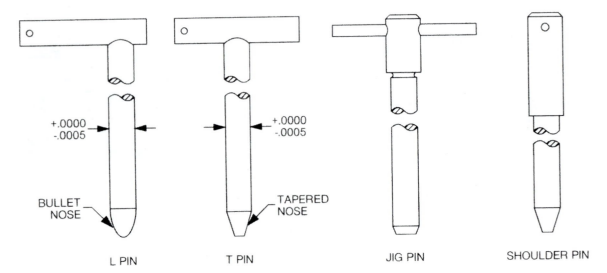

Figure 5–39 Plain alignment pins.

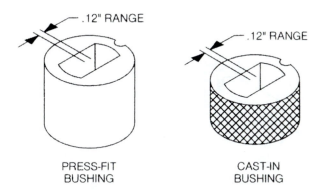

Figure 5-40 Slotted locator bushings.

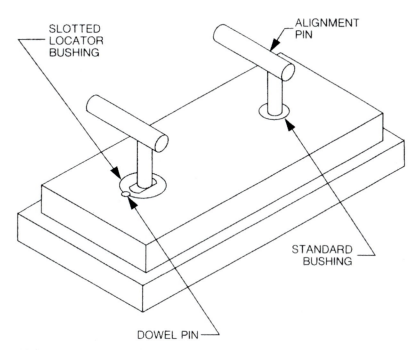

Figure 5-41 Slotted locator bushings are installed and aligned in the mounting plate with a dowel pin.

A recurring problem with any jig pin is keeping the pins and the workholder together. One method frequently used to keep the pins attached to the workholder is a cable assembly (Figure 5–42). These cables are commercially available in several lengths to suit their specific application. Likewise, different styles are available to suit the alignment pins and similar devices. Although often used to attach jig pins, other removable items, such as drill bushings or cutter-setting devices may also be attached to the tool with these cables.

Occasionally, with some tool designs, using a cable assembly may be objectionable. Depending on the application, these cables can sometimes get in the way or make loading and unloading the workpieces difficult. Here another method to keep the jig pin with the tool may be used.

The self-locking style of the jig pin, as shown in Figure 5–43, is a self-contained assembly that permits the jig pin to function as intended while staying firmly attached to the tool. As shown, these locking alignment pins are a variation of the plain "L" and

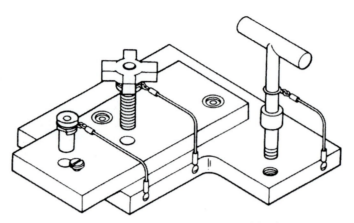

Figure 5–42 Cables attaching pins to tool body.

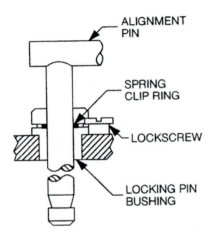

Figure 5–43 A self-locking jig pin.

"T" pin designs. The pins are furnished with a locking pin bushing. These bushings are mounted in a hole and attached to the workholder with a lockscrew. A spring clip ring in the locking pin bushing holds the alignment pin at any point along its travel by applying spring pressure on the pin. This spring clip also prevents the pin from falling out by engaging the groove end when the pin is in its fully retracted position. The pin stays in place and will not fall out even when the tool is turned upside down.

In situations in which the pins are removed from the tool but also need a self-locking feature, the quick-release style of alignment pin may be the best choice. Quick-release alignment pins have an integral self-locking element. This locking mechanism retains the pin in the hole and prevents the pin from falling out as the workholder is moved around. However, when necessary, the pins can also be removed completely from the tool body.

These quick-release alignment pins are available in several styles to suit many different applications. The simplest quick-release pin design is the detent pin (Figure 5–44). This pin has a simple spring-loaded dual ball arrangement. In use, the pin is simply pressed in and pulled out of the hole. An internal spring applies a light spring force to extend the balls. As the pin is inserted or removed, the spring collapses as the pin is moved through the mounting hole. Then, as the end of the pin exits the hole, the spring returns the locking balls to their extended position.

Another variation of this style alignment pin is the ball lock pin. As shown in Figure 5–45, this group of alignment pins includes the single-acting, double-acting, and adjustable ball lock pins. These pins are similar to the detent type, except that a ball-locking mechanism is used in place of a simple spring to extend and lock the balls. This mechanism ensures a positive lock and prevents the balls from retracting into the pin body. In use, these pins are installed and removed by pushing a button or pulling a handle to disconnect the ball-locking mechanism. The single-acting ball lock pins have a single release positions, while the double-acting ball lock pins have a fixed grip length.

For applications where the grip length of the pin must be controlled, the adjustable ball lock pin may be used. As shown in Figure 5–46, the grip length of

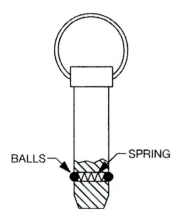

Figure 5-44 A detent pin.

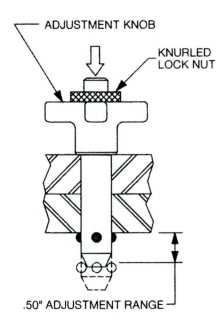

Figure 5-46 An adjustable ball lock pin.

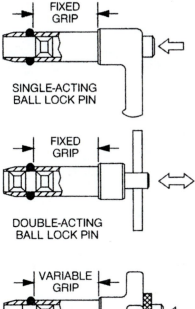

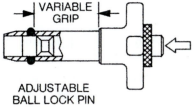

Figure 5-45 Single-acting, double-acting, and adjustable ball lock pins.

this style pin is variable and is easily adjusted by turning the adjusting knob. This design allows the adjustable ball lock pin to be used for setups where an odd-sized grip length is needed or where the pin must also act as a clamping device. To preset the grip length, simply turning the adjusting knob to the desired grip clamping device. Once the proper grip length is set, the position is fixed by turning the knurled lock nut against the adjusting knob. To use the adjustable lock pin as a clamp, simply depress the button and inset the pin in the mounting hole. Then

turn the adjusting knob to clamp the workpiece between the knob and the balls. These adjustable ball lock pins have a .50" grip length adjustment range.

Another quick-release alignment pin design intended for specialized applications is the expanding pin (Figure 5–47). The expanding pin is used for those applications where the pin diameter rather than the pin length must be variable. When engaged, the diameter of this pin expands up to .006" to lock the pin in place against the sides of the hole. As shown, these expanding pins are engaged and released with a cam lever.

Lifting Devices

Designing any jig or fixture involves many design choices and decisions. Often designers can become caught up in the design specifications and overlook the general requirements of the tool. Providing appropriate devices for handling and lifting these tools, for example, are important considerations that

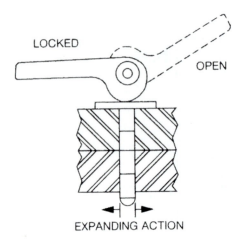

Figure 5–47 An expanding pin.

should not be neglected. Although lifting devices are usually included with larger or heavier jigs and fixtures, even smaller tools should be equipped with elements to make handling the tool easier and safer.

As a general rule, mechanical lifting devices should always be used for tools weighing over 25 to 35 pounds. Lighter tools may not require special lifting devices, but they should nevertheless be equipped with handles or handholds to make moving and positioning the jig or fixture safer. Even small tools are difficult to pick up without a secure gripping device. Without such a device, the operator is likely to grab whatever is handy to pick up the tool. Here the tool could easily slip and hurt the operator or damage the tool if it were dropped.

The handles selected should be strong enough to support the weight of the tool and large enough to provide a safe and secure grip. The handles must also be positioned at the natural balance points of the tool. Most small tools require only two handles, with one positioned on each side of the tool body. If more than two handles are needed, another type of lifting device is probably more appropriate.

The simplest handling device or small tool is the machined handhold (Figure 5–48). These handholds are usually cut directly into the base, as part of the machining operations of the tool body. When machining these handholds, the slot should be deep enough to offer a good grip and have a rounded bottom. The handholds should also have rounded or chamfered

edges to make using the handholds more comfortable. If installed handles are needed, either "D"-shaped or straight handles, as shown in Figure 5–49, may be used, depending on the type and size jig or fixture. In either case, the handles should always be placed high enough on the tool body to eliminate any chance of pinched fingers when placing the tool on a machine table.

Commercial hoist rings are another style of lifting device that is both convenient and efficient. They are also the safest handling devices for most lifting operations. However, when lifting light to medium loads, eyebolts may be a cost-effective alternative to commercial hoist rings.

When using eyebolts, a few general safety points must be considered. The lifting angle should always be as great as possible when using eyebolts (Figure 5–50). Standard eyebolts are most often solid units designed to be used under a tensile, or pulling, load. Reducing this lifting angle will often shift the load on the eyebolt. If the load is shifted too far, side-loading forces rather than tensile forces are applied. If the side-loading forces are excessive, they may cause the eyebolt to fail. To reduce any possibility of failure, both the lifting angle and the weight of the tool are important considerations. If there is any doubt about the suitability of eyebolts for any lifting task, commercial hoist rings should be used instead.

Most hoist rings are made with a swiveling lifting ring and are attached to the workholder with a screw. This swivel-type lifting ring design allows the hoist ring to maintain a tensile load on the screw while compensating for different lifting angles. Fixed mounted hoist rings are commonly used for applications where a fixed relationship between the hoist rings and the workholder is desired. This style of hoist ring is attached directly to the tool with socket-head cap screws (Figure 5–51). These hoist rings are quite compact, and the low-profile design permits them to be installed in areas where hoist rings would be too large. Although the base of the hoist rings is mounted in a fixed position, the lifting ring is free to pivot 180 degrees to align the ring to the other hoist rings and to the lifting forces.

Swivel-mounted style hoist rings are often used for heavier lifting applications where additional

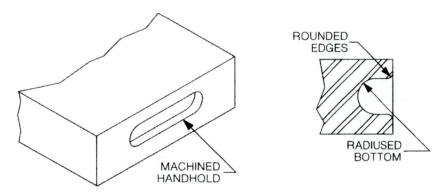

Figure 5–48 A machined handhold.

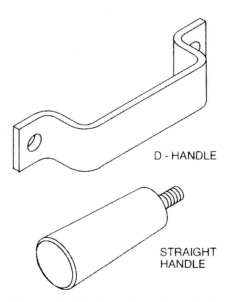

Figure 5–49 A machined handhold.

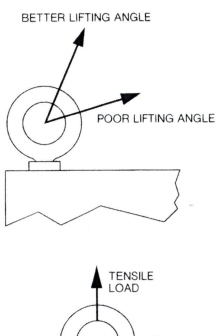

Figure 5–50 Lifting angle when using eyebolts.

movement of the hoist ring is needed. As shown in Figure 5–52, these hoist rings are mounted with a single, center-mounted socket-head cap screw. With this style of hoist ring, both the mounting base and the lifting ring are free to move. Here the lifting ring can pivot 180 degrees and the complete unit can rotate 360 degrees about the mounting screw. Lifting pins, as shown in Figure 5–53, are a variation of the standard swivel-type hoist ring. These hoist rings are well suited for medium–heavy loads where their quick-release feature is desired. These pins are mounted in a hole and have a four-ball plunger arrangement for both holding the pins in place and lifting. A quick-release button at the end of the pin makes installation

and removal quick and easy. Here the button is pushed to depress the locking balls. This allows the pin to be easily inserted or removed as necessary.

Numerous additional accessories are available to assist the tool designer in designing jigs and fixtures.

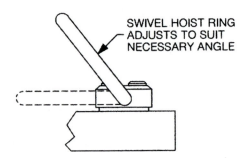

SWIVEL HOIST RING

SWIVEL HOIST RING ADJUSTS TO SUIT NECESSARY ANGLE

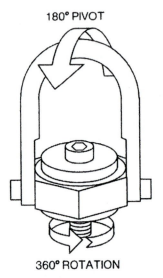

180° PIVOT

360° ROTATION

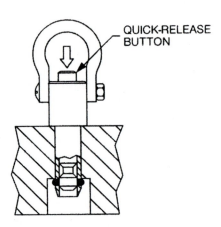

QUICK-RELEASE BUTTON

Figure 5–51 A swivel-type lifting ring design allows the hoist ring to maintain a tensile load on the screw while compensating for different lifting angles.

Figure 5–52 Swivel-mounted style hoist rings are mounted with a single, center-mounted socket-head cap screw.

Figure 5–53 Lifting pins are a variation of the standard swivel-type hoist ring.

Several accessories are used throughout the design section of this text to illustrate proper application.

SUMMARY

The following important concepts were presented in this unit:

- The three most common tool body forms are cast, welded, and built-up.
- Using preformed materials can save time and money in building jigs and fixtures.
- Several types of drill bushing are used for jigs. The most common types are renewable, press-fit, and liner bushings.
  - For the jig to perform as desired, the correct bushing must be selected.
  - To work properly, the bushing must be installed correctly.
- Set blocks are used to reference the cutter for operations with fixtures.

- The proper fasteners must also be selected and installed correctly to ensure a well-built and sturdy jig or fixture.
- Jig pins are the family of locating, alignment, and clamping pins widely used for tooling applications.
- Appropriate lifting elements and mechanical lifting devices must be used for handling tools weighing over 25 to 30 pounds.

REVIEW

1. Match the tool body type to the characteristic that best describes it: 1—cast; 2—welded; 3—built-up.

   a. Easiest to modify
   b. High strength and rigidity
   c. Could warp during fabrication
   d. Requires a pattern
   h. Longest lead time
   i. Could use wood for construction
   j. Often requires secondary machining

e. Best for vibration dampening

f. Epoxy resin can be used

g. Assembled with screws and dowels

k. Shortest time required to make

l. Could loosen during use

2. Identify the bushings shown in Figure 5–54.

a. Head-type liner bushing

b. Headless press-fit bushing

c. Slip-renewable bushing

d. Fixed-renewable bushing

e. Headless liner bushing

f. Head-type press-fit bushing

g. Serrated bushing

h. Extended-range bushing

i. Knurled bushing

j. Template bushing

3. How long should a bushing be?

4. How thick should the bushing wall be?

5. How much space should there be between the bushing and the workpiece for extremely accurate drilling?

6. What is the standard distance between the work and the bushing for general applications?

7. What happens if the bushing is too close to the workpiece?

8. What happens if the bushing is too far from the workpiece?

9. What clearance must the tool designer consider when placing bushings close to the work surface?

10. List two methods of using bushings in holes that are close together.

11. What are set blocks used for?

12. What controls the use of set blocks?

13. What additional tool must be used with set blocks? Why?

14. What size feeler gauge should be used for the best results?

15. What information should be etched on the feeler gauge?

16. What term describes the general family of locating, alignment, and clamping pins?

17. What device is used to ensure that alignment pins will fit inslightly misaligned holes?

18. The single-acting, double-acting, and adjustable are all styles of which quick-release alignment pin?

19. Mechanical lifting devices should always be used for tools that weight how much?

Figure 5–54

20. If there is any doubt about the suitability of eyebolts, what should be used for lifting a tool?

21. Answer the following questions with "true" or "false."

   A. Hex-head cap screws are the most commonly used screws for jig and fixture work.

   B. When possible, special bolts should be used to make the tool operate more efficiently.

   C. Shoulder screws are a common variation of the socket-head cap screw used for jig and fixture work.

   D. Interchangeable keys usually take more time to install than do standard plain keys.

   E. T-bolts, slot nuts, and studs are used to hold jigs and fixtures to machine tables.

   F. Jig pins are more accurate than plain dowels in aligning parts of a tool.

   G. Pull dowels are intended for use in blind holes.

   H. The five most common dowels are plain, tapered, pull, spring, and grooved.

   I. Cables keep loose parts attached to the tool to prevent loss or damage.

Considerations of Design Economics

UNIT 6

Design Economics

OBJECTIVES

After completing this unit, the student should be able to:

- Identify and define the principles of design economy.
- Complete an economic analysis of a tool design.

DESIGN ECONOMY

The demands of modern industry for maximum productivity at minimal cost are a challenge to the tool designer. In addition to developing designs for efficient and accurate jigs and fixtures, the tool designer is responsible for finding ways to keep the cost of special tools as low as possible. To do this, he must know and apply design economy.

Design economy begins with the tool designer's ideas and is carried through to the completion of the tool. Design details should be carefully studied to find ways to reduce costs and still maintain part quality. The tool designer is aided in this task by following the principles of economic design.

Simplicity

Simplicity is necessary in tool design. Design details should be made as basic and uncomplicated as possible, and every detail should be considered for possible savings in time and materials. Over-elaborate jigs and fixtures serve only to increase costs without adding significantly to accuracy or quality. Basic and simple designs minimize costs, labor, and confusion. All tool designs should be made as simple as the part design permits.

Preformed Materials

Preformed materials can greatly reduce tooling costs by eliminating many machining operations. Whenever practical, preformed materials, such as drill rods, structural sections, premachined bracket materials, tooling plate, and precision-ground flat stock, should be specified in the design.

Standard Components

Commercially available standard jig and fixture components can greatly improve tooling quality. They can

89

also achieve sizable savings in labor and materials. Standard components, such as clamps, locators, supports, drill bushings, pins, screws, bolts, nuts, and springs, should be planned into the design to reduce labor and material expenses.

Secondary Operations

Secondary operations, such as grinding, heat-treating, and some machining, should be limited to areas that are necessary for efficient tool operation. Grinding should be performed only on areas that contact either the part or the machine. Hardening operations should also be limited to areas that are subjected to wear, such as supports, locators, and moving parts. Secondary machining of surfaces that do not directly affect the accuracy of the tool should be eliminated.

Tolerance and Allowance

Generally, the tolerance of a jig or fixture should be between 20 and 50 percent of the part tolerance. Overly accurate tooling is economically wasteful and no more valuable than tooling within the required tolerance. When the tolerance applied to a tool design is unnecessarily close, the only effect on the part is higher cost.

Simplified Drawings

Tool drawings are a sizable part of the total tooling cost. Any savings gained in the drawing reduce the tool cost. The following list is a general guide to simplifying tool drawings:

- When practical, words should replace drawn details.
- Eliminate unnecessary or redundant views, projections, or details.
- When possible, replace drawn details with symbols.
- Use templates and guides to reduce drawing time.
- Standard parts should be drawn only for clarity, not for detail. Refer to these by part numbers or names.

Be careful not to oversimplify a drawing. As a rule, any shortcut that simplifies the drawing and still delivers the message is acceptable.

By applying the rules of design economy on the drawing board, the tool designer can realize substantial savings in time, labor, and materials. These economic principles are applied to all design examples and suggestions throughout the practical design units of this text.

ECONOMIC ANALYSIS

The tool designer must furnish management with an idea of how much tooling will cost and how much the production method saves over a specific run. This information is generally furnished in the form of a tooling estimate, which includes the estimated cost of the tool and projected savings over alternate methods. The estimate also includes any special conditions that may justify the cost of the tooling, such as close tolerances or high-volume production (Figure 6–1). For a valid estimate, the tool designer must accurately estimate the cost and productivity of the design in terms of materials, labor, and the number of parts per hour the tool will produce.

Estimating Tool Cost and Productivity

The simplest and most direct way to determine the cost of a tool design is to add the total costs of material and labor needed to fabricate the tool. This must be done carefully so that no part or operation is forgotten. One method is to label each part of the tool (Figure 6–2), and list the materials in a separate parts list. Then, using a cost work sheet, list each part and calculate the material and labor for each operation (Figure 6–3). The time allowed for each machining operation includes time for setup and breakdown as well as actual machining. The final expense added is the cost of designing the tool.

The next step in estimating is calculating the number of parts per hour the tool will produce. The simplest method is to divide 1 hour by the single-part time, or the time it takes to load, machine, and unload each part. Expressed as a formula, this calculation becomes:

$$Ph = \frac{1}{S}$$

```
┌─────────────────────────────────────────────────────────┐
│                  TOOLING ESTIMATE                        │
│                                                          │
│  PART _BRACKET BLOCK_____   PART NO _154-7A__          │
│  LOT SIZE _375____          REPEAT ORDER - YES ___ NO _X_ │
│  TYPE OF OPERATION REQUIRED -                            │
│      MILL 45° BEVEL ON ONE SIDE - TOL. ±.005             │
│  DESCRIPTION OF PROPOSED TOOLING -                      │
│      VISE HELD MILL FIXTURE                              │
│  ESTIMATED TOOLING COSTS - $39.00                       │
│  ESTIMATED PRODUCTION RATE - _75_ PARTS PER HOUR         │
│  ESTIMATED COST PER PART - $.62                         │
│  ESTIMATED SAVINGS OVER ALTERNATE METHOD - $422.25      │
│  ALTERNATE TOOLING METHODS -                            │
│      TOOL ROOM OR PROTOTYPE DEPT.                       │
│  REMARKS / RECOMMENDATIONS -                            │
│   TOOL ROOM OR PROTOTYPE DEPT. TOO SLOW - 25 PARTS      │
│   PER HOUR, AND TOO COSTLY - $1.85 PER PART.            │
│   RECOMMEND FIXTURE BE AUTHORIZED.                      │
│   _Sam Davis_____          _2/5_____              │
│   TOOL DESIGNER                    DATE                 │
└─────────────────────────────────────────────────────────┘
```

Figure 6-1 Tooling estimate.

Where Ph = Parts per hour
$\quad\quad\ S$ = Single-part time

Example: How many parts per hour will a jig produce if the machining time is .0167 hour and it takes .0027 hour to load the part and another .0027 hour to unload it:

$$Ph = \frac{1}{S}$$

By substituting the known values, the formula becomes:

$$Ph = \frac{1}{.0167 + .0027 + .0027}$$
$$= \frac{1}{.0221}$$
$$= 45.25 \text{ parts per hour}$$

The following chart can be used to convert standard clock time from hours, minutes, and seconds into decimal hours for easier calculation:

1 hour	= 1.0 hour
1/2 hour	= .5 hour
1/4 hour	= .25 hour
6 minutes	= .1 hour
1 minute	= .0167 hour
1 second	= .000277 hour

Calculating Labor Expense

Labor is the single most expensive factor in manufacturing. If labor expenses can be reduced, so can overall production costs. Jigs and fixtures reduce machining

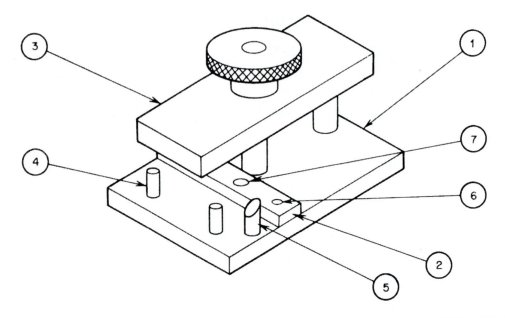

TOOLING BREAKDOWN					
P/N	QTY	DESCRIPTION	SIZE	MATL	REMARKS
1	1	BASE PLATE	$\frac{1}{2}$ X 3 X $5\frac{1}{2}$	A2 T.S.	PREC. GRND. STOCK
2	1	LOCATOR BLOCK	$\frac{1}{2}$ X $\frac{1}{4}$ X 3	" "	" " "
3	1	CLAMP ASSEMBLY	$1\frac{1}{2}$ X 4	—	COM. CLAMP # 22 - C3
4	2	DOWEL PIN	$\frac{1}{4}$ X $\frac{1}{2}$	COMM.	COM. PIN # 8721 - 1
5	1	SET / STOP PIN	$\frac{3}{8}$ X $\frac{3}{4}$	"	" " # 8721 - 5
6	2	DOWEL PIN	$\frac{1}{4}$ X $\frac{5}{8}$	"	" " # 8721 - 3
7	2	S.H. CAP SCREW	10 - 24 X $\frac{5}{8}$	"	" " # 14972

Figure 6-2 Parts list.

time in most applications and reduce or eliminate the need for skilled labor. Since special tooling transfers the required skill and accuracy from the operator to the tool, unskilled labor can produce accurate parts at a reduced wage rate. The formula to calculate the cost of labor is:

$$L = \frac{LS}{Ph} \times w$$

Where L = Cost of labor
LS = Lot size
Ph = Parts per hour
w = Wage rate

Example: Five thousand parts are to be milled using a fixture capable of producing 60 parts per hour. What is the cost of labor if the machine operator earns $6.75 per hour?

$$L = \frac{LS}{Ph} \times w$$

By substituting the known values, the formula becomes:

$$L = \frac{5000}{60} \times \$6.75$$
$$= 83.33 \times \$6.75$$
$$= \$562.50 \text{ labor expense}$$

```
┌─────────────────────────────────────────────────────────────┐
│                    COST WORK SHEET                           │
├──────┬─────────────────────────────────────┬────────┬───────┤
│ ITEM │      REQUIRED OPERATIONS            │ LABOR  │ MATL. │
│  NO  │                                     │        │       │
├──────┼─────────────────────────────────────┼────────┼───────┤
│  1   │                                     │        │ $3.75 │
│      │ LAYOUT                              │ .5 HR  │       │
│      │ DRILL AND REAM                      │ .30    │       │
│      │ DRILL AND TAP                       │ .40    │       │
│  2   │                                     │        │ $.70  │
│      │ DRILL AND CBORE                     │ .20    │       │
│      │ (DRILLED AND REAMED WITH BASE PLATE)│        │       │
│  3   │                                     │        │ $6.50 │
│  4   │ 2 EA. @ .08                         │        │ $.16  │
│  5   │                                     │        │ $.10  │
│      │ GRIND ANGLE                         │ .40    │       │
│  6   │ 2 EA. @ .045                        │        │ $.09  │
│  7   │ 2 EA. @ .05                         │        │ $.10  │
│      │          ASSEMBLY AND INSP.         │ .50    │       │
│      │          DESIGN                     │ 1.0    │       │
│      ├─────────────────────────────────────┼────────┼───────┤
│      │                                     │ 3.30HR │$11.40 │
│      ├─────────────────────────────────────┴────────┴───────┤
│      │ LABOR EXPENSE                                         │
│      │ 2.3 HR @ $8.00 = $18.40 (FABRICATION)                 │
│      │ 1.0 HR @ $9.00   $9.00 (DESIGN)                       │
│      │                 ─────                                 │
│      │                 $27.40                                │
│      │                        ┌──────────┬────────┐          │
│      │               LABOR    │  $ 27.40 │         │         │
│      │               MATERIALS│  $ 11.40 │         │         │
│      │               TOTAL    │  $ 38.80 │         │         │
└──────┴────────────────────────┴──────────┴─────────┘
```

Figure 6–3 Cost work sheet.

Calculating the Cost Per Part

A comparison of tool costs or labor expenses cannot give the tool designer enough information to determine the true economic potential of a design. For accuracy, he must calculate how much the design is worth in terms of total production and cost per part. The formula for finding this value is:

$$Cp = \frac{TC + L}{LS}$$

Where Cp = Cost per part
TC = Tool cost
L = Cost of labor
LS = Lot size

Example: What is the cost per part of a milling operation for 7000 parts when the fixture costs $55.00 and the labor expense is $784.12?

$$Cp = \frac{TC + L}{LS}$$

By substituting the known values, the formula becomes:

$$Cp = \frac{\$55.00 + \$784.12}{7000}$$
$$= \frac{\$839.12}{7000}$$
$$= \$.119, \text{ or } \$.12 \text{ per part}$$

Calculating Total Savings

To determine the most economical production method, the tool designer must compare production alternatives. This calculation can be done in two forms, depending on the situation regarding the tooling used. The first formula, which assumes that both alternatives being considered require special tooling to produce the part, is:

$$TS = LS \times (Cp\ 1 - Cp\ 2)$$

Where *TS* = Total savings
       LS = Lot size
       Cp = Cost per part

Example: A part requiring six holes is to be drilled using a jig. The first tool can produce the parts for $.19 each, the second for $.12 each. What will be the savings over a production run of 1000 parts if the second tool is used?

$$TS = LS \times (Cp\ 1 - Cp\ 2)$$

By substituting the known values, the formula becomes:

$$TS = 1000 \times (\$.19 - \$.12)$$
$$= 1000 \times \$.07$$
$$= \$70.00 \text{ total savings}$$

In the case of production alternatives, where only one method requires special tooling, the formula used is:

$$TS = LS \times (Cp\ 1 - Cp\ 2) - TC$$

Where *TS* = Total savings
       LS = Lot size
       Cp = Cost per part
       TC = Tool cost

Example: A flange-plate adapter costs $.24 per part to mill without a fixture and $.10 per part when a fixture is used. Assuming the fixture costs $128.00, how much will the fixture save over a production run of 1500 parts?

$$TS = LS \times (Cp\ 1 - Cp\ 2) - TC$$

By substituting the known values, the formula becomes:

$$TS = 1500 \times (\$.24 - \$.10) - \$128.00$$
$$= 1500 \times \$.14 - \$128.00$$
$$= \$210.00 - \$128.00$$
$$= \$82.00 \text{ total savings}$$

Calculating the Break-Even Point

The break-even point is the minimum number of parts a tool must produce to pay for itself. Any number less than this minimum results in a loss of money; any number more results in a profit. It is logical to assume that the lower the break-even point, the higher the profit potential.

The formula used to find the break-even point is:

$$BP = \frac{TC}{(Cp\ 1 - Cp\ 2)}$$

Where *BP* = Break-even point
       TC = Tool cost
       Cp = Cost per part

Example: A lathe fixture costs $150.00 to build and produces parts at a cost of $.20. How many parts must it produce to pay for itself when compared to an alternate method that requires no special tooling and is capable of making the parts at a cost of $.40 each?

$$BP = \frac{TC}{(Cp\ 1 - Cp\ 2)}$$

By substituting the known values, the formula becomes:

$$BP = \frac{\$150.00}{(\$.40 - \$.20)}$$

$$= \frac{\$150.00}{\$.20}$$

$$= 750 \text{ parts needed to break even}$$

COMPARATIVE ANALYSIS

The tool designer must consider and evaluate several options before making a tooling recommendation to management. By comparing each method, he can see the tooling requirements in terms of costs versus savings. Then the method that returns the most for each dollar spent can be selected. When preparing this comparison, the tool designer must weigh all the economic factors in relation to expenses and productivity.

Example: Five hundred guide plates must be milled to receive a locating block. The tool designer has determined three possible alternatives:

1. Have a toolmaker, who earns $12.00 per hour, mill the plates at a rate of 25 per hour.

2. Use limited tooling that costs $35.00 in the production department. The machine operator in this department, who earns $7.00 per hour, can make a part every 1.2 minutes.

3. Use a more expensive tool that costs $110.00 but is capable of producing a part every 24 seconds. This would be done in the production department, where a machine operator earns $7.00 per hour.

Which alternative should the tool designer select as the most efficient and economical?

Before a decision can be made, the tool designer must organize this information. The simplest method is shown in Figure 6–4. This comparison work sheet is first constructed by listing the alternatives across the top and the economic and productivity factors along the side. Then the known values as shown are filled in. The remaining values are calculated from the economic and productivity formulas and are used to complete the work sheet.

The first values that should be calculated are the parts per hour that the tools in alternatives 2 and 3 will produce. The formula to do this is:

$$Ph = \frac{1}{S}$$

COMPARISON WORK SHEET			
ECONOMIC & PRODUCTIVITY FACTORS	ALTERNATIVES		
	# 1	# 2	# 3
LOT SIZE	500	500	500
TOOL COST	0	$35.00	$110.00
PARTS PER HOUR	25		
LABOR/HOUR	$12.00	$7.00	$7.00
LABOR/LOT			
COST PER PART			

Figure 6–4 Comparison work sheet.

For alternative 2 the formula and calculations are:

$$Ph = \frac{1 \text{ hour}}{1.20 \text{ minutes}}$$

$$= \frac{1}{.0167 + (20 \times .000277)}$$

$$= \frac{1}{.02224}$$

$$= 44.96, \text{ or } 45 \text{ parts per hour}$$

For alternative 3 the formula and calculations are:

$$Ph = \frac{1 \text{ hour}}{24 \text{ seconds}}$$

$$= \frac{1}{(24 \times .000277)}$$

$$= \frac{1}{.006648}$$

$$= 150.42, \text{ or } 150 \text{ parts per hour}$$

The next calculation computed is the cost of labor for the entire production run. The formula is:

$$L = \frac{LS}{Ph} \times w$$

For alternative 1 the formula and calculations are:

$$L = \frac{500}{25} \times \$12.00$$

$$= \$240.00 \text{ labor cost}$$

For alternative 2 the formula and calculations are:

$$L = \frac{500}{45} \times \$7.00$$

$$= \$77.78 \text{ labor cost}$$

For alternative 3 the formula and calculations are:

$$L = \frac{500}{150} \times \$7.00$$

$$= \$23.33 \text{ labor cost}$$

The tool designer now uses this information to calculate the cost of each alternative on a per-part

basis. Many decisions will be based on these figures. The formula used to determine the cost per part is:

$$Cp = \frac{TC + L}{LS}$$

For alternative 1 the formula and calculations are:

$$Cp = \frac{\$240.00}{500}$$

$$= \$.48 \text{ cost per part}$$

For alternative 2 the formula and calculations are:

$$Cp = \frac{\$35.00 + \$77.78}{500}$$

$$= \$.226 \text{ cost per part}$$

For alternative 3 the formula and calculations are:

$$Cp = \frac{\$110.00 + \$23.33}{500}$$

$$= \$.267 \text{ cost per part}$$

The completed comparison work sheet, shown in Figure 6–5, contains enough information for the tool designer to make recommendations to management. For the tool designer to make the best possible choice, each alternative must be evaluated in terms of plus and minus factors. In evaluating the information in the comparison work sheet, the tool designer draws the following conclusions.

The first alternative saves the cost of tooling, but because of the slow production rate and high labor cost, the savings are lost. This method may be useful for a small run of fewer than 50 parts or for experimental production purposes. When cost is the only factor, the first alternative is not suitable.

The third alternative produces the parts at a higher production rate and a lower labor cost than do the other alternatives. The savings are again offset, this time by the tool cost. If the production run were greater, this method would be the least expensive. For the lot size specified, the third alternative is too costly.

The second alternative has the lowest cost per part of the three alternatives and will return the most

COMPARISON WORK SHEET			
ECONOMIC & PRODUCTIVITY FACTORS	ALTERNATIVES		
	#1	#2	#3
LOT SIZE	500	500	500
TOOL COST	0	$35.00	$110.00
PARTS PER HOUR	25	45	150
LABOR/HOUR	$12.00	$7.00	$7.00
LABOR/LOT	$240.00	$77.78	$23.33
COST PER PART	$.48	$.226	$.267

Figure 6–5 Comparing the alternatives.

for each dollar invested. For these reasons, it is the alternative the tool designer should select.

How much is actually saved? How many parts must this tool produce to pay for itself? These questions can be answered by calculating the total savings and the break-even point. The formulas for calculating the total savings are:

$$TS = LS \times (Cp\ 1 - Cp\ 2)$$
or
$$TS = LS \times (Cp\ 1 - Cp\ 2) - TC$$

To calculate the total savings between alternatives 1 and 2, the second formula is used:

$$TS = 500 \times (\$.48 - \$.226) - \$35.00$$
$$= \$92.00 \text{ saved by using alternative}$$
2 rather than alternative 1

To calculate the total savings between alternatives 2 and 3, the first formula is used:

$$TS = 500 \times (\$.267 - \$.226)$$
$$= \$20.50 \text{ saved by using alternative}$$
2 rather than alternative 3

The formula to calculate the break-even point is:

$$BP = \frac{TC}{(Cp\ 1 - Cp\ 2)}$$

To calculate the break-even point between alternatives 1 and 2, the formula and calculations are:

$$BP = \frac{\$35.00}{(\$.48 - \$.226)}$$
$$= 138 \text{ parts to break even}$$

To calculate the break-even point between alternatives 2 and 3, the formula and calculations are:

$$BP = \frac{\$35.00}{(\$.267 - \$.226)}$$
$$= 854 \text{ parts to break even}$$

SUMMARY

The following important concepts were presented in this unit:
- The following principles of economic design are an important element in keeping costs low while maintaining part quality:
  - Keep all designs simple and uncomplicated.
  - Use preformed materials where possible.
  - Always use standard components.
  - Reduce or eliminate secondary operations.
  - Do not use overly tight tolerances.
  - Simplify tool drawings.

- Performing an economic analysis helps the designer consider a variety of tooling alternatives to find the most efficient and cost-effective design. The major elements of this analysis are:
  - Estimating the tool cost and productivity.
  - Calculating the values necessary to determine the best tooling alternatives.
  - Preparing a comparative analysis of the tooling alternatives.

FORMULA SUMMARY

The information for formulas used to answer the questions at the end of the chapter has been listed together to assist you in answering those questions. Refer back to the information provided in the chapter to study the sequence of computations and determine which formula to use to complete the problem.

FORMULA SHEET

Production

Ph = Parts per hour
S = Single-part time

Includes load time and unload time

$$Ph = \frac{1}{S}$$

Labor expense

L = Cost of labor
LS = Lot size
Ph = Parts per hour
w = Wage rate

$$L = \frac{LS}{Ph} \times w$$

Cost per part

Cp = Cost per part
TC = Tool cost
L = Cost of labor
LS = Lot size

$$Cp = \frac{TC + L}{LS}$$

Total savings-economical production

TS = Total savings
LS = Lot size
Cp 1 = Cost per part—first tool
Cp 2 = Cost per part—second tool

$$TS = LS \times (Cp\ 1 - Cp\ 2)$$

Production alternatives

TS = Lot size
LS = Lot size
Cp 1 = Cost per part—first tool
Cp 2 = Cost per part—second tool
TC = Tool cost

$$TS = LS \times (Cp\ 1 - Cp\ 2) - TC$$

Break-even point

BP = Break-even point
TC = Tool cost
Cp1 = First cost per part
Cp 2 = Second cost per part

$$BP = \frac{TC}{(Cp\ 1 - Cp\ 2)}$$

REVIEW

1. List and briefly describe the six principles of economic design.
2. Using the listed alternatives, prepare a comparative analysis for the following tooling problem: A total of 950 flange plates require four holes accurately drilled 90 degrees apart to mate with a connector valve. Which of the listed alternatives is the most economically desirable?
   a. Have a machinist who earns $10.00 per hour lay out and drill each part at a rate of 2 minutes per part.
   b. Use a template jig, capable of producing 50 parts per hour and costing $18.00, in the production department, where an operator earns $6.50 per hour.
   c. Use a duplex jig, which costs $37.50 and can produce a part every 26 seconds, in the production department, where an operator earns $6.50 per hour.
3. Assuming everything to be the same as in problem 2, which alternative is the most economical for 135 parts?

UNIT 7

Developing the Initial Design

OBJECTIVES

After completing this unit, the student should be able to:

• Describe how the designs for jigs and fixtures are planned.

• List the human factors involved in tool design.

• List the safety factors related to tool design.

PREDESIGN ANALYSIS

All tool design ideas begin in the mind of the tool designer. A great deal of planning and research is needed to turn tooling ideas into practical hardware.

The first step in designing a tool is organizing all relative information. Part drawings and production plans are carefully studied to find exactly what tool is required. Preliminary plans for the tool are developed, usually by means of sketches. The tool designer must develop alternatives that are practical and cost-effective. Finally, tool drawings are made from the tools that can be built.

Overall Size and Shape of the Part

The tool designer must consider how the size and shape of the part influence the bulk and mass of the tool. For example, the mating parts shown in Figure 7–1 have the same hole patterns. The tool needed for the end cap, however, is much smaller and simpler than the tool required for the housing. In this case, a template jig could be used for the end cap and a table jig could be used for the housing.

Type and Condition of Material

The type and condition of part material directly influence how the tool is made. Parts from soft materials,

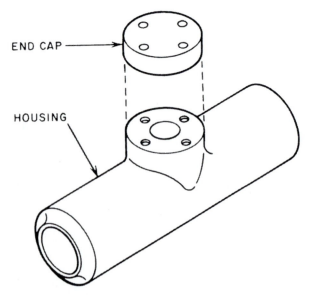

Figure 7–1 Parts having the same hole pattern may require different tools.

such as aluminum, magnesium, or plastic, are easier and faster to cut than harder materials. Since cutting forces are reduced for these materials, the design of the tool is directly affected. Reduced cutting forces allow lighter, less rigid tools, but the higher production rate requires faster tool operation.

The condition of the part material also affects how the part is held and located. Rolled or extruded bar-stock is more uniform in size than cast parts and is normally easier to locate. In addition, cast parts are sometimes more fragile than solid sections, and clamping pressure must be reduced to prevent breaking or cracking the casting.

Type of Machining Operation

The particular machining operation to be done specifies the type of tool to be made. In some cases, multipurpose tools can be designed for more than one operation, such as the drill jig/milling fixture (Figure 7–2). Normally, single-purpose tools are preferred for high-speed production.

The machining operation also determines how rigid the tool will be made. For example, a gang-milling fixture must be built stronger than a keyway fixture. A drill jig for large holes must be made stronger than a jig for small holes. As a rule, increased cutting forces require added tool strength and rigidity.

Degree of Accuracy Required

The effect accuracy has on the design is usually reflected in the tool tolerances. The general rule of tolerance is that 20 to 50 percent of the part tolerance is applied to the tool. The degree of required accuracy

determines this tolerance. Figure 7–3A shows a part that requires the slot to be within ±.001 inch of the .38-inch dimension. This is a much closer tolerance than that shown in Figure 7–3B. Here the tolerance is ±.010 inch. The slot location is much more critical for the first part than for the second part. Therefore, the tool must reflect this added precision.

Number of Pieces to be Made

The number of pieces to be made has a direct bearing on how well the tool is made. For example, a production run of 1500 parts requires a jig. The jig must not cost more to make than it saves; therefore, it must be made as simple and as inexpensively as possible. If, however, a run of 150,000 parts needed the same jig, more money could be spent to make the tool.

As a rule, larger production runs justify more detailed and expensive tooling than do smaller runs. This is because the tool will be in service longer and production speeds are generally higher. Longer production runs also require replaceable parts to be used in making the tool. Bushings are sometimes left out of short-run tools. They are included, along with liners and lock screws, in tools that are used in longer production runs. Details, such as locators and clamps, are also affected by the size of production runs.

Locating and Clamping Surfaces

The part drawing must be studied to find the best surfaces to locate and clamp the part. The order of preference is:

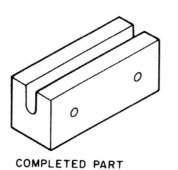

COMPLETED PART

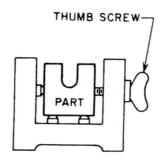

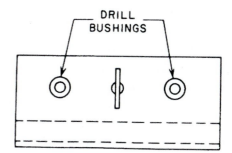

Figure 7–2 Multipurpose tooling—drill jig/milling fixture.

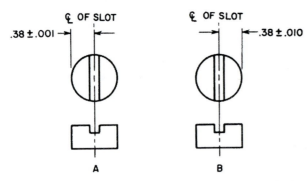

Figure 7–3 The degree of accuracy required determines how the tool should be made.

1. Holes
2. Two machined surfaces that form a right angle
3. One machined and one unmachined surface that form a right angle
4. Two unmachined surfaces that form a right angle

The prime requirement for a locating surface is repeatability. The parts must be positioned identically, within the tolerance limits, part after part.

Clamping surfaces must be rigid and capable of holding the part without bending. Bending can distort the machining operation. If the clamping surface can bend, it must be supported. If a finished surface is used to hold the part, the clamp should have a nonmar cap or pad to prevent damage to the finished surface.

Type and Size of Machine Tool

The process planning engineer normally selects the machine tool for each operation. However, if a better tool could be used, the tool designer should consult the process engineer before beginning the design. For example, when holes are drilled with a drill jig, a drill press should be used. Little is gained from using a vertical mill or a jig borer, since the accuracy is built into the jig, not into the machine tool.

Once the machine tool has been selected, the tool designer must know the size of machine elements and ranges of operation before beginning the design. This is required so the tool designer can position the details, such as clamps, locators, holddowns, slot keys, and other parts, in places where they will not interfere with the operation of the tool. The machine reference sheet provided in the operator's instruction or maintenance

book for each machine contains this information (Figure 7–4). Using this reference sheet saves many hours of measuring and checking each machine before designing a tool.

Type and Size of Cutters

Normally, the type and size of cutters are specified by the process engineer, but occasionally the tool designer may do this. Before the tool designer selects the cutters, every detail about the tools being used must be known to ensure that the part is properly referenced to the tool and that enough tool clearance is provided. Figure 7–5 shows a specifications chart. Each type and size of cutter is made to a standard size. Again, using this type of engineering data sheet saves time.

Another source of information is the industrial supplier. Suppliers know their products and can answer tooling questions or furnish information relative to most tool design problems.

Sequence of Operations

Quite often the tool designer must design more than one tool for a part. When this is the case, the sequence of operations must be determined as well as which tool to design first. For example, if a drill jig for a part is designed first, then the holes provide an excellent location for the milling fixture that is needed in the next operation.

DESIGNING AROUND THE HUMAN ELEMENT

No matter how mechanized our society becomes, there will always be a need for people in the manufacturing industries. Before deciding on the final design, the tool designer must consider the human factors in relation to the operation of the tool. Operators, setup personnel, and inspectors are all involved with the proposed jig or fixture. They must be considered in the tool design.

Design Ergonomics

Ergonomics is a science that studies the human body and uses what it learns about how the human body works to determine the best design of objects, systems, and environmental systems for human interaction. Ergonomics is applied in one form or another to anything that involves people. In the case of tool design, the ergonomic considerations include aspects of anatomy, physiology, psychology, and design. The work environment for the operator and the component must be compatible with the capabilities of the human body and mind and must meet the needs of the manufacturing environment.

It is important that the designer consider ergonomics during the design phase and make good use of information that is provided from the machining technician as well as from industrial engineers involved in basic motion and time study. The designer's job is a onetime event. Once the design is complete and any necessary revisions are completed, the design is released for build and journeys onto the production floor. But the machining technician will load and unload parts continuously for manufacture for extended periods. A design that causes repetitive motion injuries or other work-related injuries because it did not respect ergonomic design eventually results in higher product cost. The following partial list of questions provides a starting point for the tool designer to consider when planning a tool design.

- Is the operation of the tool smooth and rhythmic?
- Can both hands be used at the same time?
- Are hands clear and free of moving parts?
- Do both hands start and stop together?
- Does the intended motion minimize operator fatigue?
- Can feet be used to lessen hand and arm fatigue?
- Is the tool height appropriate?
- Are controls and clamps within easy reach of the operator?
- Are handles designed to reduce hand and finger fatigue?
- Is safety designed into the tool with respect to the operation of the supporting equipment?

This is a list of some very basic considerations. Each design problem may have specific ergonomic requirements that are particular to the type of manufacturing that it will serve. Information is readily available through a variety of resources that specifically address these concerns. Consult textbooks, the Internet, and the Occupational Safety & Health Administration (OSHA) for further references.

Safety as Related to Tool Design

The first consideration in designing any tool must be safety. No matter how fast a tool works or how much money it saves, if it is not safe to operate, it is useless. Safety must be planned into every design detail. The tool designer must always remember the person operating the tool. The following checklist should be consulted during every step of the design to ensure that the tool is completely safe to operate.

- Is the tool clear of the cutters during the loading and unloading operations?
- Are all clamps and controls located so the operator does not have to reach over the tool?
- Are any operator movements required close to a moving or revolving tool?
- Are chip guards needed to protect the operator and others nearby?
- Are all sharp edges on the tool chamfered?
- Are attached accessories (pins, feeler gauges, wrenches, etc.) far enough away to prevent tangling in the tool?
- Is the entire operation visible from the operator's position?
- Could the part be pulled from the tool?
- Is the tool body rigid enough to resist all cutting forces?
- Could the clamping device loosen during the machining cycle?

The Occupational Safety & Health Administration was created by the U.S. Occupational Safety and Health Act and is referred to as OSHA. OSHA maintains and mandates standards and conducts mandatory inspections for manufacturing facilities. Its purpose is to ensure that the working environment is safe and healthy with respect to workers' requirements. Information is available from this organization that provide the designer with additional safety resources.

DIMENSIONAL DRAWINGS

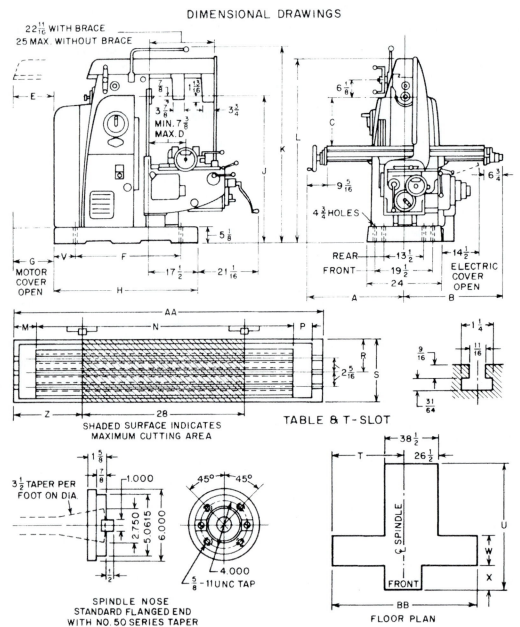

Figure 10A
Dimensional Drawing—No. 2 MI Plain and Universal Machines

SIZE	A		B		C		D		E	F	G	H	J	K
	Max.	Min.	Max.	Min.	Max.	Min.	Without Brace	With Brace						
Universal	50 13/16	22 5/16	46 1/8	17 5/8	18 3/16	0	17 3/8	14 1/4	12 1/2	35 3/4	14 1/4	47 7/8	48 1/4	64 5/8
Plain	49 5/16	20 13/16	49 5/16	20 13/16	18 13/16	0	17 3/8	14 1/4	12 1/2	35 3/4	14 1/4	47 7/8	48 1/4	64 5/8

SIZE	L	M	N	P	R	S	T	U	V	W	X	Z	AA	BB
Universal	61 3/8	3 5/8	44 1/2	2 7/16	5	10	50 13/16	83 3/16	6 3/8	20	16 3/8	13	52 3/4	96 15/16
Plain	61 3/8	3 3/4	45 9/16	2 3/4	5	10	49 5/16	83 3/16	6 3/8	20	16 3/8	11 1/2	54 7/16	98 5/8

Figure 7-4 Machine reference sheet.

MACHINE SPECIFICATIONS

GENERAL SPECIFICATIONS

	2 MI UNIV.	2 MI PLAIN
Table		
Working surface	52 3/4" x 10"	54 7/16" x 10"
Size over-all	52 3/4" x 10"	54 7/16" x 10"
T-slots, number and size	Three——11/16"	Three——11/16"
Distance between T-slots.	2 5/16"	2 5/16"
Swivels	45° R and L	. . . .
Range		
Longitudinal.	28"	28"
Cross.	10"	10"
Vertical	18"	18 1/2"
Centerline spindle to top of table │ Max	18	18 1/2"
│ Min	0	0
Spindle		
Spindle nose.	No. 50 Std.	No. 50 Std.
Hole for draw-in bolt	1 1/8"	1 1/8"
Speeds │ Number.	16	16
│ Range	25-1500 r/min	25-1500 r/min

25, 33, 43, 56, 76, 100, 130, 168, 225,
295, 385, 500, 680, 895, 1160, 1500 rpm.

Low range of spindle speeds, 20 to 1200 r/min, or high range of spindle
speeds, 33 to 2000 r/min, may be obtained at the time the order is placed.

Feed		
Number of feeds	16	16
Range │ Longitudinal and cross	1/4-30"/min.	1/4-30"/min.
│ Vertical	1/8-15"/min.	1/8-15"/min.

1 1/4, 5/16, 7/16, 5/8, 7/8, 1 1/4, 1 3/4, 2 3/8, 3 1/4, 4 1/2, 6 1/8,
8 3/8, 11 1/2, 16, 22, 30" per min. Vertical rates are 1/2 the foregoing.
An optional feed range of 1/2" to 60" per minute can be supplied at
the time the order is placed. Longitudinal, cross, and vertical feed rates
are then in the same proportion as standard.

Power Rapid Traverse		
Longitudinal and Cross.	150"/min.	150"/min.
Vertical	75"/min.	75"/min.
Dividing Head		
Size (nominal swing)	10"	. . . .
Max. length of work between centers	28"	. . . .
Overarm (Dynapoise)		
Underside to centerline of spindle.	6 1/8"	6 1/8"
Drive		
Spindle drive motor	5 hp	5 hp
Feed drive motor.	1 1/4 hp	1 1/4 hp
Floor space		
Maximum size.	97" x 84"	95 5/8" x 83 1/4"
Area	56 1/2 sq. ft.	57 sq. ft.
Shipping data (Approx.)		
Net weight	4900 lb.	4650 lb.
Gross weight, domestic.	5600 lb.	5300 lb.
Gross weight, export	5900 lb.	5600 lb.
Shipping case │ Size (length x height x width)	72" x 70" x 42"	72" x 70" x 42"
│ Volume.	123 cu. ft.	123 cu. ft.
Code Names		
Conventional Table Feed Machines	MIULL	MIPLL
Automatic Table Feed Machines	. . . .	MPIAL

Note.—General specifications for Plain Machines with and without Automatic Table Cycles.
　　　　Automatic Table Cycles not available on Universal Machines.

Figure 7-4 (continued).

ENGINEERING SPECIFICATION MODIFICATION CHART NIAGARA CUTTER

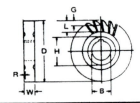

A
STAGGERED TOOTH SIDE MILLING CUTTERS

Cutter Diameter D	Diameter Range	Width Range Available W	Bore Size B	Max. Hub Dia. H	Tooth Lgt. L	Gash Depth G	No. Teeth T
2-1/8	2.00– 2.125	.170– .375	3/4	1-1/8	5/16	3/16	14
2-1/2	2.37– 2.500	.230– .500	7/8	1-1/4	5/16	7/32	16
2-3/4	2.62– 2.750	.230– .500	1	1-1/2	3/8	1/4	16
3	2.87– 3.00	.090– .250	1	1-1/2	1/2	5/32	28
3	2.87– 3.00	.170–1.250	1	1-1/2	7/16	7/32	16
3	2.87– 3.00	.170–1.250	1-1/4	1-3/4	7/16	7/32	18
3-1/4	3.12– 3.250	.220– .500	1-1/4	1-3/4	1/2	7/32	18
		.687– .750					
3-1/2	3.37– 3.50	.200–1.000	1	1-1/2	5/8	7/32	18
3-1/2	3.37– 3.50	.200–1.000	1-1/4	1-3/4	5/8	7/32	18
3-3/4	3.62– 3.750	.220– .500	1-1/4	1-3/4	5/8	7/32	18
		.687– .750					
4	3.87– 4.00	.090– .250	1	1-5/8	3/4	3/16	32
4	3.87– 4.00	.090– .250	1-1/4	1-7/8	3/4	3/16	32
4	3.87– 4.00	.170–1.000	1	1-3/4	3/4	11/32	18
4	3.87– 4.00	.170–1.500	1-1/4	1-7/8	3/4	11/32	18
4-1/4	4.12– 4.25	.220– .500	1-1/4	1-7/8	3/4	11/32	18
		.687– .750					
4-1/2	4.37– 4.50	.200– .625	1	1-5/8	7/8	3/8	18
		.687– .750					
		.900–1.000					
4-1/2	4.37– 4.50	.200– .625	1-1/4	1-7/8	7/8	3/8	18
		.687– .750					
		.900–1.000					
5	4.87– 5.00	.090– .250	1	1-5/8	7/8	7/32	36
5	4.87– 5.00	.090– .250	1-1/4	1-7/8	7/8	7/32	36
5	4.87– 5.00	.170– .500	1	1-5/8	1	3/8	24
5	4.87– 5.00	.170–1.000	1-1/4	1-7/8	1	3/8	24
5-1/2	5.37– 5.50	.220– .500	1-1/4	1-7/8	1	3/8	24
		.687– .750					
6	5.87– 6.00	.100– .250	1	1-3/4	1	7/32	40
		.312– .375					
6	5.87– 6.00	.100– .250	1-1/4	1-7/8	1	7/32	40
		.312– .375					
6	5.87– 6.00	.170– .500	1	1-5/8	1-1/8	7/16	24
6	5.87– 6.00	.170–1.250	1-1/4	1-7/8	1-1/8	7/16	24
6	5.87– 6.00	.170–1.000	1-1/2	2-1/4	1-1/8	7/16	28
7	6.87– 7.00	.100– .187	1	1-5/8	1-1/8	1/4	44
7	6.87– 7.00	.100– .187	1-1/4	1-7/8	1-1/8	1/4	44
7	6.87– 7.00	.220–1.000	1-1/4	1-7/8	1-1/4	19/32	24
7	6.87– 7.00	.220–1.000	1-1/2	2-1/4	1-1/4	19/32	28
8	7.87– 8.00	.100– .250	1-1/4	1-7/8	1-1/4	1/4	48
		.312– .375					
8	7.87– 8.00	.100– .250	1-1/2	2-1/4	1-1/4	1/4	48
		.312– .375					
8	7.87– 8.00	.170–1.000	1-1/4	2-1/2	1-1/4	19/32	28
8	7.87– 8.00	.170–1.000	1-1/2	2-1/2	1-1/4	19/32	28
9	8.87– 9.00	.220–1.000	1-1/4	2-1/2	1-3/8	19/32	28
9	8.87– 9.00	.220–1.000	1-1/2	2-1/2	1-3/8	19/32	28
10	9.87–10.00	.170– .250	1-1/4	3-1/4	1-1/2	1/4	56
		.312– .375					
10	9.87–10.00	.170– .250	1-1/2	3-1/4	1-1/2	1/4	56
		.312– .375					
10	9.87–10.00	.170–1.000	1-1/2	3-1/4	1-1/2	19/32	32
12	11.87–12.00	.220– .375	1-1/2	4	1-3/4	1/4	64
12	11.87–12.00	.280–1.000	1-1/2	4	1-3/4	11/16	36

B
STRAIGHT TOOTH SIDE MILLING CUTTERS

Cutter Dia. D	Diameter Range	Width Range Available W	Bore Size B	Max. Hub Dia. H	Tooth Lgt. L	Gash Depth G_A	No. Teeth T
2	1.87– 2.00	.170– .375	1/2	7/8	5/16	3/16	14
2	1.87– 2.00	.170– .375	5/8	1	5/16	3/16	14
2-1/2	2.37– 2.50	.055– .125	7/8	1-1/4	3/8	5/32	30
2-1/2	2.37– 2.50	.220– .500	7/8	1-1/4	3/8	3/16	16
3	2.87– 3.00	.055– .250	1	1-1/2	5/8	3/16	32
3	2.87– 3.00	.170–1.000	1	1-1/2	1/2	1/4	20
4	3.87– 4.00	.055– .250	1	1-5/8	5/8	3/16	36
4	3.87– 4.00	.055– .250	1-1/4	1-7/8	5/8	3/16	36
4	3.87– 4.00	.200–1.000	1	1-5/8	3/4	9/32	24
4	3.87– 4.00	.200–1.000	1-1/4	1-7/8	3/4	9/32	24
5	4.87– 5.00	.055– .250	1	1-5/8	3/4	3/16	44
5	4.87– 5.00	.055– .250	1-1/4	1-7/8	3/4	3/16	44
5	4.87– 5.00	.220–1.000	1	1-5/8	1	3/8	26
5	4.87– 5.00	.220–1.000	1-1/4	1-7/8	1	3/8	26
		.055– .250					
6	5.87– 6.00	.312– .375	1	1-5/8	7/8	3/16	48
		.055– .250					
6	5.87– 6.00	.312– .375	1-1/4	1-7/8	7/8	3/16	48
		.440– .500					
6	5.87– 6.00	.687– .750	1	1-1/2	1-1/8	3/8	30
6	5.87– 6.00	.220–1.000	1-1/4	1-7/8	1-1/8	3/8	30
7	6.87– 7.00	.100– .187	1	1-5/8	1	3/16	52
7	6.87– 7.00	.100– .187	1-1/4	1-7/8	1	3/16	52
7	6.87– 7.00	.220–1.000	1-1/4	1-7/8	1-1/4	13/32	32
8	7.87– 8.00	.100– .125	1	1-5/8	1-1/8	7/32	56
		.100– .250					
8	7.87– 8.00	.312– .375	1-1/4	1-7/8	1-1/8	7/32	56
		.100– .250					
8	7.87– 8.00	.312– .375	1-1/2	2-1/4	1-1/8	7/32	56
8	7.87– 8.00	.220–1.000	1-1/4	1-7/8	1-1/4	7/16	34
9	8.87– 9.00	.220–1.000	1-1/4	2-1/2	1-1/4	1/2	36
		.170– .250					
10	9.87–10.00	.312– .375	1-1/4	3-1/4	1-1/4	7/32	64
		.170– .250					
10	9.87–10.00	.312– .375	1-1/2	3-1/4	1-1/4	7/32	64
10	9.87–10.00	.220–1.000	1-1/2	3-1/4	1-1/2	1/2	38
12	11.87–12.00	.220– .375	1-1/2	3-1/4	1-3/8	1/4	72
12	11.87–12.00	.220–1.000	1-1/2	4	1-3/4	5/8	42

STANDARD KEYWAYS AND COLLAR DIAMETERS

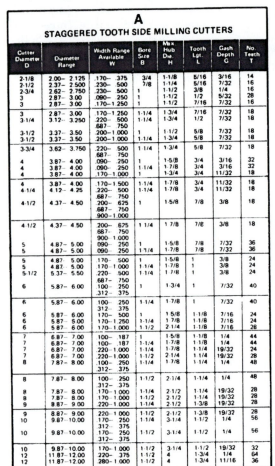

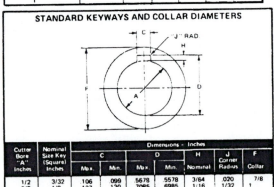

Cutter Bore "A" Inches	Nominal Size Key (Square) Inches	Dimensions - Inches						
		C		D		H	J Corner Radius	F
		Max.	Min.	Max.	Min.	Nominal		Collar
1/2	3/32	.106	.099	.5678	.5578	3/64	.020	7/8
5/8	1/8	.137	.130	.7085	.6985	1/16	1/32	1
3/4	1/8	.137	.130	.8325	.8225	1/16	1/32	1-1/8
7/8	1/8	.137	.130	.9575	.9475	1/16	1/32	1-1/4
1	1/4	.262	.255	1.1140	1.1040	3/32	3/64	1-1/2
1-1/4	5/16	.325	.318	1.3950	1.3850	1/8	1/16	1-7/8
1-1/2	3/8	.410	.385	1.6760	1.6660	5/32	1/16	2-1/4
1-3/4	7/16	.473	.448	1.9580	1.9480	3/16	1/16	2-1/2
2	1/2	.535	.510	2.2080	2.1980	3/16	1/16	2-3/4
2-1/2	5/8	.660	.635	2.7430	2.7330	7/32	1/16	3-1/2
3	3/4	.785	.760	3.2750	3.2650	1/4	3/32	4
3-1/2	7/8	.910	.885	3.9000	3.8900	3/8	3/32	5
4	1	1.035	1.011	4.4000	4.3900	3/8	3/32	5-1/2
4-1/2	1-1/8	1.160	1.135	4.9630	4.9530	7/16	1/8	6
5	1-1/4	1.285	1.260	5.5250	5.5150	1/2	1/8	6-1/2

For intermediate size bore use the Keyway for the next larger size bore listed.

STEP-MILLING INTERLOCK OPTIONS

OPTION	CUTTER DIAMETER INTERLOCK RANGES (INCHES)	HOLE SIZE	NO. OF TEETH
1	2¾,3	1	16
2	3½,4,4½	1	18
3	3,3¼,3½,3¾,4,4¼,4½	1¼	18
4	5,6	1	24
5	5,5½,6,7	1¼	24
6	8,9	1¼	28
7	6,7,8,9	1½	28

SIDE MILLING CUTTER DEPTH OF RECESS (R)

2" Through 6" Dia.		7" Through 12" Dia.	
Width	Recess Depth	Width	Recess Depth
3/16– 7/32	1/32	1/4 –5/16	1/32
1/4 –11/32	3/64	3/8	3/64
3/8 – 7/16	1/16	7/16	1/16
1/2 – 9/16	3/32	1/2 –9/16	3/32
5/8 & Wider	1/8	5/8 & Wider	1/8

CHECK YOUR MILLING CUTTER PRINTS WITH ENGINEERING SPECIFICATION CHARTS A & B TO ELIMINATE SPECIAL MILLING CUTTERS.

Figure 7–5 Engineering specifications for cutters.

PREVIOUS MACHINING OPERATIONS

This phase of design is closely related to the sequence of operations. The tool designer must know what operations, if any, take place before the operation being planned. In this way, locators and clamps can be positioned to take advantage of the existing machined surfaces. This is important when more than one person is designing tools for the same part.

DEVELOPING TOOLING ALTERNATIVES

Every tool design problem has an almost unlimited number of possible solutions. The tool designer must find the solution that is the fastest, most economical, and most accurate.

When developing tooling alternatives, the tool designer must keep speed, accuracy, and economy in mind at all times. Often, while developing designs, a combination of ideas is better to work with than limiting the tool to only one possible design. For example, if a jig can be a template or a plate type, both should be incorporated into the final design.

As outlined in Unit 1, the tool designer must answer the following questions before any design can be selected.

- Should special tooling be used or should existing tooling be modified?
- Should multiple-spindle or single-spindle machines be used?
- Should the tool be capable of more than one operation?
- How should each operation be checked?
- Should special gauges be made?
- Will the savings justify the cost of the tool?
- Is there enough leverage provided on handheld jigs to prevent spinning?
- Has every possible detail been studied to protect the operator?

This general checklist should be used to evaluate the safety of each design. As a rule, if there is any doubt about the safety of using a tool, redesign the tool until it is considered safe to use.

NOTE TAKING

After studying all the preliminary information, the tool designer can begin the actual tool design. During work on the design, all the necessary data are kept close at hand by taking notes. The designer should record any particular point that might be useful later. For example, if during the analysis the designer thinks of a good way to hold the part, the idea should be jotted down. If notes are not taken, good ideas can be lost. No one can possibly remember every detail without some sort of written reminder. It does not matter how a note is recorded, just get it down on paper before it is forgotten.

Engineering grid paper and isometric drawing grid paper are essential elements of the tool designer's note-taking tools. Quick sketches of concepts or ideas may say more than notes or fragmented sentences when they are revisited at a later time. Grid paper can provide a quick reference to scale and size and can assist the designer in making a more accurate sketch of an idea. CAD systems are an excellent tool, and when they are used to produce a quick sketch, they provide a concept that can be retrieved later for a formal design.

SUMMARY

The following important concepts were presented in this unit:
- To determine the best possible tool design, a pre-design analysis should be done to evaluate the workpiece and the operations to be performed. This analysis should include:
  - Overall size and shape of the part
  - Type and condition of workpiece material
  - Type of machining operations required
  - Degree of accuracy required
  - Number of pieces to be made
  - Locating and clamping surfaces
  - Type and size of machine tools
  - Type and size of cutters
  - Sequence of operations
- Safety is a primary concern in the design of any jig or fixture.

- Developing tooling alternatives is the best way to find the exact design for any particular part.
- Note taking is an important part of any evaluation and helps the designer recall important data that might otherwise be forgotten.

REVIEW

1. Briefly describe the principal factors that must be analyzed with regard to the following areas:
   a. Overall size and shape of the part
   b. Type and condition of material
   c. Type of machining operation
   d. Degree of accuracy required
   e. Number of pieces to be made
   f. Locating and clamping surfaces
   g. Type and size of machine tool
   h. Type and size of cutters
   i. Sequence of operations
   j. Previous machining operations
2. List six human factors that must be considered when designing a tool.
3. List ten safety factors that must be considered during the design of any tool.
4. Conduct an Internet research of ergonomics and write a paragraph that describes its benefits to the tool design process.
5. Conduct an Internet research of OSHA and write a paragraph that describes the benefits that this organization provides to the tool designer.

UNIT 8

Tool Drawings

OBJECTIVES

After completing this unit, the student should be able to:

- Identify the types of tool drawings.
- Specify methods to simplify tool drawings.
- Identify dimensional forms.
- Specify the rules of metric dimensioning.

TOOL DRAWINGS VERSUS PRODUCTION DRAWINGS

Tool drawings are used to transfer detailed instructions from the tool designer to the toolmaker. The form and specifications of these drawings are normally established within each company to meet particular needs. However, there are standards and conventions that all companies follow.

Tool drawings differ from standard production drawings in the amount of detail shown. Toolmakers are highly skilled technicians. Therefore, they require less detailed information on drawings. For example, if the tool designer specifies two .25-inch dowel pins to be located on 2.000-inch centers, she does not need to specify the hole size on a tool drawing. Following standard toolmaking practices would automatically

make the hole smaller to allow for a press fit on the dowels. Likewise, the tool designer eliminates much of the drawn details found on production drawings and replaces them with word descriptions.

Tool drawings fall into two general types: assembly and detailed. Some tool drawings are made in the assembly form only because of the simplicity of the tool. However, larger, more complex tools require detailed drawings to describe each part.

Assembly Drawings

Assembly tool drawings show the entire tool in its completed form with all parts in their proper place (Figure 8–1). This tool is simple enough to show every detail without an individual drawing of each part having to be made.

Detailed Drawings

Detailed tool drawings are used to show tools with many parts that must be drawn separately to show true sizes and shapes. Figure 8–2 shows a tool that is too complex for an assembly drawing. Here a detailed drawing must be used.

In either case, an assembly drawing is first drawn in orthographic or isometric form. If necessary, balloon references are used to identify each of the parts

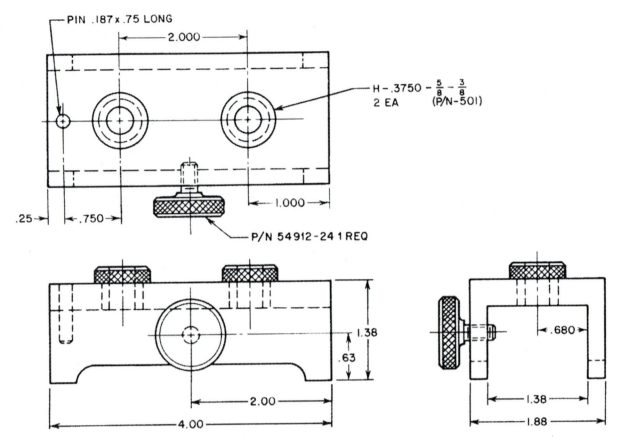

Figure 8-1 Assembly drawing.

for the detailed drawings. Using assembly drawings, along with detailed drawings, allows the toolmaker to see how each part is located in the final tool. It also permits much easier assembly (Figure 8–3).

SIMPLIFIED DRAWINGS

The simplified drawing practices outlined in Unit 6 assist the designer by reducing the amount of time it takes to make a tool drawing. The following are points to remember in making tool drawings:

- When practical, words should replace drawn details.
- Eliminate all unnecessary or redundant views, projections, or details.
- When applicable, symbols are used in place of drawn details.
- When possible, templates and guides are used to reduce drawing time.

- Standard parts should be drawn only for clarity, not for detail. Refer to these parts by number and name.
- CAD drafters use vendor libraries of standard components, eliminating the time-intensive manual methods of drawing templates. It is now possible to provide a more complete drawing because these libraries can be accessed through vendor-supplied CDs or through vendor web sites.
- Font selection in the CAD software package replaces lettering guides and expedites the process of adding notes. Geometric dimensioning and tolerancing symbols are also available in CAD software.

When using the simplified form of drawing, the designer must not oversimplify. As a rule, any shortcut that simplifies a drawing and still delivers the intended message is acceptable.

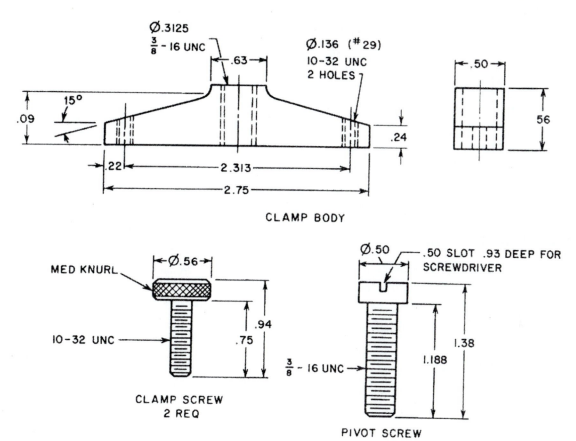

Figure 8-2 Detailed drawing.

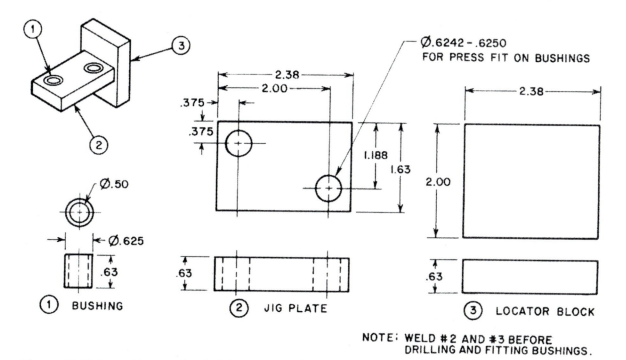

Figure 8-3 Assembly and detailed drawings used together.

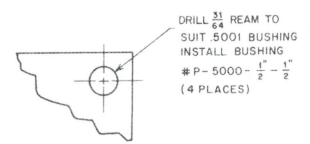

Figure 8-4 Using words to identify size.

Using Words on Drawings

Words can save countless hours of drawing time. To be effective, however, they must clarify, not complicate, a drawing. The bushing holes in the jig plate in Figure 8–4 are identified by words rather than by the numerical sizes. This is quicker than making a drawing of the actual bushings.

Reducing the Number of Views

Standard drafting practices usually require three views of an object. If one or two views totally describe the object, the third view should be eliminated. The set block in Figure 8–5 shows how one drawn detail and a word description, ".63 thick," can save drawing the second and third views. When reducing the number of views, the designer must make sure that none of the necessary information is omitted.

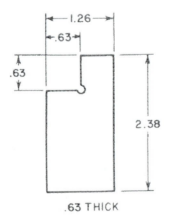

Figure 8-5 Reducing the number of views.

Figure 8-6 Screw thread symbols.

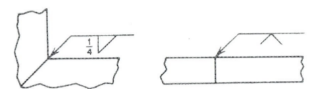

Figure 8-7 Welding symbols.

Symbols

When appropriate, symbols should replace drawn details. Symbols, such as those for the screw threads in Figure 8–6, should replace the actual drawn threads. Likewise, welding symbols are used in Figure 8–7. Symbols contain more specific information than drawn details do and they take less time to draw.

Using Templates, Guides, and CAD Libraries

Templates save hours of drawing time by allowing the designer to trace a printed drawing of a tooling detail or component. Templates also detail size differences in various models of standard parts. Figure 8–8 shows available lengths of jig feet and rest buttons. The designer selects the size desired and draws only that portion of the template. Templates are also available in scale. The most common scales are full, one-half, and one-quarter.

Drawing guides are available in several types and sizes. Common figures, such as circles, hexagons, and squares, can be drawn accurately and quickly with these guides. Lettering guides can also help the designer make crisp, sharp letters (Figure 8–9).

CAD software for design and drafting has replaced most of the drafting boards in the United States today. Although the board may be used to introduce the student to drafting conventions, stu-

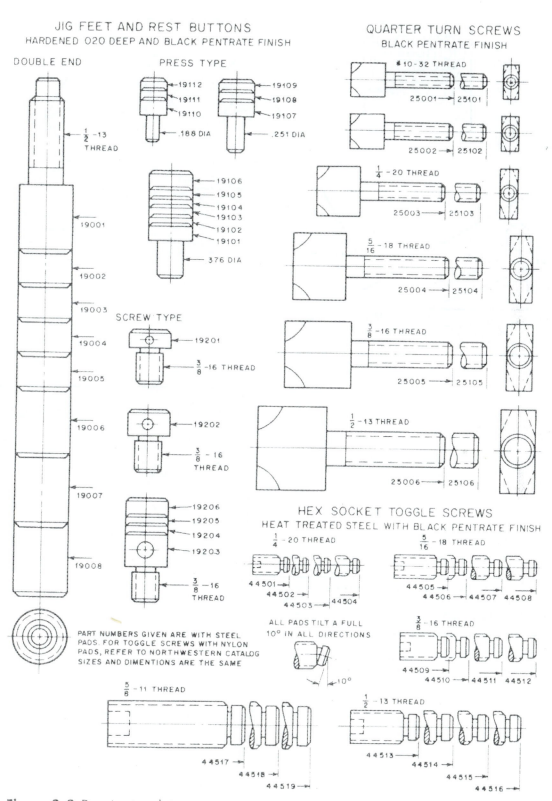

Figure 8-8 Drawing templates.

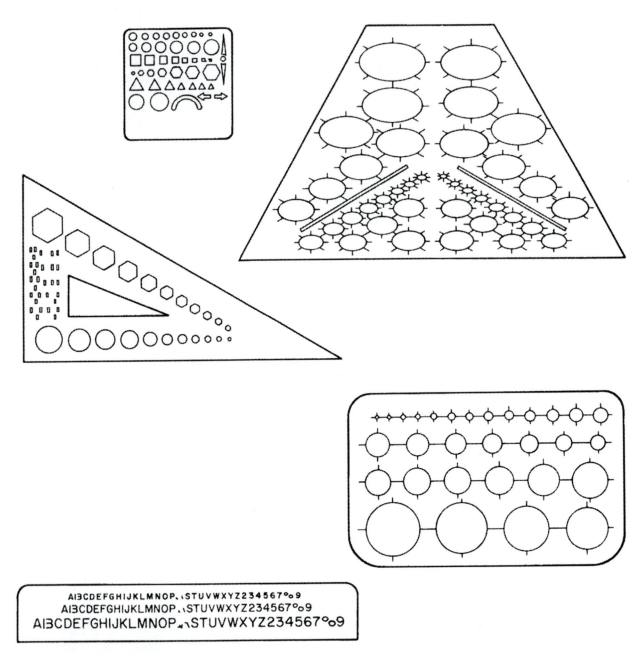

Figure 8–9 Drawing and lettering guides.

dents must also acquire CAD skills in order to be marketable as designers or entry-level drafters. In keeping with this progression, many vendors of components for jigs and fixtures began developing libraries for their component parts. They have gratefully made these libraries available to business, industry, schools, and colleges at no cost. Component libraries can also be found on vendor web sites and can be easily downloaded.

Describing Standard Parts

Standard parts such as bolts, screws, pins, washers, drill bushings, and clamp assemblies should be described by name, part number, and size, rather than by drawing the entire detail (Figure 8–10). Drawing the entire detail in such cases serves no useful purpose and only adds time and cost in making the drawing.

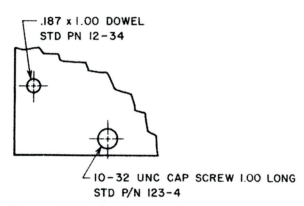

.187 x 1.00 DOWEL
STD PN 12-34

10-32 UNC CAP SCREW 1.00 LONG
STD P/N 123-4

Figure 8-10 Identifying standard parts.

MAKING THE INITIAL DRAWING

The first step in making a tool drawing is sketching. Sketches allow the designer to see ideas and designs in graphic form before the tool drawing is begun.

When sketching the initial design, the designer must make sure the sketch is as close as possible to the actual drawing. All machine factors, such as tool clearances, methods of mounting the tool, and table sizes, must be considered. To ensure that these areas have been thought about, draw all relevant machine factors into the sketch. The milling fixture in Figure 8-11 shows the relationship of both the machine arbor and the table to the fixture. By doing this, the designer determines the approximate size of the cutter at the beginning, thereby avoiding a costly mistake. The tool designer must solve all the problems on the drawing board, not on the shop floor.

CAD systems can eliminate the potential for machine interferences graphically. Solid models provide the ability to rotate the model with the jig or fixture in place and to identify any clearance problems with the machine tool early in the process. Concurrent engineering makes it possible to share the electronic data, providing the design to the team for review and evaluation. Feedback from the team can translate into design changes that head off any problems early in the design process. The electronic file is protected, allowing the designer the ability to be the only one to make changes. The efficiencies of the CAD tool are a requirement to do business globally and to remain competitive.

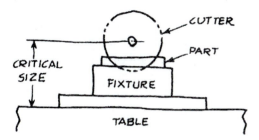

Figure 8-11 Showing tool relationships.

Developing the Sketch

The first step in sketching any tool design is drawing the part. The flange in Figure 8-12 requires a hole to be tapped in the hub. After the part is sketched, the appropriate locating point is selected and a locator is drawn in (Figure 8-13). Since the tapped hole is positioned between two of the flange holes, a secondary locator must be drawn to maintain this relationship. A diamond pin is used, in this case to simplify loading and unloading.

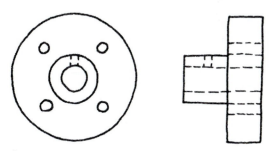

Figure 8-12 Flange plate.

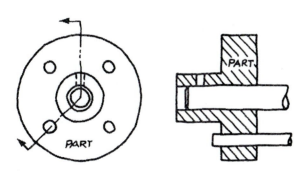

Figure 8-13 Selective locating points.

To ensure that the hole is parallel to the base, the tool body is drawn as shown in Figure 8–14. The simplest form of tool body for this fixture is an angle plate, which provides both a horizontal base and a vertical flange to mount the locators.

The final step in sketching this tool is putting the tap in place. If the angle-plate flange is too long, as in Figure 8–15, placing the tap in the drawing will call attention to it. Also notice that the tap will hit the larger locator. The designer can still make any necessary changes before tool construction is begun.

Once the sketch is complete, the tool designer begins the actual tool drawing. For this fixture, the drawing should be an assembly type because of the simplicity of the tool (Figure 8–16). Details, such as chamfer edges, clearance hole for the tap, and specific mounting instructions for the locators, are specified in the final tool drawing. Normally omitted are additional details, such as table location and tool location.

Sketches allow the tool designer to try different ideas while the tool is still on the drawing board. Many design errors can be changed or adjusted before tool production begins.

The CAD environment provides a friendly approach to the sketching process and can save design time once the final concept is determined. Sketching techniques in CAD allow the designer to develop the concept in much the same manner as a manual sketch and see the geometry change to match the dimension. The outcome is a sketch that can be developed into the final concept. Electronically, many possible concepts may be developed and archived fro tracking purposes. It is also possible to create your own libraries of parts for future use in similar designs. It is a simple process to copy a component from one drawing and insert it into another. Smart designers will create their own libraries for future reference. Vendor libraries are used for component selection.

DIMENSIONING TOOL DRAWINGS

Dimensions describe the overall limits or size of a part or detail. Only those dimensions that are absolutely necessary to manufacture or inspect a part are included in tool drawings. Dimensions that do not directly aid in the manufacture or inspection of the part must be omitted for the sake of clarity.

There are two methods of describing sizes on a drawing: limit dimensioning and basic size dimensioning. Each manufacturing company decides which method better suits its needs. The intent in any case is

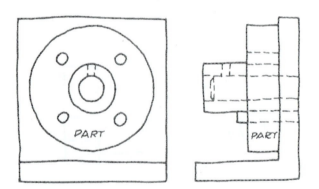

Figure 8–14 Angle-plate tool body.

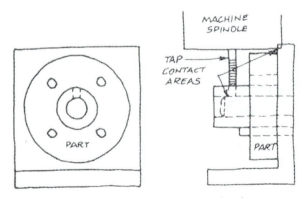

Figure 8–15 Placing the tool in the sketch.

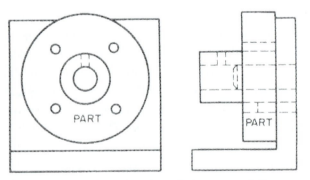

Figure 8–16 Completed tool drawing of flange-tapping fixture.

to establish acceptable dimension variations in a way that will not confuse shop personnel.

Limit Dimensions

Limit dimensions relate the upper and lower limits of size to each detail (Figure 8–17). On older drawings, external features were shown with the larger limit value on top and the smaller value below; internal features were shown in just the opposite way: the smaller value on top and the larger value below. However, with newer drawings, the larger value is shown on top and the smaller value is shown below, regardless of the type of detail or feature being dimensioned. When limit dimensions are written horizontally, the smaller value is shown to the left and the larger value is shown to the right, separated by a short dash. The main advantage of this dimensioning form is that it eliminates the need to calculate the upper and lower size limits before making the part.

Basic Size Dimensions

Basic size dimensions are used to describe the basic size and the allowable variation from that size (Figure 8–18). The two types of basic size dimensions are unilateral and bilateral.

Unilateral dimensions specify tolerance in only one direction (Figure 8–19A). The tolerance value can be plus or minus of the basic size, but not both.

Bilateral dimensions specify tolerance in two directions. If the plus and minus values are the same, the tolerance is considered an equal bilateral tolerance (Figure 8–19B). If one value is larger, the tolerance is an unequal bilateral tolerance (Figure 8–19C).

Designating Critical Dimensions

On parts where several dimensions must be shown, it is sometimes difficult for the designer to relate to the toolmaker which size is critical and which is not. One

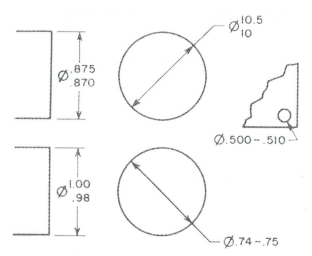

Figure 8–17 Limit dimensions.

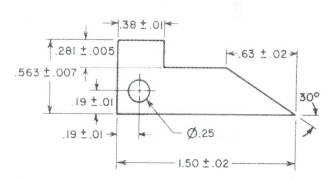

Figure 8–18 Basic size dimensions.

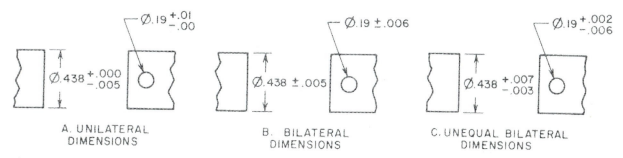

Figure 8–19 Tolerance variations.

method is to designate the critical size area directly on the drawing (Figure 8–20). Another method is to apply tolerances only to critical dimensions. The toolmaker automatically assumes that untoleranced dimensions are controlled by the general tolerances in the title block or in specific notes.

Designating Specific Fits

Standard combinations of internal and external diameters have been established to allow designers to specify exactly what amount of clearance or interference they desire between mating parts.

MILLIMETER AND INCH DIMENSIONING

The two primary systems of measurement used in manufacturing are the SI (Systeme Internationale d'Unites) and the U.S. Customary. The metric system, or more appropriately the SI system, was internationally adopted in 1954. It is rapidly becoming the world standard of measurement. Many U.S. companies have begun to change over completely to the SI metric system.

The first major step in making this changeover is to develop all new drawings in the standard SI unit, the millimeter. Several types of drawing have evolved to make the transition between systems as smooth as possible. They include dual-dimensioned drawings, drawings with millimeter dimensions and inch cross-references, and straight SI drawings with millimeter dimensions.

Dual-Dimensioned Drawings

Dual-dimensioned drawings are a compromise between U.S. Customary and SI drawings to permit a gradual transition to the SI (Figure 8–21). In this style, the primary system or design units are normally placed on top for vertical dimensions or to the left for horizontal dimensions. It is relatively easy to determine which system is the design or primary system, since those dimensions tend to be rounded more.

The following are the main advantages of dual-dimensioning:

- It permits gradual transition from the U.S. Customary to the metric system.
- It allows parts to be made without hazardous conversions on the shop floor.

Here are the main disadvantages:

- It prolongs eventual changeover to total SI drawings.
- It takes longer to dimension drawings.
- It creates a cluttered appearance that could cause misreadings.

SI Drawings with U.S. Customary Cross-References

This style of drawing is an improvement over dual-dimensioning in that only one dimension is given for each value. The conversion values are given in a chart on the drawing sheet (Figure 8–22).

The advantages of this system:

- Drawings are not cluttered with excessive dimensions.
- These drawings are easier to interpret than dual-dimensioned drawings.
- When no longer needed, the conversion chart can be removed, leaving a drawing completely in SI units.

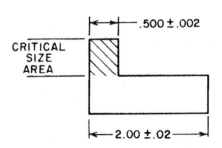
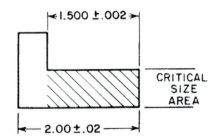

Figure 8–20 Specifying critical dimensions.

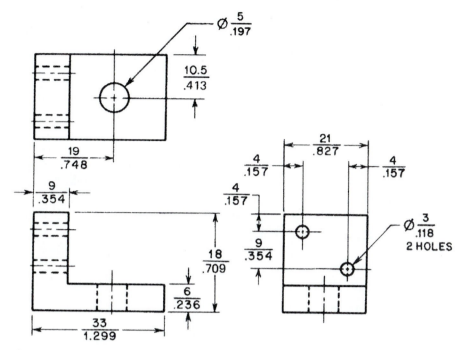

Figure 8-21 Dual-dimensioned drawing.

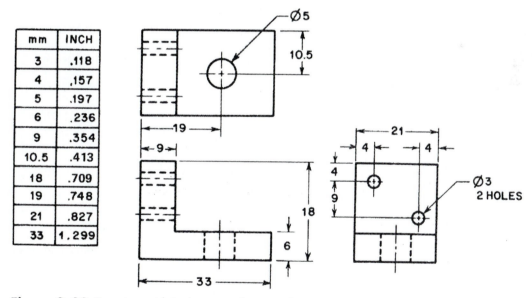

Figure 8-22 Drawing with inch cross-reference chart.

The disadvantages:

- It prolongs eventual changeover to total SI drawings.
- It takes longer to draw because of the added conversion chart.

Straight SI Drawings

Straight SI drawings are drawn with dimensional values expressed solely in millimeters (Figure 8–23). They are fundamentally the same as U.S. Customary

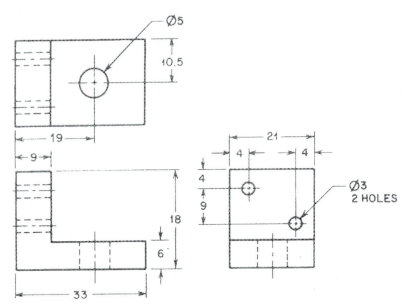

Figure 8-23 Straight SI drawing.

drawings except that the unit of measurement is in millimeters. The advantages of straight SI drawings are:

- There is no misinterpretation due to more than one dimensioning system.
- They require the shortest drawing time of the three methods discussed.

These are the disadvantages:

- Total conversion to metric tooling and machinery must be complete.
- There is the possibility of error due to conversions made on the shop floor to compensate for inch-type measuring tools.

The newer machine tools provide the ability to switch from English to metric very easily. Toolroom equipment can be retrofitted or may already have digital readouts that are capable of switching from one system to another. Programmable equipment will run on either system, eliminating the need to make conversions. Company culture determines the choice of systems. Globally competitive designers and manufacturers have gravitated to the metric system for business reasons.

Rules for Using SI Drawings

While SI metric drawings are virtually identical in form to inch-type forms, there are differences in dimensional rules. Listed next are the primary differences and rules the tool designer must remember:

- All units on SI drawings are expressed in millimeters (mm).
- Where inch signs (") are omitted on inch-type drawings, the SI unit (mm) abbreviation is also omitted.
- All fractional millimeter values must be preceded by a zero (for example, 0.15 mm).
- Neither commas nor spaces are used with values over 1000.
- Meaningless zeros are omitted from SI numerical values. For example, one half of a millimeter is written as 0.5, not 0.500.
- Units of angularity still use the degree increment and symbol (°).
- Thread values are expressed in their original form; they are not converted. For example, 1/2-20 UNF does not become M12.7-1.27.

Standard Drafting Sheets

Specific size drafting sheets have been developed to standardize drawings throughout the manufacturing community. By standardizing sizes, problems in use and storage have been eliminated. Figure 8–24 shows the standard sheet sizes for U.S. Customary and SI metric systems.

Reduction and Enlargement Ratios

Quite often parts cannot be drawn to their actual size. Either the part is too large and must be reduced to fit on the paper, or it is too small and must be enlarged. In these cases, drawings are made to a specific scale. Standard scale ratios used with inch and millimeter drawings are shown in Figure 8–25.

Tolerance and Allowance Comparison

Regardless of the system used to make a part, there must be a specific relationship in actual size values for given degrees of manufacturing precision. A comparison of typical inch and millimeter tolerance values is shown in Figure 8–26.

Projection

One important difference between international and North American drawings is the method of projection. SI drawings made in North America differ from those drawn in Europe in that the projections and arrangement of views are not the same. European drawings use the first-angle projection as a basis for all mechanical drawings (Figure 8–27). While the views are actu-

INCH STANDARD	MILLIMETER STANDARD
A – 8 $\frac{1}{2}$ X 11	A 6 – 105 X 148
B – 11 X 17	A 5 – 148 X 210
C – 17 X 22	A 4 – 210 X 297
D – 22 X 34	A 3 – 297 X 420
E – 34 X 44	A 2 – 420 X 594
	A 1 – 594 X 841
	A 0 – 841 X 1189

Figure 8–24 Standard drafting sheets.

INCH STANDARD		MILLIMETER STANDARD
$\frac{1}{16}$" = 1"		1 : 100
$\frac{1}{8}$" = 1"		1 : 50
$\frac{1}{4}$" = 1"		1 : 20
$\frac{1}{2}$" = 1"	REDUCTION	1 : 10
$\frac{3}{4}$" = 1"		1 : 5
1" = 1"		1 : 1
2" = 1"		2 : 1
3" = 1"		5 : 1
5" = 1"	ENLARGEMENT	10 : 1
10" = 1"		20 : 1
ETC.		ETC.

Figure 8–25 Reduction and enlargement ratios.

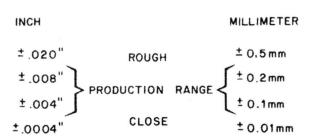

Figure 8-26 Tolerance and allowance comparison.

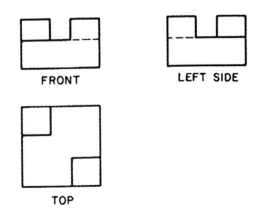

Figure 8-27 First-angle projection.

ally identical, the placement and arrangement differ from the third-angle projection, shown in Figure 8–28, which is common in the United States and Canada.

GEOMETRIC DIMENSIONING AND TOLERANCING

In addition to the dimensioning methods already discussed, a new form of dimensioning is currently finding wide acceptance in industry for tooling and production drawings—geometric dimensioning and tolerancing. This system uses the function of the particular part feature as a guide for dimensioning rather than simply stating a size and tolerance value. Thus, not only the size limitations of a particular feature but also its geometric characteristics can be controlled using this dimensioning method.

Feature Control Symbols

The primary symbols used in geometric dimensioning and tolerancing are the *feature control symbols* (Figure 8–29). These are actually a combination of symbols, letters, and numbers that are contained in a frame. Depending on the application, a feature control symbol will normally consist of a *geometric characteristic symbol*, a *tolerance value*, and a *datum reference*. This sequence of information is called the international sequence and represents the new method of constructing feature control symbols. The older method, called the American sequence, used the geometric characteristic symbol, datum reference, and tolerance value sequence. It was replaced by the international sequence in 1982, but you may still see it on many prints.

Geometric Characteristic Symbols

The basic geometric characteristics and their symbols are shown in Figure 8–30. These characteristics and symbols, as shown, are grouped into five basic cate-

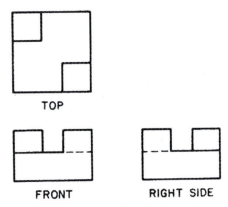

Figure 8-28 Third-angle projection.

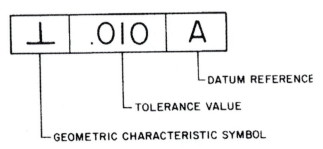

Figure 8-29 Basic feature control symbol.

gories identified as form, orientation, profile, location, and runout. Additionally, each characteristic is designated as applicable to individual features, related features, or both individual and related features. Individual features are those that are not dependent on another surface or area to establish the designated form. Thus, a flat surface is flat only to itself and has no relationship to another surface. Likewise, the other form characteristics—straightness, circularity, and cylindricity—are individual controls. Related controls, such as orientation, location, and runout, are dependent on other surfaces or areas to establish their form. Parallelism, for example, must be related, since a second surface must be used to establish a parallel surface. Profile may be either individual or related, depending on the application. The exact meaning and interpretation of each geometric characteristic symbol is shown in Figure 8–31.

Tolerance Values

The tolerance value shown in the feature control symbol shows the total amount a part feature may vary from the stated dimension.

Datum Reference

The datum reference indicates a *datum*—a specific surface, line, plane, or feature that is assumed to be perfect and is used as a reference point for dimensions or features. Datums are identified by letters and are shown on the part with *datum feature symbols* that refer to the datum reference in the feature control symbol (Figure 8–32). The letter *B* in the feature control symbol refers to datum *B* on the part. The datum feature symbol is shown as a letter with a dash on either side contained in a box. All letters in the alphabet except *I*, *O*, and *Q* may be used to identify datums. *I*, *O*, and *Q* are not used because they closely resemble numbers. On large prints, where more than 23 datums are required, double letters (*AA*, *AB*, *AC*, etc.) may also be used.

Datum references can be shown as a single datum or as multiple datums. When only one letter is shown in the feature control symbol, it means the feature is related only to that single datum. When a feature must be related to more than one datum, multiple datums are specified (Figure 8–33). Datums are

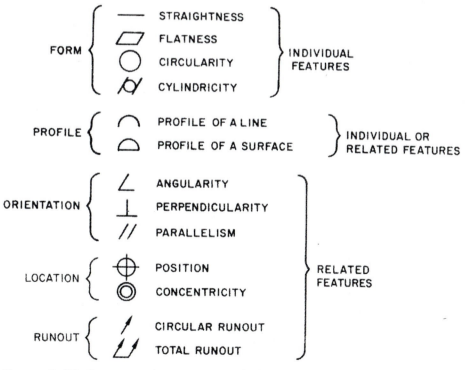

Figure 8–30 Geometric characteristic symbols.

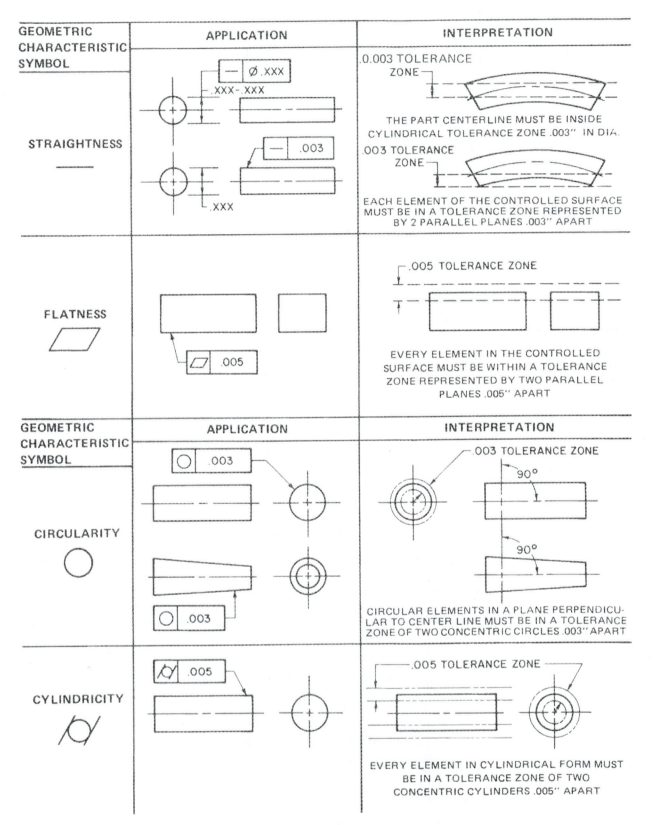

Figure 8-31 Interpreting geometric characteristic symbols.

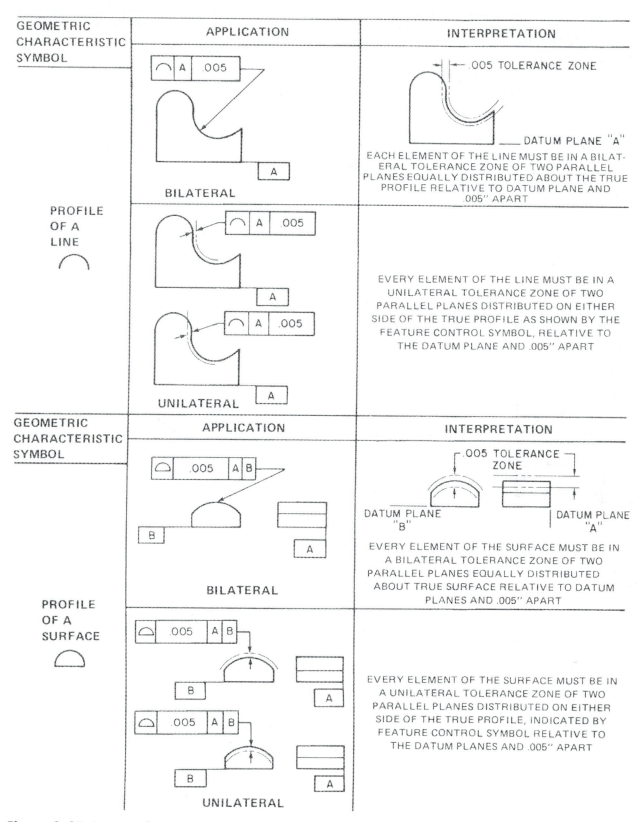

GEOMETRIC CHARACTERISTIC SYMBOL

APPLICATION

INTERPRETATION

PROFILE OF A LINE

.005 TOLERANCE ZONE

DATUM PLANE "A"

EACH ELEMENT OF THE LINE MUST BE IN A BILATERAL TOLERANCE ZONE OF TWO PARALLEL PLANES EQUALLY DISTRIBUTED ABOUT THE TRUE PROFILE RELATIVE TO DATUM PLANE AND .005" APART

BILATERAL

EVERY ELEMENT OF THE LINE MUST BE IN A UNILATERAL TOLERANCE ZONE OF TWO PARALLEL PLANES DISTRIBUTED ON EITHER SIDE OF THE TRUE PROFILE AS SHOWN BY THE FEATURE CONTROL SYMBOL, RELATIVE TO THE DATUM PLANE AND .005" APART

UNILATERAL

GEOMETRIC CHARACTERISTIC SYMBOL

APPLICATION

INTERPRETATION

PROFILE OF A SURFACE

.005 TOLERANCE ZONE

DATUM PLANE "B"

DATUM PLANE "A"

EVERY ELEMENT OF THE SURFACE MUST BE IN A BILATERAL TOLERANCE ZONE OF TWO PARALLEL PLANES EQUALLY DISTRIBUTED ABOUT TRUE SURFACE RELATIVE TO DATUM PLANES AND .005" APART

BILATERAL

EVERY ELEMENT OF THE SURFACE MUST BE IN A UNILATERAL TOLERANCE ZONE OF TWO PARALLEL PLANES DISTRIBUTED ON EITHER SIDE OF THE TRUE PROFILE, INDICATED BY FEATURE CONTROL SYMBOL RELATIVE TO THE DATUM PLANES AND .005" APART

UNILATERAL

Figure 8–31 (continued).

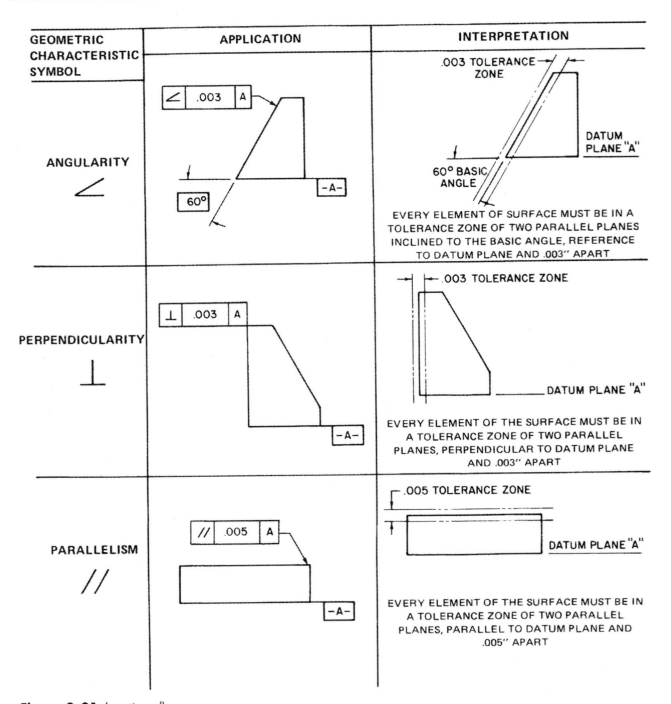

Figure 8–31 (continued).

arranged in their order of precedence; a primary datum is shown first, a secondary datum is shown next, and a tertiary datum is shown last. If two datum surfaces, such as two primary datums, have equal values, they are indicated by a dash between the datum reference letters (Figure 8–34).

SUPPLEMENTARY SYMBOLS

Supplementary symbols, or modifying symbols, are used in addition to the basic elements in the feature control symbol to further define and clarify the meaning and intent of the basic feature control symbol.

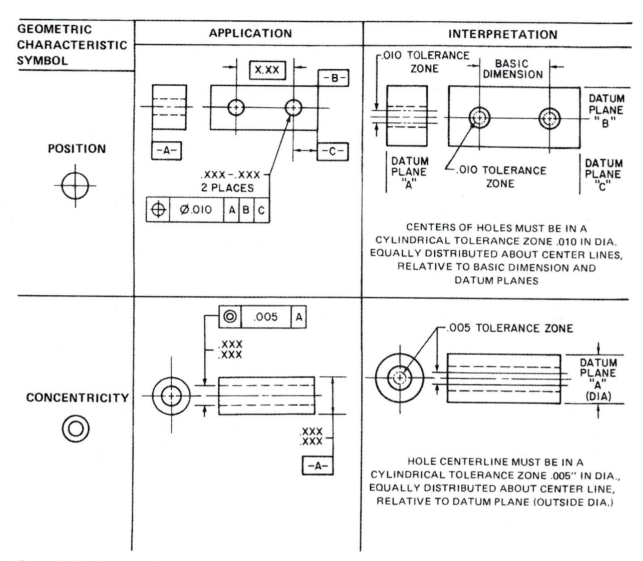

GEOMETRIC CHARACTERISTIC SYMBOL	APPLICATION	INTERPRETATION

POSITION

CENTERS OF HOLES MUST BE IN A CYLINDRICAL TOLERANCE ZONE .010 IN DIA. EQUALLY DISTRIBUTED ABOUT CENTER LINES, RELATIVE TO BASIC DIMENSION AND DATUM PLANES

CONCENTRICITY

HOLE CENTERLINE MUST BE IN A CYLINDRICAL TOLERANCE ZONE .005" IN DIA., EQUALLY DISTRIBUTED ABOUT CENTER LINE, RELATIVE TO DATUM PLANE (OUTSIDE DIA.)

Figure 8–31 (continued).

The symbols commonly used with geometric dimensioning and tolerancing are shown in Figure 8–35.

Maximum Material Condition

The *maximum material condition*, or MMC, is the condition in which the part feature has its maximum amount of material within the specified limits of size. This is the largest size of an external feature and the smallest size of an internal feature (Figure 8–36).

The MMC modifier is shown as a circled *M* (Ⓜ) and is used only on part features that vary in size.

This modifier may be used with either the tolerance value, the datum reference, or both (Figure 8–37A). In use, it means the specified tolerance value applies only at the MMC size of the feature.

Least Material Condition

The *least material condition*, or LMC, is the opposite of maximum material condition (MMC). The LMC shows the least material within the specified limits of size. This value is the smallest size of an external feature or the largest size of an internal feature (Figure 8–36.)

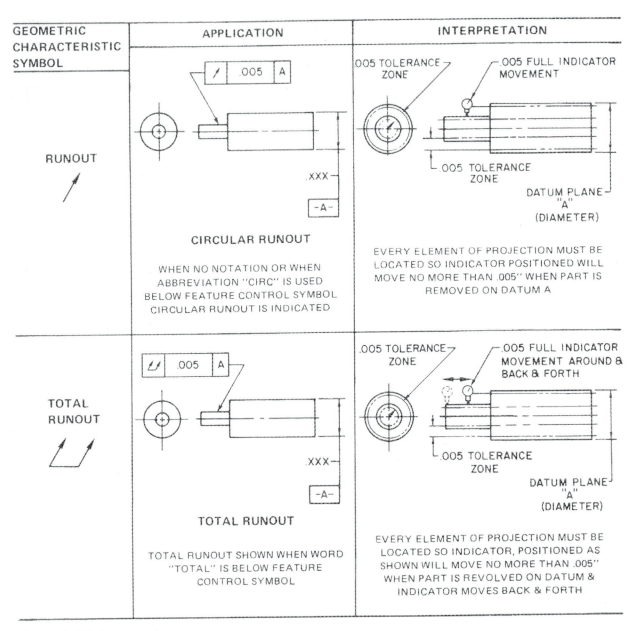

GEOMETRIC CHARACTERISTIC SYMBOL	APPLICATION	INTERPRETATION
RUNOUT	CIRCULAR RUNOUT WHEN NO NOTATION OR WHEN ABBREVIATION "CIRC" IS USED BELOW FEATURE CONTROL SYMBOL CIRCULAR RUNOUT IS INDICATED	EVERY ELEMENT OF PROJECTION MUST BE LOCATED SO INDICATOR POSITIONED WILL MOVE NO MORE THAN .005" WHEN PART IS REMOVED ON DATUM A
TOTAL RUNOUT	TOTAL RUNOUT TOTAL RUNOUT SHOWN WHEN WORD "TOTAL" IS BELOW FEATURE CONTROL SYMBOL	EVERY ELEMENT OF PROJECTION MUST BE LOCATED SO INDICATOR, POSITIONED AS SHOWN WILL MOVE NO MORE THAN .005" WHEN PART IS REVOLVED ON DATUM & INDICATOR MOVES BACK & FORTH

Figure 8–31 (continued).

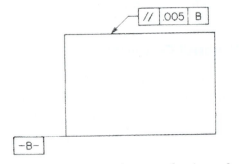

Figure 8–32 Application of a datum feature symbol.

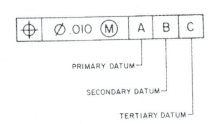

Figure 8–33 Specifying multiple-part datums.

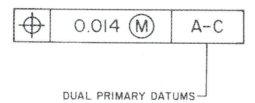

DUAL PRIMARY DATUMS

Figure 8-34 Specifying a dual primary datum.

SUPPLEMENTARY SYMBOLS	
Ⓜ	MAXIMUM MATERIAL CONDITION (MMC)
Ⓛ	LEAST MATERIAL CONDITION (LMC)
Ⓢ	REGARDLESS OF FEATURE SIZE (RFS)
Ⓟ	PROJECTED TOLERANCE ZONE
∅	DIAMETER
S∅	SPHERICAL DIAMETER
R	RADIUS
SR	SPHERICAL RADIUS
( )	REFERENCE DIMENSION
☐	BASIC DIMENSION

Figure 8-35 Supplementary symbols.

The LMC modifier is shown as a circled L (Ⓛ) and, like the MMC modifier, is used only on features that can vary in size. This modifier may also be used with either the tolerance value, the datum reference, or both (Figure 8–37B). When used, it means that the specified tolerance value applies only at the LMC size of the feature.

Regardless of Feature Size

The *regardless of feature size*, or RFS, modifier, like the other modifiers, is used only on those features that can vary in size. It is shown as a circled S (Ⓢ) and can be applied either to the tolerance value, the datum reference, or both (Figure 8–37C). When used, the RFS modifier means that the stated tolerance value applies regardless of the actual size of the feature.

The current standard does not use the RFS symbol. RFS is assumed if no other modifier is used. If there are no other modifiers in the feature control frame, the designer is allowing the default status of RFS. It should be mentioned here that part geometries using RFS result in a greater cost to manufacture. Regardless of feature size indicates that variability must be built into the production tooling. Instead of using a solid locator for a bore that has been toleranced at RFS, a collet adapter might be used. The collet will provide the adjustment for size that is needed to meet the variable design requirements. Maximum material condition and least material condition are more economical modifiers, since both could be designed and built to a specified size.

Projected Tolerance Zone

The *projected tolerance zone*, shown as a circled P (Ⓟ), is used to specify the height of a mating part that is to be assembled with the toleranced feature, usually a hole, to ensure the proper alignment of the mat-

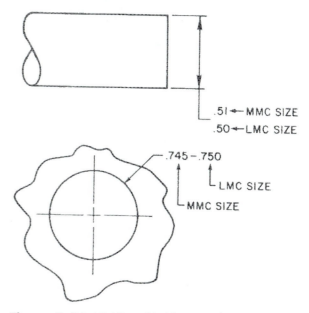

.51 ← MMC SIZE
.50 ← LMC SIZE

.745 – .750
LMC SIZE
MMC SIZE

Figure 8-36 MMC and LMC sizes of a part.

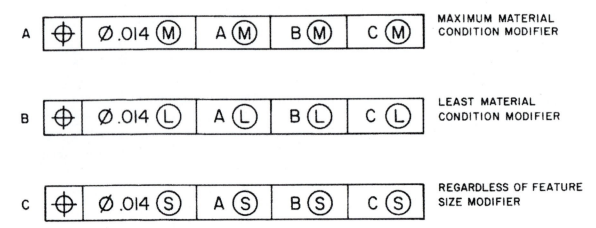

Figure 8–37 Positioning modifiers in a feature control symbol.

ing holes. This modifier is positioned below the feature control symbol, as shown in Figure 8–38.

Diameter

The diameter symbol, shown as a circle with a 60-degree slash, indicates a diameter. In a feature control symbol, it is positioned before the tolerance value. This symbol is also used to denote diameters elsewhere on the print. When used, it is placed before the dimensional value (Figure 8–39A).

Spherical Diameter

When a spherical diameter must be specified, the symbol used is an *S* with a diameter symbol (Figure 8–39B). Here again, the symbol precedes the dimensional size of the feature.

Radius

A radius is specified on a print with the letter *R* (Figure 8–39C). This symbol is also positioned before the dimensional value.

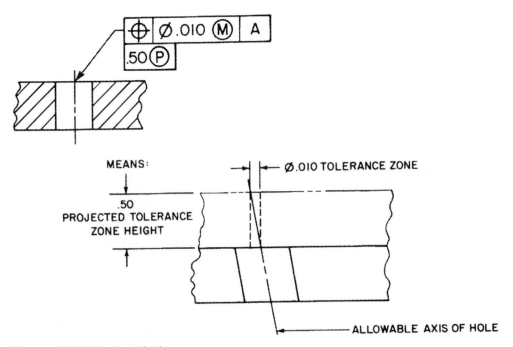

Figure 8–38 Projected tolerance zone.

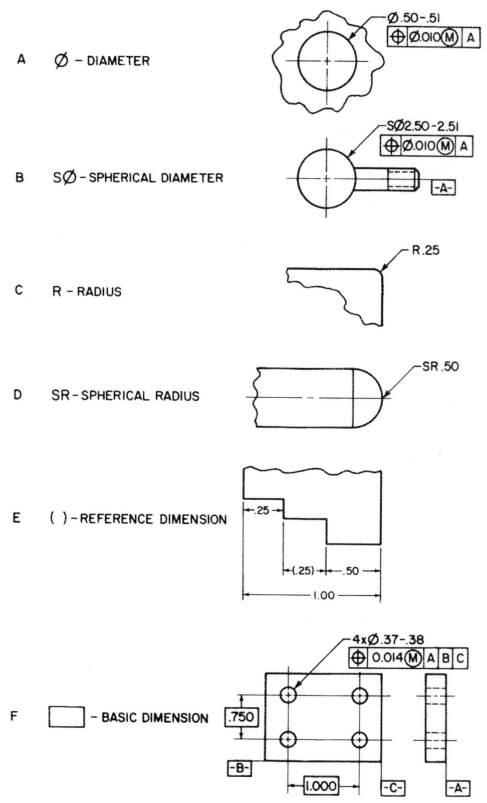

A   Ø – DIAMETER

B   SØ – SPHERICAL DIAMETER

C   R – RADIUS

D   SR – SPHERICAL RADIUS

E   ( ) – REFERENCE DIMENSION

F   ☐ – BASIC DIMENSION

Figure 8-39 Applying supplementary symbols to specify dimensional control.

Spherical Radius

When a spherical radius must be specified, the symbol *SR* precedes the dimensional size of the feature (Figure 8–39D).

Reference and Basic Dimensions

Reference and basic dimensions are treated the same way in geometric dimensioning and tolerancing as they are in standard dimensioning and tolerancing. A reference dimension, shown in parentheses ( ) in Figure 8–39E is used for information and convenience. A basic dimension, shown by a framed number in Figure 8–39F is used as a base dimension. It represents the theoretically exact, or perfect, feature size. A basic dimension forms the basis for tolerances applied to other features or in feature control symbols.

Basic dimensions determine theoretically perfect locations for features of size, such as holes, grooves, keys, and slots. Basic dimensions originate from datums.

GEOMETRICALLY DIMENSIONED AND TOLERANCED TOOL DRAWINGS

Tool drawings constructed following geometric dimensioning and tolerancing practices resemble the drawing shown in Figure 8–40. The specific units used with the print are of minimal concern with regard to the feature control symbols, since both inch dimensions and millimeter dimensions readily lend themselves to this dimensioning and tolerancing system.

COMPUTERS IN TOOL DESIGN

Computers are rapidly replacing drawing boards as the preferred method of preparing engineering drawings. Virtually every area of design has been affected by the computer. Now a workpiece design can go from inception through final documentation without ever being near a drawing board. Like other areas, tool design has been affected by the computer.

Today's typical CAD (computer-aided design) system is a relatively small unit that occupies very little space compared to older systems. Although many organizations use large mainframe computers, the microcomputer, shown in Figure 8–41, is rapidly becoming a standard part of many drafting departments. A typical CAD system used for preparing engineering drawings is shown in Figure 8–42. This type of system uses a variety of different computer programs to perform a wide variety of tasks. A CAD software package is used to prepare the part drawings, while a fixturing software program is used to add the fixturing components and elements to the engineering drawing to design the workholder. Finally, a CAM (computer-aided manufacturing) program is used to generate the codes needed to machine the parts.

Designing Jigs and Fixtures on the Computer

The application of CAD in tool design has dramatically reduced the time necessary to construct the required tool drawings. With these systems, the designer needs only to retrieve a copy of the engineering drawing on the screen, with the fixturing software, the workholder is constructed around the drawn object. Figure 8–43 shows this process. The part drawing is first retrieved on the computer screen, as shown at view A. The fixture is then built around the workpiece and shown in as many views as needed to completely describe it (view B and view C). Even the computer codes and tool paths can be shown using a suitable CAM program, as shown at view D and view E. Showing the proposed path of the cutter can help the tool designer see where any possible problems might arise between the cutter and the fixture.

The software programs normally used for tool design are composed of a series of *component libraries*. These libraries contain a wide variety of drawn components such as clamps, supports, locators, and similar commercial components that are drawn to a specified

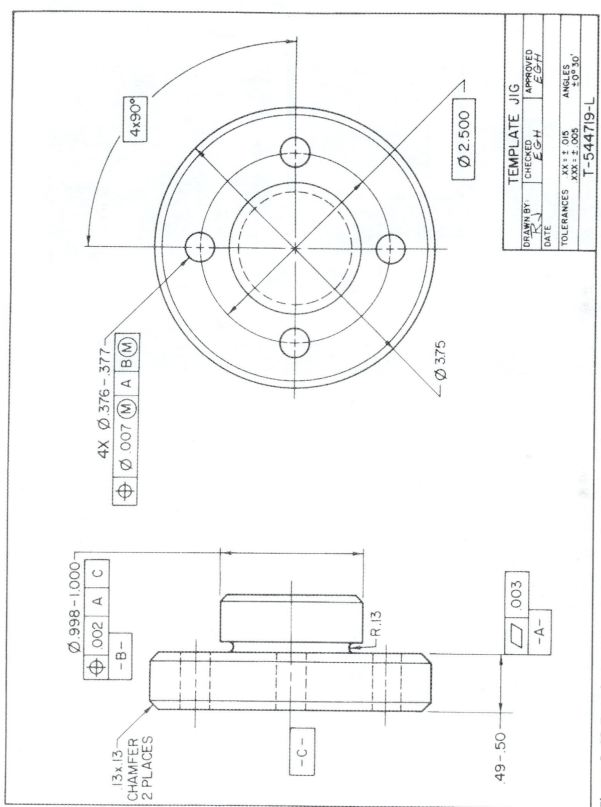

Figure 8-40 Typical part print using geometric dimensioning and tolerancing.

133

Fig. 8–41 A typical microcomputer used for CAD/CAM operations and for designing the required fixturing. (Courtesy of Hewlett-Packard Company)

Fig. 8–42 A typical CAD workstation. (Courtesy of Hewlett-Packard Company)

scale size and are contained in the fixturing software. A typical component library is shown in Figure 8–44. In use, the component is simply called up from the proper library and placed on the drawing where it is required. In most cases where commercial components are required for the workholder, the time needed to complete a tool drawing has been reduced from days to hours. Even when special parts are required for the jig or fixture, the CAD system can greatly reduce the time required to make the tool drawings. Together, these CAD, fixturing, and CAM programs have dramatically reduced the time needed to design and build the necessary fixturing for any job.

SUMMARY

The following important concepts were presented in this unit:

- Tool drawings differ from production drawings in both size and complexity.

- Tool drawings may be prepared as either assembly drawings or detailed drawings.

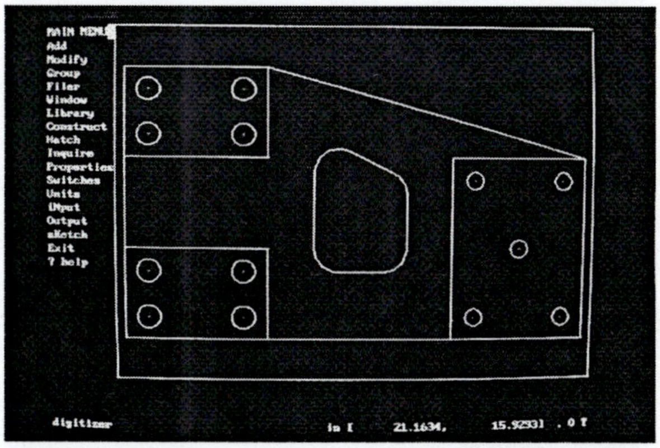

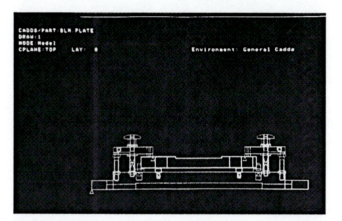

Figure 8–43A The part is retrieved on the screen.

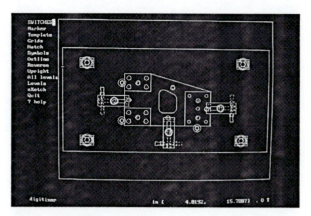

Figure 8–43B The fixture is designed around the part.

Figure 8–43C Additional views are used to clarify the design.

```
%
N0 G49 G90
N1 G00 X11.4692 Y9.5619 T1 M06
N2 M03 S2000 F2.0
N3 G43 H22 Z0.1 M08
N4 G82 P100 Z0.10 R0.1
N5 Y10.5619
N6 X13.4692
N7 Y9.5619
N8 Y13.5619
N9 Y14.5619
N10 X11.4692
N11 Y13.5619
N12 X18.4692 Y12.5619
N13 X20.4692
N14 X19.4692 Y11.0619
N15 X18.4692 Y9.5619
N16 X20.4692
N17
N18 G80 M09
N19 G00 G91 G28 Z0
```

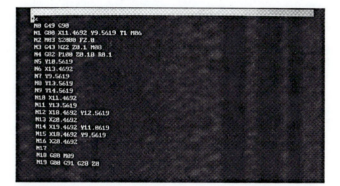

Figure 8–43D The necessary computer code for the machine tool.

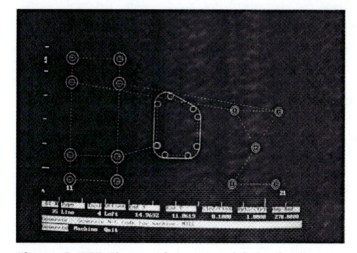

Figure 8–43E Showing the cutter path helps the designer spot any problems between the cutter and the fixture.

Figure 8–43 The process used to design a workholder on a computer (*Courtesy of Jergens, Inc.*).

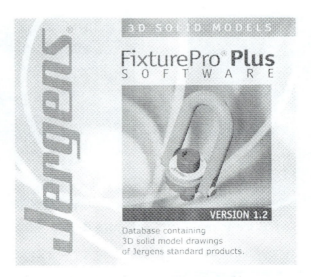

Figure 8–44 A fixture component library can save many hours of drawing standard fixturing components (*Courtesy of Jergens, Inc.*).

- To reduce the time and expense of preparing these drawings, the designer should simplify the tool drawings as much as possible. These are a few ways to do this:
  - Use words to replace drawn details.
  - Reduce the number of views.
  - Use symbols when appropriate.
  - Use templates and guides.
  - Use part numbers for standard parts.
- Each design should start as a sketch.
  - Sketching helps to formulate the design.
  - Sketches help to reduce problems and show relationships.
- The dimensioning method and placement of the dimensions can help to show any critical relationships.
- Inches and millimeters are the two units of measurement used to describe linear dimensions. Some drawings are made in inch units, while others are drawn in millimeters. Still others are shown with both types.
- Geometric dimensioning and tolerancing is finding wide acceptance as a method of dimensioning tool drawings.
- Computers are finding wider application in preparing tool drawings. In addition to being used for CAD/CAM, fixture-component libraries are helping the tool designer by reducing the need to draw standard tooling components.

REVIEW

1. Briefly describe the following types of tool drawings:
   a. Assembly
   b. Detailed
2. List five methods of simplifying tool drawings.
3. Identify the following dimensional forms with one or more of the sample dimensions on the right:
   a. Vertical limit dimension
   b. Horizontal limit dimension
   c. Unilateral dimension
   d. Unequal bilateral dimension
   e. Equal bilateral dimension

   1. $.438^{+.000}_{-.006}$
   2. $2.75 \pm .01$
   3. $1.000^{+.001}_{-.000}$
   4. $.375$
      $.370$
   5. $.250 - .252$
   6. $.375 - .370$
   7. $.750^{+.003}_{-.005}$
   8. $.50^{+.02}_{-.01}$
4. List six rules for dimensioning SI metric drawings.

5. Identify the following symbols:
   a. ▱
   b. ——
   c. //
   d. ⌰
   e. ⊥
   f. ○
   g. ∠
   h. ⌀
   i. ◠
   j. ╱
   k. ⊕
   l. ◠
   m. ◎

6. Draw the symbols for the following modifiers:
   a. Least material condition
   b. Regardless of feature size
   c. Projected tolerance zone
   d. Maximum material condition

7. Identify the maximum material condition and least material condition for the following part:

8. Identify the parts of the feature control symbol shown here:

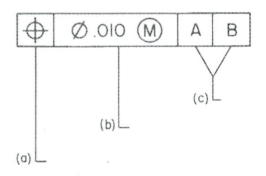

9. Identify the five basic categories of feature characteristics controlled with geometric dimensioning and tolerancing.

10. Where are the supplementary symbols that are used to control a dimension positioned with reference to the size value in a dimension?

11. Answer the following questions concerning geometric dimensioning and tolerancing using Figure 8–45.
   a. What is the largest diameter of this part?
   b. What type of control is specified for the outside diameter?
   c. Which feature establishes datum B?
   d. How is datum A established?
   e. Which two features establish the positional relationship of the four mounting holes?
   f. What type of dimension is used to show the diameter of the bolt circle for the mounting holes?
   g. How is the small shoulder diameter related to the center hole diameter
   h. What does the symbol (Ⓜ) mean in the feature control symbol for the mounting holes?
   i. How should the hole size specification for the mounting holes be read?
   j. What are the MMC and LMC sizes of the center hole?

12. What does CAD mean? What does CAM mean?

13. Where are the fixturing components located in a fixturing software program?

14. How does CAD make tool design faster and easier?

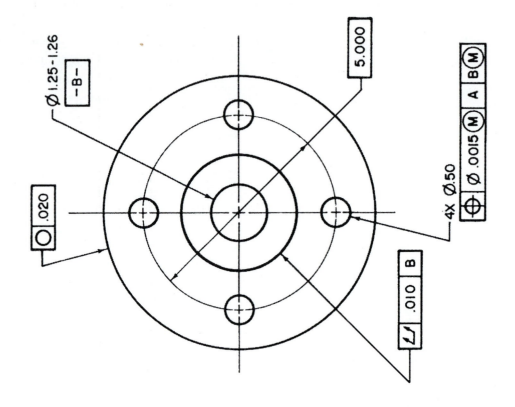

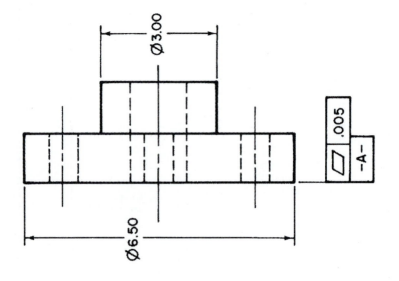

Figure 8–45

Designing and Constructing Jigs and Fixtures

UNIT 9

Template Jigs

OBJECTIVES

After completing this unit, the student should be able to:

- Analyze part data to determine suitable tool designs.
- Specify locating and supporting methods and the tools to suit a sample part.
- Design a suitable template jig for a sample part.
- Complete a tool drawing of a proposed tool to accomplish specific tasks.

TEMPLATE JIGS

Template jigs are the simplest and most basic jig type used in production. They can serve as layout guides for locating holes and contours or as a jig for low-speed production of accurate workpieces.

Template jigs are generally made without clamps and depend on pins, nests, or part shapes to reference them to the part. Since templates are the simplest form of jig, they are used extensively in low-volume production such as prototype work or short-run jobs. Cost is another attractive feature of template jigs. Although templates lack the refinements of other types of jig, they cost much less.

The main disadvantage in using template jigs is that they are not as foolproof as other types of jigs. Unless the operator is careful, many parts could be inaccurately machined.

VARIATIONS OF TEMPLATE JIGS

To accommodate the many types of parts, there are several forms and variations of the template jig. While the same basic construction principles are used, the jigs are made to suit each part.

Layout Template

The *layout template* is used as a rapid reference tool for laying out several identical parts (Figure 9–1). Layout templates can be used for locating holes, contours, and external part details. When mating parts must be laid out, the second part is laid out by turning over the template (Figure 9–2).

The material used for templates is normally determined by the projected tool life. Templates for one-time use can be made from plastic or soft aluminum sheet. Templates intended for longer production runs should be made of harder materials, such as hardened tool steel. In either case, the thickness of material should be sufficient to maintain the proper relationship between template elements. A thickness range of .050 inch to .200 inch is adequate for most applications.

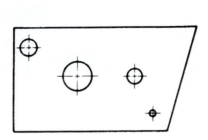
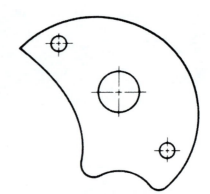

Figure 9–1 Layout template.

Flat-Plate Template Jigs

Flat-plate template jigs are used to locate holes on flat surfaces (Figure 9–3). This template is normally located with pins referenced from the edge or from other holes. The plate thickness, which should be specified, normally depends on the diameter of the hole to be drilled. The general rule is that the minimum plate thickness equals one to two times the tool diameter.

Circular-Plate Template Jigs

Circular-plate template jigs are used for cylindrical workpieces (Figure 9–4). They are usually located on a cylindrical portion of the part. While any hole pattern can be machined with these jigs, they are generally used for round hole patterns. Since clamps are not normally built into a template jig, jig pins should be used to maintain the correct hole alignment. When using a

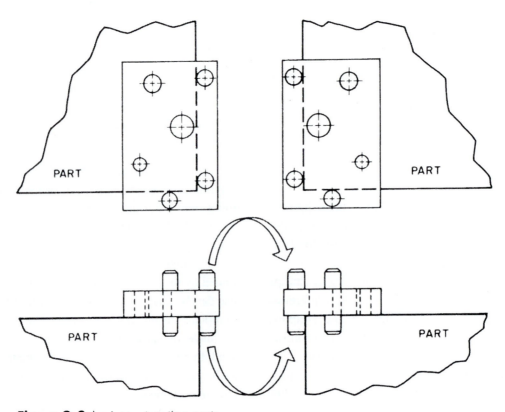

Figure 9–2 Laying out mating parts.

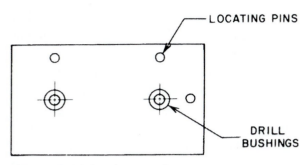

Figure 9–3 Flat-plate template jig.

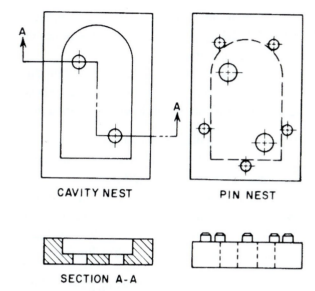

Figure 9–5 Nesting template jigs.

jig pin for this application, the first hole is drilled and the jig pin is inserted in it. The jig pin will then accurately align the jig as well as the part and will ensure the accurate location of the remaining holes.

Nesting Template Jigs

Nesting template jigs use a cavity nest or a pin nest to locate the workpiece (Figure 9–5). They can accommodate almost any form or shape part; the only restriction is the complexity of the cavity. The more detailed the cavity, the more expensive the jig. For this reason, cavity nests are normally limited to symmetrical shapes such as rounds, squares, or rectangles. When a nest is

desired for nonsymmetrical shapes, a pin nest should be specified to keep costs to a minimum.

DESIGN PROCEDURES

Once the tool designer decides that a template jig is the best choice for a particular job, the design process

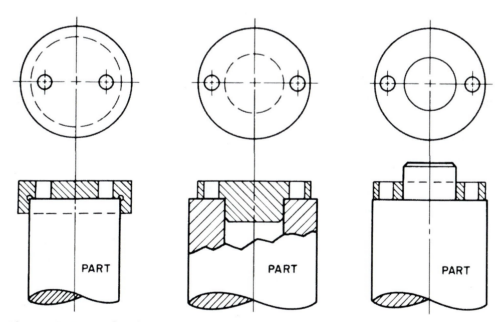

Figure 9–4 Circular-plate template jig.

begins. Following the planning processes outlined in Units 1 and 7, the tool designer assembles and evaluates all the necessary data.

An examination of the part drawing and production plan for the butt plate, part number 2164, shown in Figure 9–6, discloses the following information:

- The part is a flat disc, 2.56 inches in diameter and .75 inch thick, with a 1.000-inch hole in its center.
- The material specified is 1020 steel.
- The only operation required of the jig is to drill two holes .19 inch in diameter, 1.770 inches apart.
- The blank part received for drilling is faced, drilled, and reamed to the specific dimensions.

Using this information as a starting point, the tool designer begins designing the template jig.

Locating the Part

The first area the designer must consider is location. Since this part has a 1.000-inch center hole, it should be used to locate the part. When a center hole cannot be used to locate a cylindrical part, a nest can be used. However, in no case should a nest and a pin locator be used together for a part of this type. This results in redundant location, which can cause serious inaccuracies.

When determining the size and position of locators, be sure to maintain the required part tolerance. Locators must be positioned in reference to the part dimensions. For example, if the butt plate, part number 2164, was dimensioned as shown at view A of Figure 9–7, it could not be located the same as it would be if dimensioned as at view B. In these cases, the manner in which the part is dimensioned indicates the critical portions of the part.

Since the butt plate is dimensioned from its center hole, the hole should be selected to locate this part. When calculating the size of the locator and its relationship to the position of the drill bushings, the designer must first determine the size of the pin (Figure 9–8). The limits of size of the hole are .996 inch to 1.004 inches, or a .008-inch total possible deviation. The center position of the .19-inch holes must be controlled between .879 and .891 from center, a difference of .012 inch. Using the smallest hole size

of .996 inch as a guide, a pin of .995 inch could be specified, allowing a .001-inch clearance. Assuming the tool designer allows the toolmaker a tolerance range of ±.0005 inch, the largest this pin could be is .9955 inch and the smallest is .9945 inch.

Using these figures, the tool designer then calculates the worst possible conditions that could exist between the part and the locator. If the largest hole size (1.004 inches) and the smallest pin size (.9945 inch) are compared, as shown in Figure 9–9, the maximum center distance between the holes is .88975 inch. The minimum possible dimension is .88025 inch. Comparing these values against the part limits of .891 inch and .879 inch shown in the part drawing, these tool tolerances easily meet the required precision.

Examining the opposite condition, that of the largest locator and the smallest hole, as shown in Figure 9–10, the maximum center distance is .8855 inch, while the minimum distance is .8845 inch. Again, these values fall well within the desired precision requirements for the tool.

Since either of these conditions, although possible, seldom occurs, the tool designer should also calculate the mean or average values. As shown in Figure 9–11, the mean value of the hole is 1.000 inch and the mean size of the locator is .995 inch. This permits a maximum difference of .8875 inch and a minimum difference of .8825 inch.

Locating the Bushings

Once the locator pin size and tolerance have been determined, the bushing location and allowable tolerance must be calculated. Referring to the part drawing shown in Figure 9–6, the following data are gathered:

- The first hole is positioned .885 ±.006 inch from the center hole's centerline.
- The second hole is positioned 1.770 ±.012 inches from the centerline of the first hole.

Using these facts, the following calculations must be made:

- Maximum allowable distance between hole centerlines

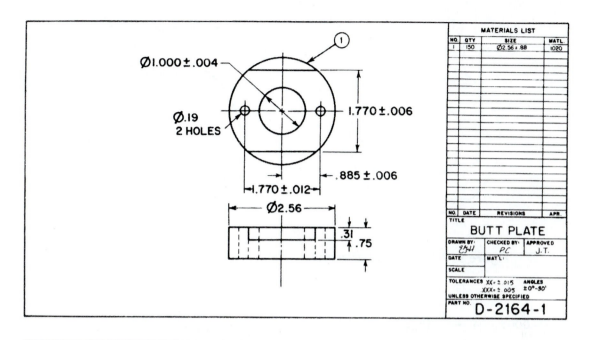

MATERIALS LIST			
NO.	QTY	SIZE	MATL.
I	150	Ø2.56 x .88	1020

NO.	DATE	REVISIONS	APR.

TITLE

BUTT PLATE

DRAWN BY:	CHECKED BY:	APPROVED
ghl	P.C.	J.T.
DATE	MAT'L:	
SCALE		

TOLERANCES XX = ± .015   ANGLES
XXX = ± .005   ± 0° - 30'
UNLESS OTHERWISE SPECIFIED

PART NO.

D-2164-1

PRODUCTION PLAN

P/N 2164	PART NAME BUTT PLATE		QUANTITY 150	ORDER NO. 1234
DRAWING NO. D-2164-1	PROCESS PLANNER D.E. LIBBY	REVISION NO	DATE	PAGE 1 OF 1

OPR. NO.	DESCRIPTION	DEPT.	MACH. TOOL
1.	CUTOFF 2.56 IN BAR STOCK TO .88 IN THICK	#68 CUTOFF P.M.	ABRASIVE CUTOFF MACH. #68-121
2.	FACE TO .75 IN THICK	#27 LATHE	TURRET LATHE #27-62
3.	DRILL AND REAM Ø1.000 ±.004 IN HOLE THRU	#27 LATHE	TURRET LATHE #27-62
4.	DRILL Ø.19 IN HOLES THRU 2 PLACES	#66 DRILLING	DRILL PRESS #66-09
5.	MILL 1.770 ±.006 x .31 IN STEP	#37 MILLING	HORZ. #37-10
6.	DEBURR	#7 FINISHING	TUMBLER #7-1207
7.	INSPECT	#7 FINISHING	NONE

OPR. NO.	TOOL DESCRIPTION	SIZE	SPECIAL TOOL NO.
7.	FUNCTIONAL GAUGE	————	FIXTURE 1-2164-3
5.	SIDE MILLING CUTTERS	4.00 x .50 x 1.00	FIXTURE T-2164-2
4.	DRILL	Ø.19	JIG T-2164-1
3.	REAMER	Ø1.000	NONE
2.	DRILL	Ø.969	NONE
1.	CUTOFF WHEEL	10.00 x .125 x 1.00	NONE

Figure 9–6 Part drawing and production plan.

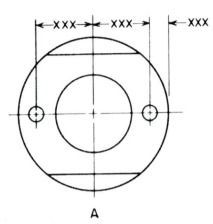

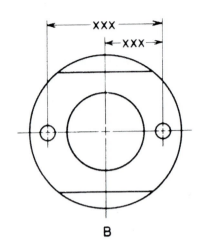

Figure 9–7 Dimensional differences.

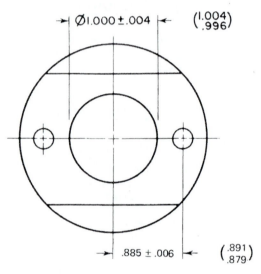

Figure 9–8 Size limits.

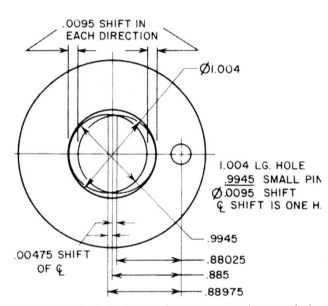

Figure 9–9 Calculating locator size—largest hole, smallest locator.

- Tolerance values that will ensure the desired precision

Figure 9–12 illustrates one method of determining these values. In the first case, the centerline of the part and the tool is shifted to the maximum allowable value of .00475 inch. The nominal size of .885 inch is then added to the offset. This value, .88975 inch, is then subtracted from the largest allowable size of .891 inch, yielding a maximum deviation of .00125 inch. In the second calculation, the part is shifted to the maximum amount allowed in the opposite direction. The offset is then subtracted from .885 inch; the difference is .88025 inch. The minimum allowable size, .879 inch, is then subtracted from the calculated value, resulting in maximum deviation of .00125 inch in the opposite direction.

In other words, the first bushing must be placed within a ±.00125-inch tolerance range to properly locate the hole in the part (Figure 9–13). Allowing the toolmaker a ±.001-inch tolerance permits a built-in ±.00025-inch wear allowance, which will lengthen the tool service life.

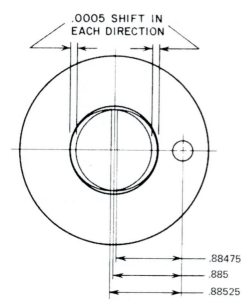

Figure 9-10 Calculating locator size—largest locator, smallest hole.

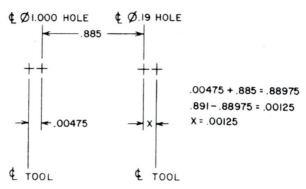

Figure 9-12 Calculating bushing locational tolerance—maximum.

Once the position of the first bushing has been decided, the locational tolerance of the second bushing must be specified. Following the general rule of tool tolerance, the tool designer should specify the center-to-center distance between the holes as 1.770 ±.006 inch, or 50 percent of the part tolerance.

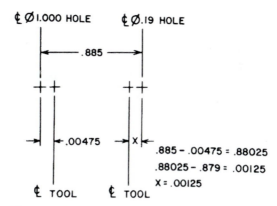

Figure 9-13 Calculating bushing locational tolerance—minimum.

Initial Jig Design

After calculating the locator and bushing values, the designer is ready to plan the rest of the tool. The first step in this initial design is rough-sketching the part. Since the butt plate is a flat disc, only two views need to be sketched (Figure 9-14).

Starting with the top view, the designer sketches in the rough outline of the jig plate (Figure 9-15). To avoid confusion, draw the part in red and the tool sketch in black. This allows for easy identification of the part at all times and prevents confusion where several lines lie close together. Once the jig plate is sketched in, the designer adds the dimensions.

Onto the front view of the part, the tool designer sketches in the front view of the jig plate. In this view, the designer must decide how long the locator pin should be. To avoid sticking and jamming, make the locator no longer than one-half the part thickness. In

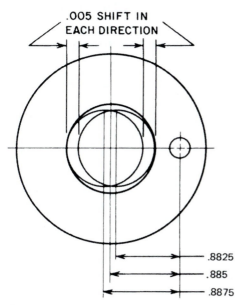

Figure 9-11 Calculating locator size—mean size, locator and hole.

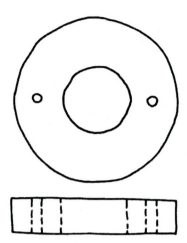

Figure 9-14 Sketching the part.

this case, the pin should be .38 inch long. To maintain the proper alignment between the holes, a pin must be placed in the first hole after it is drilled. The designer should specify a standard jig pin to keep the tool cost as low as possible.

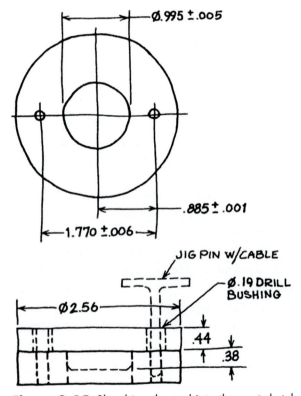

Figure 9-15 Sketching the tool into the part sketch.

Completing the Tool Drawing

Once the initial sketch has been drawn and the tool designer is satisfied that the tool will perform the desired function, the tool drawing is started. Since the initial sketch gives only very general information, the tool designer must make many other determinations to complete the more detailed and specific tool drawing. In addition to all dimensional values and tolerances, the tool drawing must include any special instructions the toolmaker will need to fabricate the tool (Figure 9–16).

TOOL DESIGN APPLICATION

1. Analyze the part drawing and production plan in Figure 9–17 to determine all relevant information necessary to design an appropriate template jig.
2. Compare similarities and differences between the assigned part and the example part.
3. Prepare an initial sketch to calculate necessary dimensions.
4. Design the tool drawing, including all necessary data needed to build the tool.

Suggestions for Design

- Select the locational method and make calculations by following the procedure outlined in the example problem.
- Specify all relevant data and sizes on the initial sketch.
- Maintain tolerance values specified on the part drawing.
- Analyze prior operations to determine locational surfaces and work to be performed.

SUMMARY

The following important concepts were presented in this unit:
- Template jigs are the most basic type of jig.
  - Template jigs do not normally have a built-in clamp.
  - Jig pins are often used to maintain correct alignments.

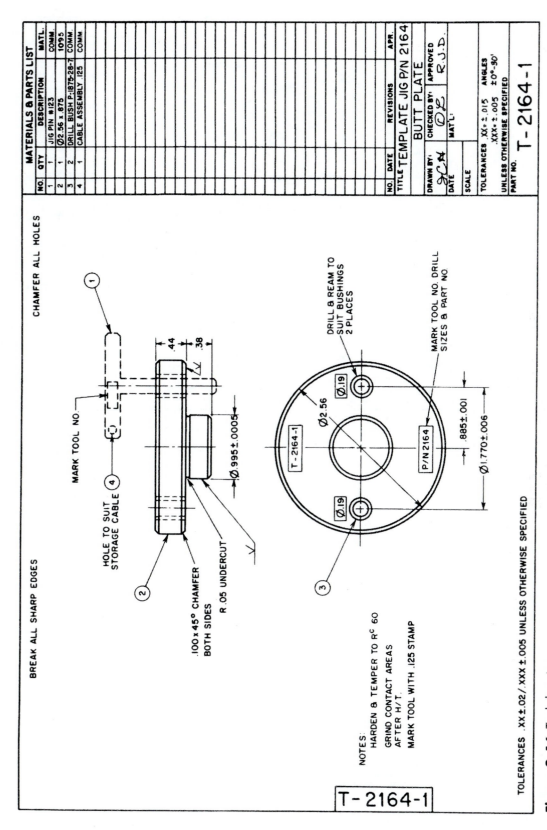

Figure 9-16 Tool drawing.

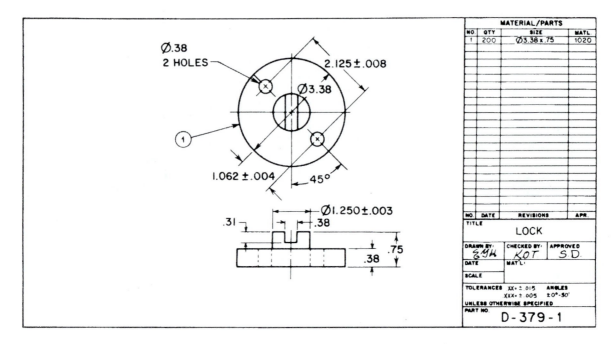

NO.	QTY	SIZE	MATL
1	200	Ø3.38 x .75	1020

NO.	DATE	REVISIONS	APR.

TITLE

LOCK

DRAWN BY:	CHECKED BY:	APPROVED
GMH	KOT	S.D.
DATE	MAT'L	
SCALE		

TOLERANCES XX= ±.015   ANGLES
              XXX= ±.005   ±0°-50'
UNLESS OTHERWISE SPECIFIED

PART NO.

D-379-1

PRODUCTION PLAN

P/N	379	PART NAME	LOCK	QUANTITY	200	ORDER NO	529-77

DRAWING NO	D-379-1	PROCESS PLANNER	R.J. LA RUE	REVISION NO		DATE		PAGE 1	OF 1

OPR NO.	DESCRIPTION	DEPT.	MACH. TOOL
1.	CUTOFF 3.38 BAR STOCK TO .88 THICK	#68 CUTOFF RM.	HORZ. BANDSAW #68-20
2.	FACE TO .75 THICK	#27 LATHE	TURRET LATHE #27-63
3.	TURN .38 SHOULDER TO Ø1.250±.003	#27 LATHE	TURRET LATHE #27-63
4.	DRILL Ø.38 HOLE THRU 2 PLACES	#66 DRILLING	DRILL PRESS #66-12
5.	MILL .38 x .31 SLOT IN BOSS	#37 MILLING	HORZ. MILL #37-10
6.	DEBURR	#7 FINISHING	TUMBLER #7-12071
7.	INSPECT	#7 FINISHING	NONE
7.	FUNCTIONAL GAUGE		FIXTURE 1-379-3
5.	MILLING CUTTER	4.00 x .375 x 1.000	FIXTURE T-379-2
4.	DRILL	.375	JIG T-379-1
OPR NO.	TOOL DESCRIPTION	SIZE	SPECIAL TOOL NO

Figure 9-17 Part drawing and production plan.

- Before the design is started, the tool drawing must be studied to obtain all necessary information about the part.
- Careful calculations must be made to determine the exact size and location of the locators and bushings.
- Each design should start as a sketch.
  - Sketching helps to formulate the design.
  - Sketches help to reduce problems and show relationships.
  - Once the design is set and the problems have been solved, the final design drawing is prepared.

REVIEW

1. What are the most common forms of template jig?
2. How are template jigs usually referenced to the workpiece?
3. What are the main advantages of template jigs?
4. What is the main disadvantage of template jigs?
5. What are the two common forms of nest?
6. Which nest is better for nonsymmetrical workpieces?
7. Why is it a bad practice to use both a nest and a pin locator to locate a part?
8. As a general rule, what percentage of part tolerance is applied to the tool?
9. What can a designer do to avoid confusion in the initial design sketches?
10. Why should the locator be only half as long as the part thickness?

UNIT 10

Vise-Held and Plate Fixtures

OBJECTIVES

After completing this unit, the student should be able to:

- Analyze part data to determine suitable tool designs.
- Specify locating, supporting, and clamping methods, and details to suit specific sample parts.
- Analyze requirements and calculate and design a cam-action clamp to hold a workpiece in a fixture.
- Design a suitable vise-held fixture and complete the required tool drawings.
- Design a suitable plate fixture and complete the required tool drawings.

Vise-held fixtures and plate fixtures are two of the most basic fixture types. Each is made from a single-base plate on which is mounted a variety of details for locating, supporting, and clamping the part. Since these two fixtures are very similar in design and construction, both are discussed in this unit.

VISE-HELD FIXTURES

Vise-held fixtures, as shown in Figure 10–1, are basically small, lightweight, simple plate fixtures. They are generally mounted in machine vises or chucks rather than directly on machine tables. Vise-held fixtures are ideal for small parts that require only light machining. Since the tool lacks many refinements, the cost of tooling is reduced considerably.

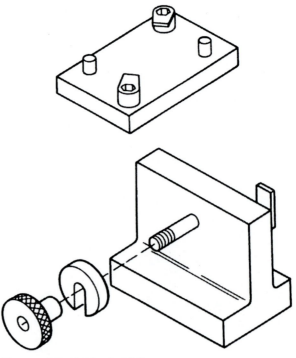

Figure 10–1 Vise-held fixtures.

DESIGNING A VISE-HELD FIXTURE

The first step in designing a vise-held fixture is the assembling and analyzing of production data. After the tool designer has decided that a vise-held fixture is the most efficient and cost-effective tool to use for a specific job, the tool design begins.

An examination of the part drawing and production plan for the butt plate, as shown in Figure 10–2, discloses the following information:

- The part is a flat disc that is 2.56 inches in diameter and .75 inch thick.
- The part has three holes, one 1.000-inch hole and two .19-inch holes 1.770 inches apart.
- The material specified is 1020 steel.
- The operation required is milling two flats 1.770 inches apart and .31 inch deep, parallel within one-half degree.
- The size of the production run is 150 pieces.
- The blank received for milling is turned, faced, bored, and drilled.

Using this information, the tool designer begins designing a suitable vise-held fixture.

Locating the Part

This is the first area the designer must consider. The main point to keep in mind here is repeatability. Locating points must be selected that will permit accurate machining of each part in the run, regardless of any "in-tolerance" size variations. Since the butt plate was located by its center hole for the drilling operation, the center hole should again be used for the milling operation.

Using the same locating point on a part for every operation ensures that all details will be correctly related to each other. Figure 10–3 illustrates one possible condition that can result if two different locating points are used for machining the same part.

Critical details are normally identified on part drawings by tighter tolerances or by how dimensions are related to the part. As a general rule, when no machined detail such as a hole or machined corner is available, the tool designer should use the same point to initially locate the part that the drafter used to

dimension it. Once the first detail has been machined, that detail is used to locate the part for the remaining operations (Figure 10–4).

To locate and position the butt plate accurately, the designer should use the 1.000-inch center hole as the primary locator and one of the .19-inch holes as a secondary locator. In this case, the secondary locator, in addition to restricting the radial movement of the part around the primary locator, sets the proper relationship and position of the part in reference to the milling cutters (Figure 10–5).

Secondary locators should not be confused with redundant locators. *Secondary locators* are used to supplement primary locators. Each secondary locator restricts a separate direction of movement. *Redundant locators* use more than one locator to restrict the same direction of movement (Figure 10–6). If, in addition to the center hole, both .19-inch holes are used to locate the butt plate, the locators will be redundant. If, however, both .19-inch holes are used and the center hole is not used, the locators will not be redundant.

Positioning the locators on the vise-held fixture is basically the same process as positioning the first bushing for the template jig. Since the primary locator on the template jig was .995 ±.0005 inch in diameter, the primary locator on the vise-held fixture should be the same size. The secondary locator on the vise-held fixture should duplicate the position of the first bushing on the template jig—that is, .885 ±.001 inch from the primary locator (Figure 10–7). It is logical to assume that if a part was drilled using this spacing, the same part should fit over locators of the same spacing.

To prevent jamming and permit easy loading and unloading, the primary locator should engage only one-half the thickness of the part. The secondary locator should be relieved (Figure 10–8). Note the position of the relieved locator. The full diameter of the pin is placed to resist the radial movement of the part around the primary locator. One other method to ensure easy loading is to make the secondary locator .060 inch shorter than the primary locator, which allows the part to be placed on the primary locator first and revolved until it drops over the secondary locator. This is simpler than trying to place the part over two pins at the same time, which will usually jam the part between them.

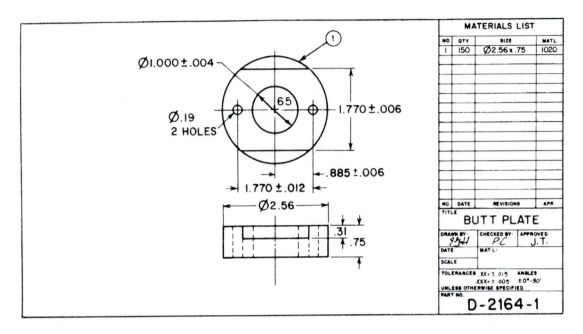

MATERIALS LIST			
NO.	QTY	SIZE	MATL
I	150	Ø2.56 x .75	1020

NO. | DATE | REVISIONS | APR

TITLE

BUTT PLATE

DRAWN BY: *8JH* | CHECKED BY: *PC* | APPROVED *J.T.*

DATE | MAT'L:

SCALE

TOLERANCES XX= ± .015 ANGLES
XXX= ± .005 ± 0°-30'
UNLESS OTHERWISE SPECIFIED

PART NO.

D-2164-1

PRODUCTION PLAN

P/N 2164	PART NAME BUTT PLATE		QUANTITY 150	ORDER NO. 1234
DRAWING NO. D-2164-1	PROCESS PLANNER D.E. LIBBY	REVISION NO.	DATE	PAGE 1 OF 1

OPR. NO.	DESCRIPTION	DEPT.	MACH. TOOL
1.	CUTOFF 2.56 BAR STOCK TO .88 THICK	#68 CUTOFF P.M.	ABRASIVE CUTOFF MACH. #68-121
2.	FACE TO .75 THICK	#27 LATHE	TURRET LATHE #27-62
3.	DRILL AND REAM Ø1.000 ±.004 HOLE THRU	#27 LATHE	TURRET LATHE #27-62
4.	DRILL Ø.19 HOLES THRU 2 PLACES	#66 DRILLING	DRILL PRESS #66-09
5.	MILL 1.770±.006 x .31 STEP	#37 MILLING	HORZ. #37-10
6.	DEBURR	#7 FINISHING	TUMBLER #7-1207
7.	INSPECT	#7 FINISHING	NONE

7.	FUNCTIONAL GAUGE		FIXTURE 1-2164-3
5.	SIDE MILLING CUTTERS	4.00 x .50 x 1.00	FIXTURE T-2164-2
4.	DRILL	Ø.19	JIG T-2164-1
3.	REAMER	Ø1.000	NONE
2.	DRILL	Ø.969	NONE
1.	CUTOFF WHEEL	10.00 x .125 x 1.00	NONE
OPR. NO.	TOOL DESCRIPTION	SIZE	SPECIAL TOOL NO.

Figure 10–2 Part drawing and production plan.

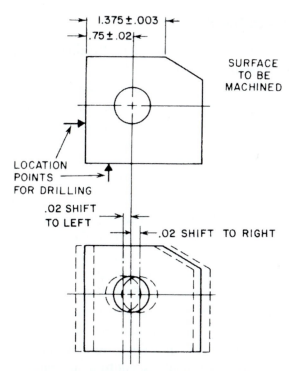

SURFACE TO BE MACHINED

1.375 ±.003

.75 ±.02

LOCATION POINTS FOR DRILLING

.02 SHIFT TO LEFT

.02 SHIFT TO RIGHT

POSSIBLE DIFFERENCES BETWEEN PARTS IF HOLE IS USED TO LOCATE PART FOR MILLING RATHER THAN ORIGINAL LOCATING SURFACES

Figure 10-3 Possible error caused by using different locating points.

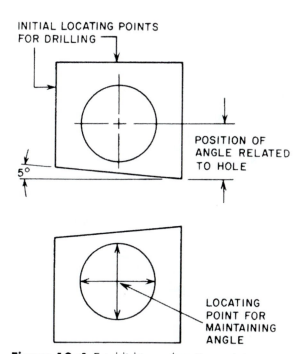

INITIAL LOCATING POINTS FOR DRILLING

POSITION OF ANGLE RELATED TO HOLE

5°

LOCATING POINT FOR MAINTAINING ANGLE

Figure 10-4 Establishing a locating point.

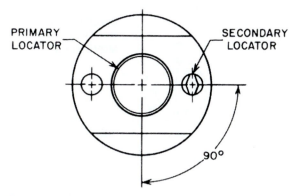

PRIMARY LOCATOR

SECONDARY LOCATOR

90°

Figure 10-5 Secondary locator.

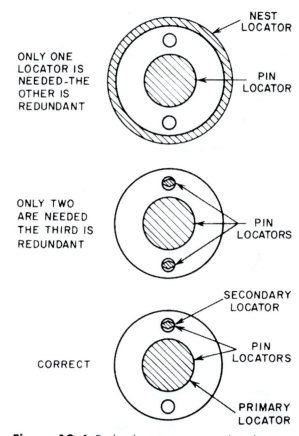

NEST LOCATOR

ONLY ONE LOCATOR IS NEEDED-THE OTHER IS REDUNDANT

PIN LOCATOR

ONLY TWO ARE NEEDED THE THIRD IS REDUNDANT

PIN LOCATORS

SECONDARY LOCATOR

CORRECT

PIN LOCATORS

PRIMARY LOCATOR

Figure 10-6 Redundant versus secondary locators.

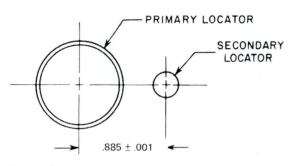

PRIMARY LOCATOR

SECONDARY LOCATOR

.885 ± .001

Figure 10-7 Positioning the secondary locator.

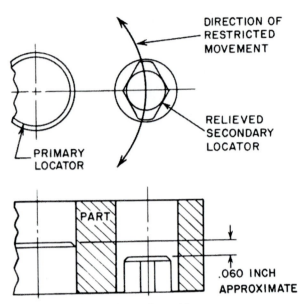

Figure 10–8 Using a relieved locator.

Supporting the Part

Special supports are not required, since the part is completely machined and its thickness is sufficient to resist bending. In this case, to reduce costs the part is supported by the base of the fixture. Solid support buttons can be used, but any benefit is offset by the cost of the buttons and the time to install them.

Clamping the Part

Clamp type, style, pressure, and location are all important factors in selecting a clamp. In the case of the butt plate, the clamp should be quick-operating and capable of moving clear off the part for faster loading and unloading. Using these requirements as a guide, the designer chooses a cam-action strap clamp similar to the clamp shown in Figure 10–9.

Once all the tool details have been selected, they must be placed in proper relation to each other to make sure the tool will work. The best way to do this is by sketching, Figure 10–10. Once the sketch is complete and the designer is confident that the tool will perform as intended, the size of the tool base must be decided. In the case of this fixture, the base must be 2.75 inches wide and 6.00 inches long. To properly hold and support the part and tool details, it should be at least 1.00 inch thick. To keep the cost down, the base plate should be cut from the standard cast section (Figure 10–11).

Locating the Cutters

The milling operation involved in fabricating the butt plate requires parallel shoulders; therefore, the best

Figure 10–9 Cam-action strap clamp.

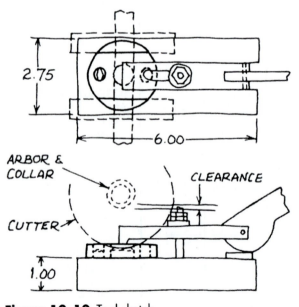

Figure 10–10 Tool sketch.

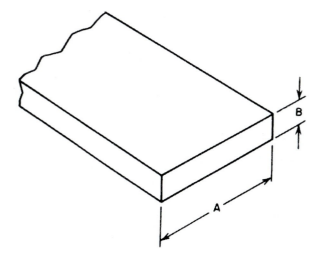

Figure 10-11 Standard commercial cast section.

FLAT SECTIONS		
PART NO.	- A -	- B -
CL — 1 — CF	3	.625
CL — 2 — CF	4	.750
CL — 3 — CF	5	.750
CL — 4 — CF	6	1.000
CL — 5 — CF	8	1.250

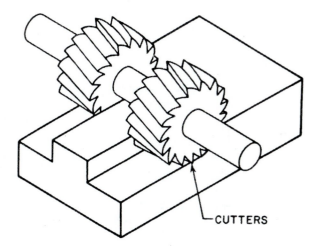

Figure 10-12 Straddle milling.

machining method is straddle milling. *Straddle milling* machines both sides at the same time (Figure 10–12).

The size of the set block used to locate the cutters is calculated by first determining the maximum and minimum allowable sizes on the butt plate. Once this is done, the size of the feeler gauge must be determined. In this case, a .125-inch feeler gauge is selected. Finally, the feeler gauge size is subtracted from the minimum butt plate size to determine the size of the set block.

Care must be used when specifying the locational and size tolerances of the set block. Since the tolerance for the part is ±.006 inch, the total allowable error in both size and position of the set block should be held to less than ±.003 inch.

Again considering the extreme permissible dimensions, the tolerance for location should be held to ±.0015 inch from the centerline of the fixture. The size tolerance should be 1.514 ±.0015 inches. Using these conditions

affects the part size by only .003 inch, as shown in Figure 10–13, which is well within the part tolerance.

While the set block shown can be used on either side to set the cutters, most set blocks are made to be used only on one side. In these cases, the set block is used to locate the position of one cutter accurately. The spacing collar between the cutters then accurately positions the second cutter. When the set block shown in Figure 10–13 is used, a .125-inch feeler gauge is used to set both the horizontal and the vertical positions on the same cutter. The dual locating surfaces are provided on this set block to permit either cutter to be set to the fixture.

When constructing the tool drawing, as shown in Figure 10–14, two points must be noted on the drawing. First, the exact size of the collar that separates the cutters is 1.770 inches. Second, the tool operator must be instructed to grind the cutters together on the same arbor to ensure that their size is identical.

Adding the size of the cutters as a third variable into the relationship of the cutters might cause a slight problem. However, either the size of the feeler gauge or the size of the collar can be adjusted to suit the specific size requirements, if necessary.

PLATE FIXTURES

Plate fixtures, as shown in Figure 10–15, are the most versatile and common form of fixture used today.

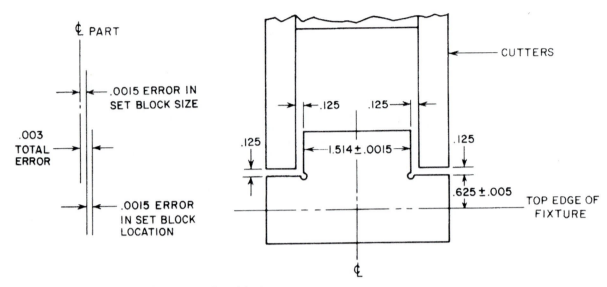

Figure 10–13 Size and position of set block.

There is almost no limit to the variety of work that can be performed with them. The basic plate fixture is a simple plate on which the part is mounted, located, and held with a variety of tooling details. While the basic appearance and style of a plate fixture are similar to those of a vise-held fixture, the main difference between the two is the workpiece size. Plate fixtures are designed to machine larger, heavier parts than vise-held fixtures are.

DESIGNING A PLATE FIXTURE

The first step in designing a plate fixture is assembling and analyzing all relevant data concerning the part to be machined. Once a plate fixture has been determined to be the most economic and efficient fixture to use for the operation, the designer begins to analyze specific information about the part to form design ideas.

The following information was assembled from an analysis of the part drawing and production plan for the housing cover (Figure 10–16).

- The part is a rough casting approximately 5.00 inches wide, 8.25 inches long, and 3.50 inches thick.
- The wall thickness is approximately .40 inch thick with supporting webs .31 inch thick.

- The part has no prior machining.
- The material is cast aluminum.
- The operation required is to gang-mill the surface that contacts the housing base.
- The production run is 75 pieces.
- The blank is received in its cast condition. The only operation that has been performed is the removal of mold flash.

Using this information, the designer begins to design the plate fixture.

Locating the Part

Consulting the part drawing, the designer notes that the step to be milled is dimensioned from the long side. Using this side as the secondary locating point, the designer specifies two locators to be used against this surface. The short side then becomes the tertiary, or third, locating point. One locator is used here (Figure 10–17). The primary locating point for this fixture, as with any other plate fixture, is the base.

Since the housing cover is a rough casting, solid locators may prove to be impractical to use because of surface irregularities. Therefore, adjustable locators are often preferred.

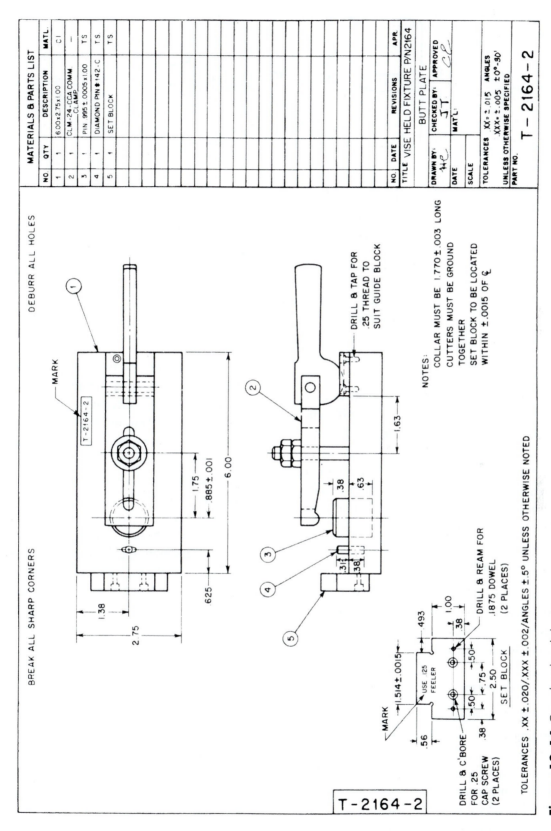

BREAK ALL SHARP CORNERS

DEBURR ALL HOLES

MATERIALS & PARTS LIST

NO.	QTY	DESCRIPTION	MATL.
1	1	6.00×2.75×1.00	CI
2	1	CLM-24-CCA COMM CLAMP	—
3	1	PIN .995±.0005×1.00	TS
4	1	DIAMOND PIN ∅.142-C	TS
5	1	SET BLOCK	TS

NO.	DATE	REVISIONS	APR.

TITLE VISE HELD FIXTURE P/N 2164
BUTT PLATE

DRAWN BY: HC	CHECKED BY: JT	APPROVED CL
DATE		MAT'L
SCALE		

TOLERANCES .XX±.015 ANGLES
.XXX±.005 ±0°-30'
UNLESS OTHERWISE SPECIFIED

PART NO. **T-2164-2**

NOTES:
COLLAR MUST BE 1.770±.003 LONG
CUTTERS MUST BE GROUND
TOGETHER
SET BLOCK TO BE LOCATED
WITHIN ±.0015 OF ℄

DRILL & TAP FOR
.25 THREAD TO
SUIT GUIDE BLOCK

MARK

T-2164-2

1.75
.885±.001
6.00
625
1.38
2.75

1.63

.38
.63

.31
.38

DRILL & C'BORE
FOR .25
CAP SCREW
(2 PLACES)

MARK

1.514±.0015
USE .125
FEELER
SET BLOCK

.56
.493
.38
1.00
.50
.75
2.50
.50
.38

DRILL & REAM FOR
.1875 DOWEL
(2 PLACES)

TOLERANCES .XX ±.020/.XXX ±.002/ANGLES ±.5° UNLESS OTHERWISE NOTED

T-2164-2

Figure 10-14 Completed tool drawing.

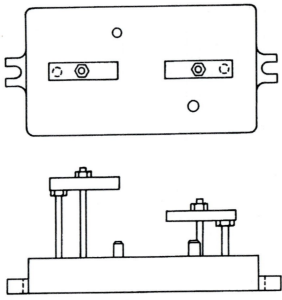

Figure 10–15 Plate fixtures.

Supporting the Part

Since the primary locating point for this part is its cast irregular top, solid supports may not provide sufficient support. A combination of fixed and adjustable supports should be used (Figure 10–18). In most cases, once the adjustable locator has been set, there should be no reason to readjust it during the run. If this problem does arise, allowances for any movement must be considered when setting the cutters.

Clamping the Part

Clamping is performed by four cam-action strap clamps positioned two on each side (Figure 10–19). Cam clamps were again selected for their fast action and easy removal from the part.

Locating the Cutters

In the case of the housing cover, the tool location for the gang milling can be a tricky task. Two separate planes must be located in reference to each other (Figure 10–20). Since the part is located by adjustable locators, establishing an accurate set block location is almost impossible. As this is the case, no set block should be specified for this tool. To accurately position the cutters, the machine operator should use the trial cut method to locate the cutters properly.

One point should always be kept in mind when specifying set blocks for a fixture. Several variables could seriously limit the effectiveness of any set block. A few examples are bent machine arbors, improperly ground cutters, or irregular workpiece surfaces, as is the case with the housing cover. Although set blocks are a good feature on a tool, they also have limitations.

Once the designer is satisfied that the tool will work, the next step is pulling all the scattered parts into a design sketch. In addition to the parts already covered, the designer specifies the fixture keys that will be used to locate the tool on the machine table.

When the sketch is complete, the information is transferred to the final tool drawing (Figure 10–21). Here, savings are gained by using part numbers that replace complicated drawn details. In the final tool drawing, all standard parts are referred to by balloon references rather than by dimensions. The references can then be checked against the material list.

CALCULATING CAM CLAMPS

In the two fixtures previously discussed in this unit, a commercial cam-action clamp was specified for holding the part in place. Occasionally commercial clamps may not fit a particular need. The tool designer must design a special cam to meet this problem.

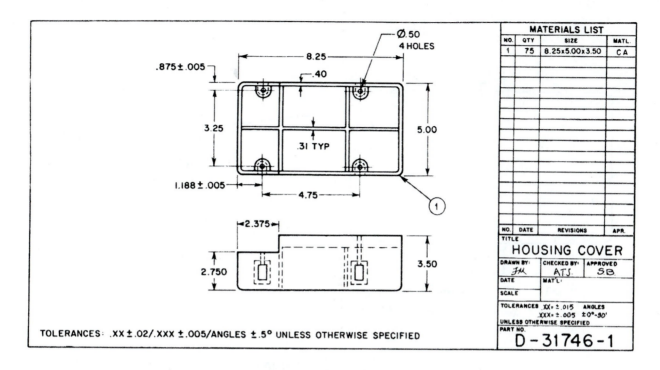

TOLERANCES: .XX ±.02/.XXX ±.005/ANGLES ±.5° UNLESS OTHERWISE SPECIFIED

MATERIALS LIST

NO.	QTY	SIZE	MATL.
1	75	8.25x5.00x3.50	C A

NO.	DATE	REVISIONS	APR.

TITLE
HOUSING COVER

DRAWN BY: *JM*	CHECKED BY: ATS.	APPROVED SB
DATE	MAT'L:	
SCALE		

TOLERANCES .XX= ±.015   ANGLES
.XXX= ±.005   ±0°-30'
UNLESS OTHERWISE SPECIFIED

PART NO.
D - 31746 - 1

PRODUCTION PLAN

P/N 31746	PART NAME HOUSING COVER		QUANTITY 75	ORDER NO 77-1234
DRAWING NO D-31746-1	PROCESS PLANNER R.J. PIERCE	REVISION NO	DATE	PAGE 1 OF 1

OPR. NO.	DESCRIPTION	DEPT.	MACH. TOOL
1	GANG-MILL MATING SURFACE TO 3.50 INCH WITH 2.750 x 2.375 INCH STEP	#37 MILLING	HORZ. MILL #37-51
2	DRILL FOUR Ø.50 MOUNTING HOLES	#66 DRILLING	RADIAL DRILL #66-14
3	DEBURR & SANDBLAST	#7 FINISHING	SANDBLASTER #7-4
4	INSPECT	#7 FINISHING	NONE

OPR. NO.	TOOL DESCRIPTION	SIZE	SPECIAL TOOL NO.
4	FUNCTIONAL GAUGE		INSP. FIXTURE 1-31746-3
2	DRILL	(1) .50	DRILL JIG T-31746-2
1	MILLING CUTTERS (4) INTERLOCKING GANG CUTTERS	(2) 6.00x4.00x1.00 (2) 3.00x5.50x1.00	MILL FIXTURE T-31746-1

Figure 10–16 Part drawing and production plan.

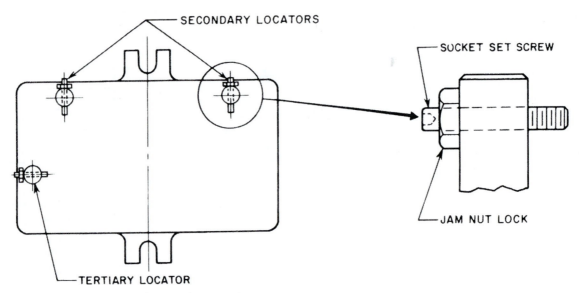

Figure 10-17 Secondary and tertiary locators.

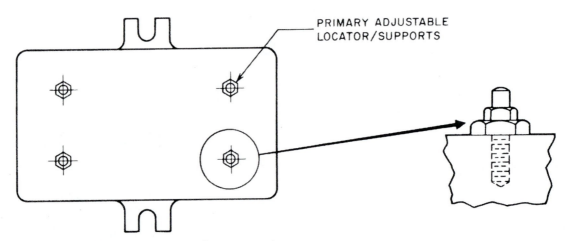

Figure 10-18 Primary locators that support the part.

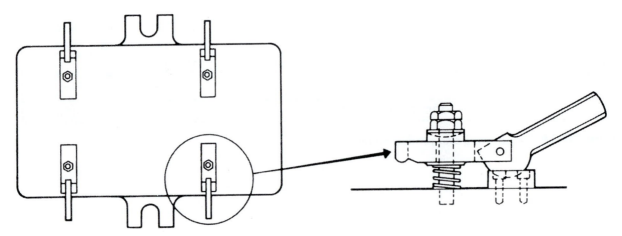

Figure 10-19 Clamping the part.

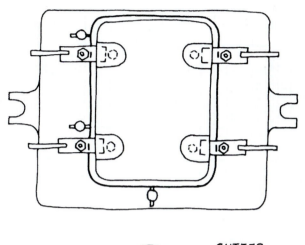

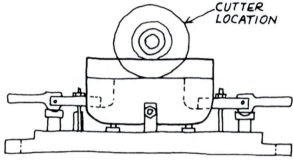

CUTTER LOCATION

Figure 10–20 Tool sketch.

Cam clamps are made in two general forms: eccentric and spiral. These clamping devices are designed to produce a specific amount of movement called *rise*. When the cam is turned through a specified arc, the arc is called the *throw*. The rise is the amount of movement perpendicular to the cam face that takes place when the cam is moved through the arc of engagement, or throw. Once this predetermined rise is reached, the cam is held in place by friction between the tool and the cam face.

Spiral cams are the most preferred type for tooling applications, since they are normally smaller, require less actuating pressure, and hold better. *Eccentric cams*, however, are easier to make. The following is one method a designer could use to calculate the required values for each type of cam.

Eccentric Cams

Eccentric cams, as shown in Figure 10–22, are not as widely used for tooling applications as are spiral cams. They require over 160 percent of the actuating torque of a similarly sized spiral cam. In addition, eccentric cams must have a radius over one and a half times as large as that of a spiral cam in order to stay locked while in use.

The first calculation that should be made to construct an eccentric cam is the amount of eccentricity. *Eccentricity* is the amount of difference between the centerlines of the pivot point and the cam radius (Figure 10–23). It is the feature of the cam that produces the rise. The formula to calculate the eccentricity is:

$$E = \frac{R}{(1 - \mathrm{Cos}\ \angle)}$$

Where E = Desired eccentricity
     R = Rise, expressed in millimeters
      or inches

Cos $\angle$ = Cosine of the angle of throw

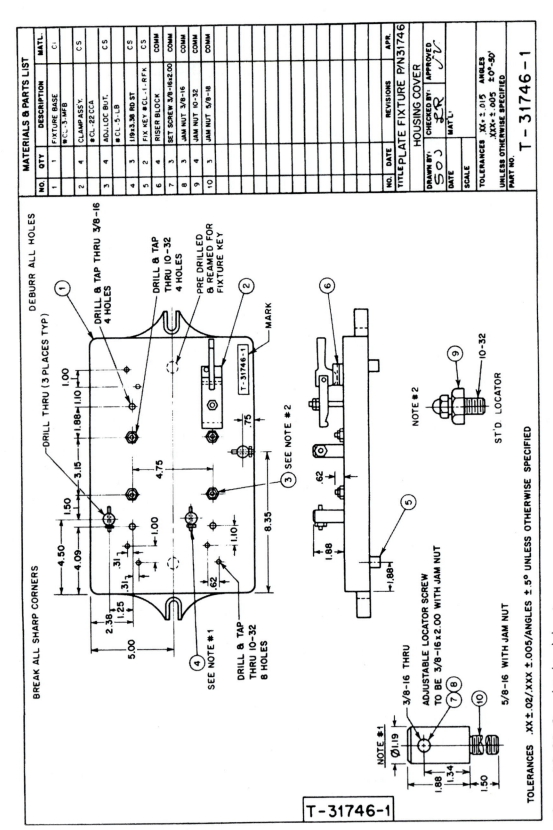

Figure 10-21 Completed tool drawing.

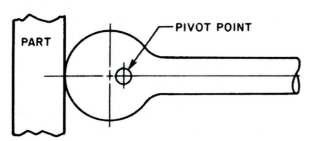

Figure 10–22 Eccentric cam.

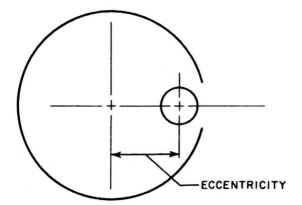

Figure 10–23 Eccentricity.

Example 1: What is the required eccentricity of an eccentric cam needed to produce a .25-inch rise with a 70-degree throw?

$$E = \frac{.25}{(1 - .34202)}$$
$$E = .37995 \text{ or } .380 \text{ inch}$$

The next calculation is the size of the required radius needed to actuate the cam (Figure 10–24).

$$Rd = E(Cos\angle + \frac{Sin}{Cf})$$

Where *Rd* = Radius required
  E = Eccentricity
  Cos ∠ = Cosine of the angle of throw
  Sin ∠ = Sine of the angle of throw
  Cf = Coefficient of friction (0.10 generally accepted value)

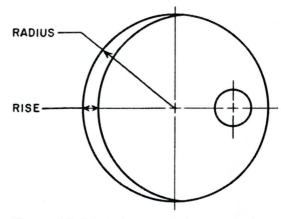

Figure 10–24 Radius required to actuate the cam.

Example 2: What is the radius required for the cam in Example 1?

$$Rd = .380 \, (.3402 + \frac{.93969}{0.10})$$
$$Rd = 3.700 \text{ inch}$$

An eccentric cam is constructed by first drawing the pivot point (Figure 10–25). The eccentric center is then offset by the amount calculated. Finally, the required radius is drawn using the eccentric point as the center.

Spiral Cams

Spiral cams, as shown in Figure 10–26, are the most popular form of cam-action clamp. This is because of

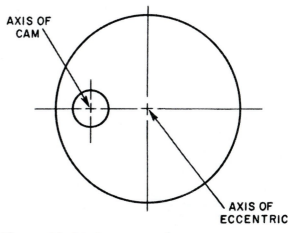

Figure 10–25 Constructing the eccentric cam.

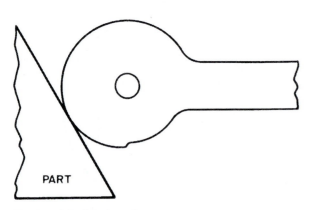

Figure 10–26 Spiral cam.

the superior hold and locking characteristics and smaller size requirements inherent in the spiral design.

The spiral cam is generated from a base circle that represents the approximate midpoint of the cam's throw (Figure 10–27). The outer and inner circles represent the limits of the cam rise.

The amount of rise a spiral cam generates is directly proportional to the length of throw and the radius of the base circle. As a general rule, there should be a .001-inch rise per inch of radius per degree of throw. The cam throw normally should be from 60 to 90 degrees.

$$R = .001 \times Rd \times T$$

Where R = Rise
Rd = Radius (base circle)
T = Throw in degrees

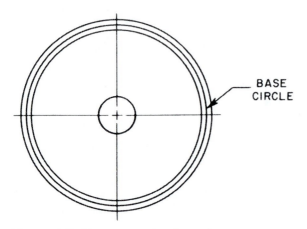

Figure 10–27 Base circle of spiral cam.

Example 1: What is the rise of a spiral cam if the cam throw is 90 degrees and the radius is 1.20 inches?

$$R = .001 \times 1.20 \times 90$$
$$R = .108 \text{ inch}$$

The amount of rise is then divided in half and added or subtracted to the base circle to find the dimensions of the outer and inner circles.

$$Or = Rd + \frac{R}{2}$$

$$Ir = Rd - \frac{R}{2}$$

Where Or = Outer circle radius
Ir = Inner circle radius
Rd = Radius (base circle)
R = Rise

Example 2: What are the outer and inner circle radii for the cam in Example 1?

$$Or = 1.20 + \frac{.108}{2}$$
$$Or = 1.254 \text{ inches}$$
$$Ir = 1.20 - \frac{.108}{2}$$
$$Ir = 1.146 \text{ inches}$$

Once the correct circle sizes are calculated and drawn, the limits of the throw should be drawn in (Figure 10–28). To allow for the possibility of inaccurate location when the pivot pin is drilled, an additional 5 to 10 degrees should be added to each end of the throw angle.

The next step is dividing the throw angle with radial lines. The number of divisions depends on how accurate the cam outline should be. For most construction drawings, 15-degree spacing is sufficiently accurate. When the radial lines are completed, the area between the outer circle and the inner circle is divided into the same number of divisions as there are radial line divisions. Each radial line and rise division line is numbered (Figure 10–29). Arcs are then constructed between the points where the lines intersect. Smoothing the lines into one continuous curve is the final step in constructing the spiral cam contour.

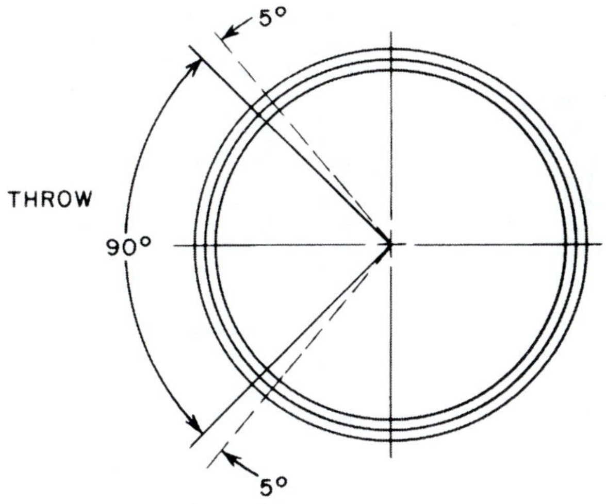

THROW

90°

5°

5°

Figure 10–28 Limits of throw.

TOOL DESIGN APPLICATION

1. Analyze the part drawings and production plans for the lock (Figure 10–30), and the adapter ring (Figure 10–31) to determine all information necessary to design an appropriate vise-held fixture and plate fixture.
2. Compare similarities and differences between assigned parts and example parts.
3. Prepare initial sketches to calculate necessary dimensions.
4. Construct the tool drawing, including all necessary data needed to build the tools.

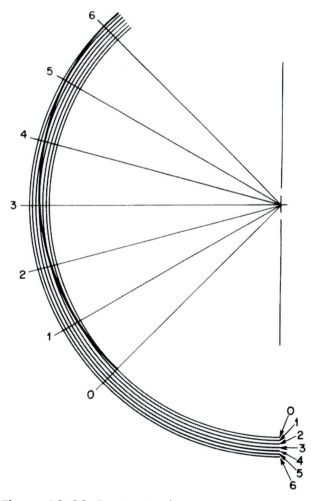

Figure 10–29 Constructing the contour.

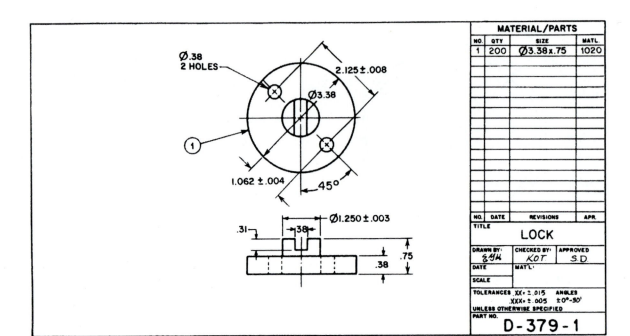

Figure 10–30 Part drawing and production plan.

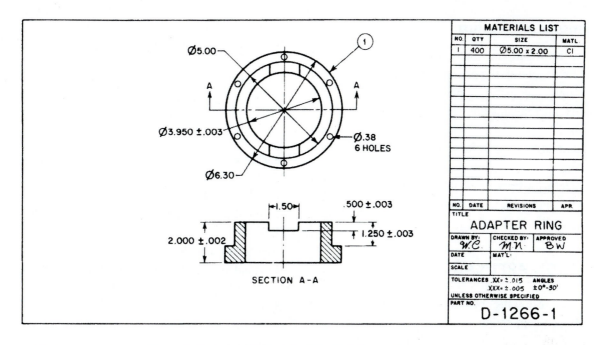

Figure 10–31 Part drawing and production plan.

The drawing portion includes:

MATERIALS LIST			
NO.	QTY	SIZE	MATL
I	400	Ø5.00 x 2.00	CI

TITLE
ADAPTER RING

DRAWN BY: W.C.   CHECKED BY: MN   APPROVED: BW

TOLERANCES .XX= ±.015   ANGLES
.XXX= ±.005   ±0°-50'
UNLESS OTHERWISE SPECIFIED

PART NO.
D-1266-1

Ø5.00
Ø3.950 ±.003
Ø.38 6 HOLES
Ø6.30

SECTION A-A
2.000 ±.002
1.50
.500 ±.003
1.250 ±.003

PRODUCTION PLAN				
P/N 1266	PART NAME ADAPTER RING		QUANTITY 400	ORDER NO 5217-77
DRAWING NO D-1266-1	PROCESS PLANNER JOE PETERS	REVISION NO. DATE		PAGE 1 OF 1

OPR. NO.	DESCRIPTION	DEPT.	MACH. TOOL
1.	TURN ROUGH CASTING TO Ø5.00	#27 LATHE	TURRET LATHE #27-4
2	BORE HOLE TO Ø3.950 ±.003	#27 LATHE	TURRET LATHE #27-4
3.	FACE SHOULDER TO .75 THICKNESS	#27 LATHE	TURRET LATHE #27-4
4.	GANG MILL Ø5.00 SIDE TO 2.00±.002 THICK WITH 1.50x5.00±.003 SLOT	#37 MILLING	HORZ. MILL #37-1
5.	DRILL 6 HOLES THRU Ø.38	#66 DRILLING	SEN. DRILL #66-8
6.	DEBURR	#7 FINISHING	BENCH
7.	INSPECT	#7 FINISHING	NONE

OPR. NO.	TOOL DESCRIPTION	SIZE	SPECIAL TOOL NO.
7.	FUNCTIONAL GAUGE	—	FIXTURE I-1266-3
5.	DRILL	(1) Ø.38	TABLE JIG T-1266-2
4.	MILLING CUTTERS	(2) 2.00x3.00x1.00 (1) 1.50x4.00x1.00	FIXTURE T-1266-1

Suggestions for Design

- Select the locational method and make calculations by following the procedures outlined in previous units and example problems.
- Specify all relevant data and sizes on the initial sketch.
- Maintain tolerance values specified on the part drawing.
- Analyze prior operations to determine locational surfaces and work to be performed.

CAM DESIGN APPLICATION

Analyze the following problems and construct an outline of each cam shape.

1. Construct an eccentric cam with a .276-inch rise and a 90-degree throw.
2. Construct a spiral cam with a base circle diameter of 2.80 inches and an 80-degree throw.

SUMMARY

The following important concepts were presented in this unit:

- Vise-held and plate fixtures are two of the most basic fixture types.
  - A plate fixture is the most common form of fixture.
  - A vise-held fixture is a small, lightweight plate fixture that is designed to be mounted in a vise.
- Before the design is started, the tool drawing must be studied to obtain all necessary information about the part.
- Careful calculations must be made to determine the exact size and location of the locators used to position the part.
- Cutter location and referencing are accomplished using a set block.
- Commercially available components should be used where possible.
- Each design should start as a sketch.

  - Sketching helps to formulate the design.
  - Sketches help to reduce problems and show relationships.
  - Once the design is set and the problems solved, the final design drawing is prepared.

REVIEW

1. What is a vise-held fixture?
2. What parts are best suited for a vise-held fixture?
3. What is the first step in designing any special workholding tool?
4. What is the main point to consider when selecting a locator for a vise-held fixture?
5. How can the designer ensure that each part detail maintains the correct relation to the other part details?
6. How are critical details identified?
7. If no machined detail is on a part, how should a designer select locational surfaces?
8. What is the difference between secondary and redundant locators?
9. Why should the secondary locator be relieved?
10. What are the important factors to consider when selecting a clamp?
11. What is straddle milling?
12. What is a set block used for?
13. What is used with a set block?
14. Why is it necessary to specify the spacing collar size?
15. Why should cutters be ground together?
16. How can the size of the cut be adjusted if necessary?
17. What is the main difference between the vise-held and plate fixtures?
18. What is a secondary function of locators?
19. How can rough cast surfaces be located?
20. Should adjustable locators be set for each part?
21. Why can't set blocks be used for every part?
22. What variables must be considered before choosing to use set blocks?
23. How can the cutters be set if no set block is used?
24. What are the two general types of cam clamp used for jigs and fixtures?
25. Which cam clamp type is better, and why?

26. What is the specific amount of movement in a cam called?

27. What is the arc of movement called?

28. Which style of cam is easiest to make?

29. What determines the rise of a spiral cam?

30. How can locational inaccuracies be allowed for when designing a spiral cam?

31. What is the general rule for rise on a spiral cam?

32. What is the normal range of cam throw?

UNIT 11

Plate Jigs

OBJECTIVES

After completing this unit, the student should be able to:

- Analyze part data to determine suitable jig types to perform specified tasks.
- Specify locating, supporting, and clamping methods, and details to suit sample parts.
- Design two plate-type jigs to suit specified sample parts, and construct the required tool drawings.

PLATE JIGS

Plate jigs are basically modified or improved template jigs. While performing the same locating function as templates, they also incorporate a means of securing the part. This added feature, clamping, is what allows plate jigs to locate parts with a high degree of accuracy and repeatability.

Plate jigs are actually a family of jigs rather than a single type of tool. Each variation of the basic plate jig has at least one distinctive feature that separates it from the others. However, all variations have the same plate as a main structural member. This plate, from which the tool gets its name, generally provides the

mounting points for locating, supporting, and clamping the part as well as positioning the drill bushing.

Plain Plate Jigs

The *plain plate jig,* as shown in Figure 11–1, is the simplest and most basic form of this jig. It uses a flat plate as its only structural member, and all details are attached and referenced to this plate. The main advantages to using a plain plate jig are minimal design and fabrication time, minimal cost, and simplicity.

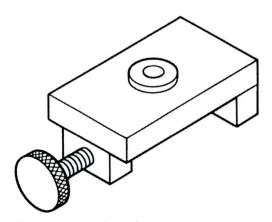

Figure 11–1 Plain plate jig.

Table Jigs

The *table jig,* as shown in Figure 11–2, is basically a plain plate jig with legs. Its main purpose is holding irregular or nonsymmetrical workpieces that cannot be held in other plate jig forms. With this jig, the part is referenced by the surface being machined rather than the opposite side. This surface relationship can be seen in Figure 11–3. Table jigs can accommodate almost any shape workpiece. Their only limitations are the size of the part and the availability of clamping surfaces. One other important point to consider is the tool thrust. The part is clamped between the jig plate and the clamping device. Therefore, the tool thrust is directed toward the clamps rather than toward the solid parts of the jig. Clamping devices must be selected to resist this thrust.

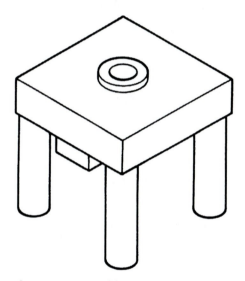

Figure 11-2 Table jig.

Sandwich Jigs

The *sandwich jig,* as shown in Figure 11–4, is almost identical to a plain plate jig. The only difference is the backup plate found on the sandwich jig. This backup plate allows the jig to hold very thin parts that could bend or distort under tool pressure. Another use of a sandwich jig is as a combination jig (Figure 11–5). One side of the tool is used to locate the part for drilling (Figure 11–5A). The opposite side is used for reaming or tapping the part (Figure 11–5B).

Leaf Jigs

The *leaf jig,* as shown in Figure 11–6, is actually a modification of the sandwich jig. Rather than using pins to locate the two members and screws to hold them together, this jig uses a hinge leaf with a cam-type latch. The part is loaded by placing it on the lower section where the locators are normally positioned. The leaf, which carries the bushings, is then lowered and latched (Figure 11–7).

DESIGNING A PLATE JIG

Designing a plate-type jig requires the same initial process required in designing any other special tool. First, all necessary data are gathered and analyzed to find the most suitable and cost-effective design. Normally, the designer has a specific design in mind before this analyzing process begins. What is being done is fitting the specifics of the part to what the designer already envisions.

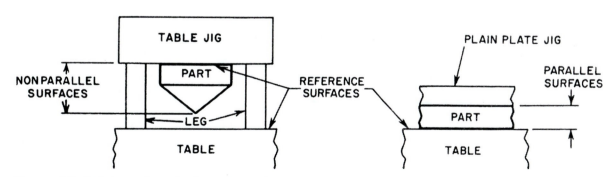

Figure 11-3 Relationship of referencing surfaces.

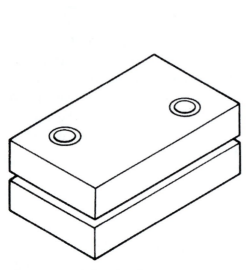

Figure 11-4 Sandwich jig.

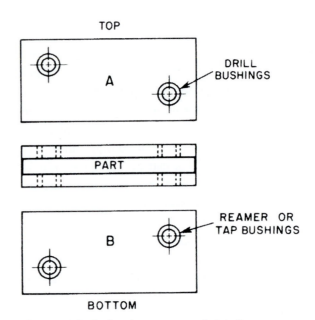

Figure 11-5 Combination sandwich jig.

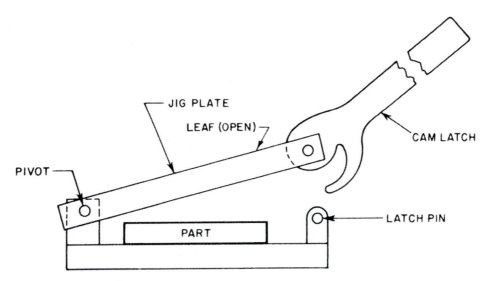

Figure 11-6 Leaf jig.

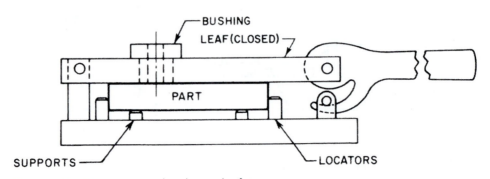

Figure 11-7 Positioning details in a leaf jig.

In the case of the housing cover, as shown in Figure 11–8, the first idea that should be checked is using a plain plate jig. Analyzing the data from the part drawing and production plan reveals the following information about this part:

- The part is a rough casting approximately 120 millimeters wide, 210 millimeters long, and 90 millimeters thick, with a 70 × 60-millimeter step on one end.
- The wall thickness is approximately 10 millimeters with supporting webs 8 millimeters thick.
- The material is cast aluminum.
- The run is seventy-five pieces.
- The operation to be performed is drilling four 12-millimeter holes.
- The blank is received as cast with one side machined.

Using this information, the designer checks the possibilities of using a plain plate jig. Since there is no requirement for a backup plate, the sandwich and leaf jigs can be eliminated, and since the unmachined side of the housing provides a sufficiently accurate supporting surface, the table-type jig is not required. Eliminating the other three types leaves the plain plate jig as the best choice. While it is possible to use any of the four types, from the standpoint of simplicity, efficiency, and economy, the plain plate jig should be selected.

Locating the Part

With a plain plate jig, the location of the jig on the part is important because the part is generally larger than the tool. In the case of the housing cover, the ideal place to locate the jig is on the step formed by the two horizontal surfaces (Figure 11–9). Using this step as the primary locating surface, the designer must then select a secondary locator. The part is a rough casting. Since the critical hole dimensions are between the hole centers rather than from an edge sight, secondary locators are the best choice (Figure 11–10). The machined surface is considered the tertiary location point.

Supporting the Part

Plate jigs do not normally include any device for supporting the part; the part generally supports the tool.

However, when large holes are drilled above 6 millimeters, the tool cannot be held by hand, so some device must be used to hold the part to the machine table. The machine specified to drill the holes in the housing cover is a radial drill press. With this machine, the part can be clamped in one position while the machine moves from hole to hole. The clamping arrangement shown in Figure 11–11 is one method of holding the part in place while machining the holes. It is important to note that the plate jig is first placed on the housing cover and clamped. Then the entire unit is clamped on the machine table.

Clamping the Part

As the step was selected as the primary locator, the clamp must be aligned to hold the jig against this step. The clamp arrangement shown in Figure 11–12 satisfies this requirement. The jig is first placed against the step, which locates the tool in one direction. The sight locators are used to locate the tool in the other direction. Once the tool is in position, the clamp screws are tightened.

Locating the Bushings

Locating the drill bushings requires calculations using the locating step as a starting point. The distance between the holes is 80 millimeters in one direction and 120 millimeters in the other direction (Figure 11–13). Starting with the 120-millimeter dimension, the designer determines by calculation that there are 30 millimeters from the step to the first hole and 90 millimeters from the step to the second hole. Using these values, the designer then applies 50 percent of the tolerance to each hole. The values then become 30 ± 0.05 and 90 ± 0.05 millimeters. Assuming the extreme permissible dimensions, the values then become 30.05 and 90.05 millimeters. Added together, the overall size becomes 120.1 millimeters, which is within the specified tolerance. To keep the tolerance closer and allow for some wear in the bushings, the designer could specify a closer tolerance. However, because the size of this run is only seventy-five pieces, wear is not a critical factor. Another factor that could affect the selection of locational tolerance is the size tolerance of the bushing mount holes. If slip-renewable bushings were specified

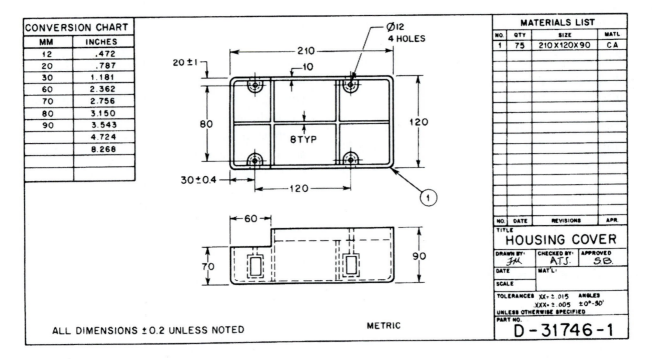

CONVERSION CHART	
MM	INCHES
12	.472
20	.787
30	1.181
60	2.362
70	2.756
80	3.150
90	3.543
	4.724
	8.268

ALL DIMENSIONS ± 0.2 UNLESS NOTED

METRIC

Ø12 4 HOLES

210
20 ±1
10
80
120
8 TYP
30 ±0.4
120

60
70
90

MATERIALS LIST

NO.	QTY	SIZE	MATL
1	75	210 X 120 X 90	C A

NO.	DATE	REVISIONS	APR.

TITLE
HOUSING COVER

DRAWN BY:	CHECKED BY:	APPROVED
JM	A.T.S.	S.B.
DATE	MAT'L:	
SCALE		

TOLERANCES XX= ±.015 ANGLES
.XXX= ±.005 ±0°-30'
UNLESS OTHERWISE SPECIFIED

PART NO.
D - 31746 - 1

PRODUCTION PLAN

P/N 31746	PART NAME **HOUSING COVER**		QUANTITY 75	ORDER NO. 77-1234
DRAWING NO. D-31746-1	PROCESS PLANNER R.J. PIERCE	REVISION NO	DATE	PAGE 1 OF 1

OPR. NO.	DESCRIPTION	DEPT.	MACH. TOOL
1	GANG-MILL MATING SURFACE TO 90mm WITH 70mm X 60mm STEP	# 37 MILLING	HORZ. MILL # 37-51
2	DRILL FOUR 12mm MOUNTING HOLES	# 66 DRILLING	RADIAL DRILL # 66-14
3	DEBURR & SANDBLAST	# 7 FINISHING	SANDBLASTER # 7-4
4	INSPECT	# 7 FINISHING	NONE
4	FUNCTIONAL GAUGE	——	INSP. FIXTURE 1-31746-3
2	DRILL	(1)12	DRILL JIG T-31746-2
1	MILLING CUTTERS (4) INTERLOCKING GANG CUTTERS	(2)100X100X25 (2)140X35X12	MILL FIXTURE T-31746-1

OPR. NO.	TOOL DESCRIPTION	SIZE	SPECIAL TOOL NO.

Figure 11-8 Part drawing and production plan.

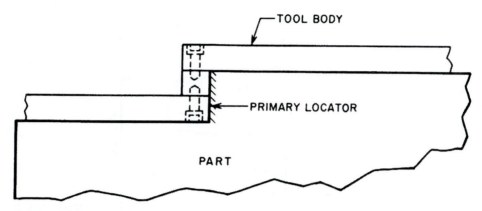

Figure 11-9 Primary locator.

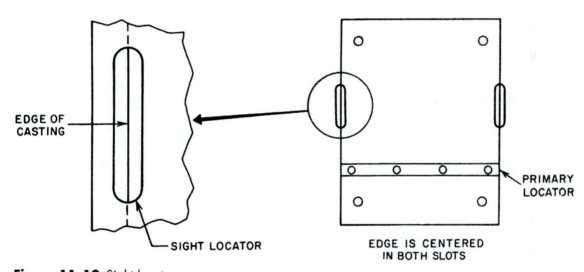

Figure 11-10 Sight locators.

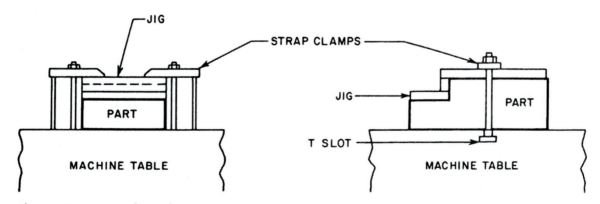

Figure 11-11 Auxiliary clamping.

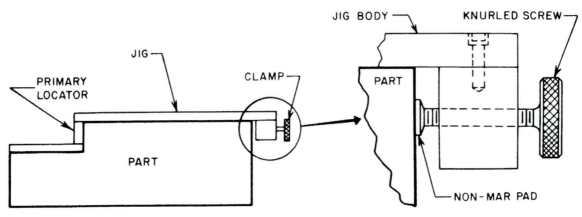

Figure 11-12 Clamping the jig to the part.

for the jig, then a slightly closer locational tolerance would be desired to offset any error in size tolerance. In the case of this plate jig, press-fit head-type bushings are specified, so this factor needs no consideration.

Initial Design Sketch

After each separate area has been considered, the tool design is sketched using all the information gathered in the previous steps. The first area to sketch is the main plate. This plate should be at least 15 millimeters thick to resist distortion. The step is constructed by attaching both bushing plates to a 5 × 15-millimeter block. Both plates are doweled and bolted to this block through the top plate. Another strip of 15 × 15-millimeter stock is attached to the opposite side to act as the clamp block (Figure 11–14).

The sight locators are made in the form of slots 30 millimeters long and 10 millimeters wide, milled on 120-millimeter centers. After the rough design

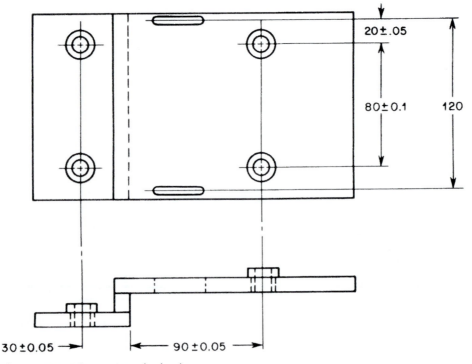

Figure 11-13 Locating the bushings.

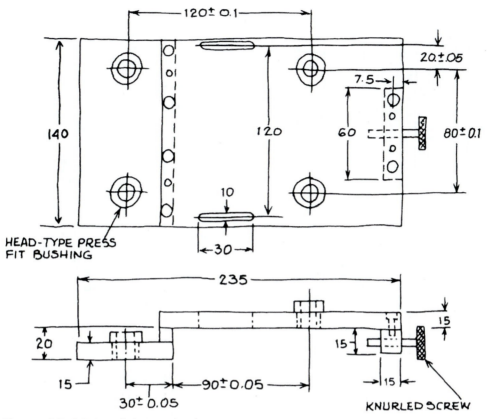

Figure 11-14 Initial design sketch.

sketch is made, the completed tool drawing is constructed (Figure 11–15).

DESIGNING A TABLE JIG

The basic process of designing a table jig is almost identical to the procedure used for the plate jig. Using the part drawing and production plan for the valve body, as shown in Figure 11–16, the designer determines the following information:

- The part is cylindrical in shape, 2.375 inches long, with an outside diameter of 1.500 inches.
- Each end is drilled and tapped on 1-8 UNC-2B.
- One hole is drilled and tapped in the side to 5/8-18 UNF-2B.
- The material specified is brass.
- The production run is 100 pieces.
- The operation required for this jig is drilling and tapping the hole in the side.
- The blank received has been turned and faced with both end holes drilled and tapped.

Using this information, the designer decides that a table jig is best suited to drill and tap this part.

Locating the Part

Locating the valve body in this tool requires the use of a vee-block locator. Since the part is to be drilled from the top, the part and the vee block must be located with the bushing hole through the vee block, as in Figure 11–17. An additional pin is located in the end of the vee block to position the hole in reference to its length. The vee block is the primary locator, and the adjustable-stop pin is the secondary locator.

Supporting and Clamping the Part

When table-type plate jigs are used, the supporting and clamping functions are both performed by the clamping device. The clamp used on the valve body is a strap clamp with a hand knob (Figure 11–18). The legs selected for the table jig are 5.000 inches long and double-ended. This is to allow clearance for

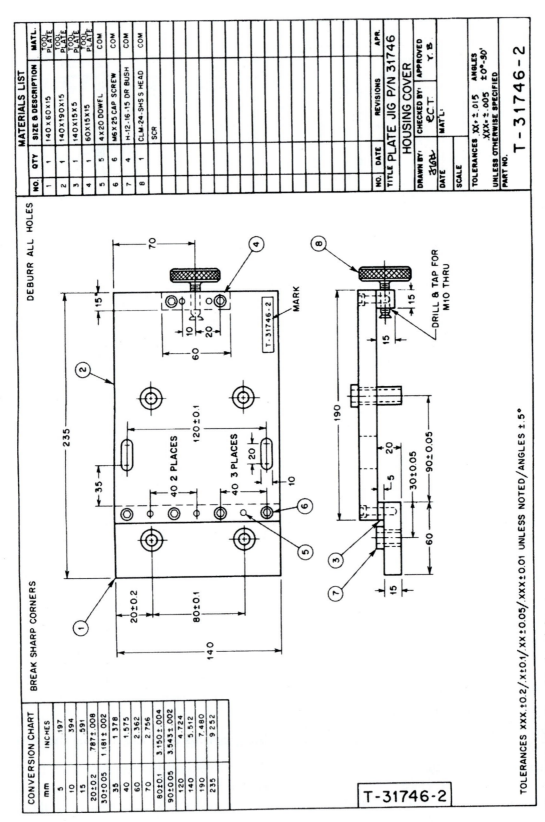

Figure 11-15 Completed tool drawing.

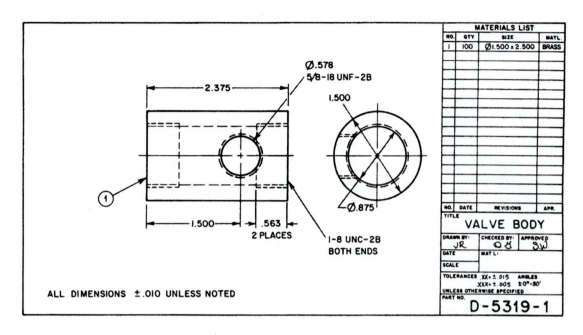

ALL DIMENSIONS ±.010 UNLESS NOTED

	MATERIALS LIST		
NO.	QTY	SIZE	MATL.
I	100	Ø1.500 x 2.500	BRASS

NO.	DATE	REVISIONS	APR.

TITLE **VALVE BODY**

DRAWN BY: JR   CHECKED BY: OS   APPROVED SW
DATE   MAT'L:
SCALE

TOLERANCES .XX = ±.015   ANGLES
.XXX = ±.005   ±0°-30'
UNLESS OTHERWISE SPECIFIED
PART NO. **D-5319-1**

PRODUCTION PLAN					
P/N 5319	**PART NAME** VALVE BODY		**QUANTITY** 100	**ORDER NO** 77-1234	
DRAWING NO. D-5319-1	**PROCESS PLANNER** D. LONGO	**REVISION NO.**	**DATE**	**PAGE** 1	**OF** 1

OPR. NO.	DESCRIPTION	DEPT.	MACH. TOOL.
1	CUTOFF Ø1.500 BAR TO 2.500 LG.	#68 CUTOFF RM.	HORZ. B'SAW #68-5
2	FACE & DRILL Ø.875 THRU	#27 LATHE	TURRET LATHE #27-12
3	TAP 1-8 UNC-28, .563 INCH DEEP EACH END	#27 LATHE	TURRET LATHE #27-12
4	DRILL Ø.578 THRU ONE SIDE	#66 DRILLING	UPRIGHT DRILL #66-8
5	TAP 5/8-18 UNF-2B	#66 DRILLING	UPRIGHT DRILL #66-8
6	DEBURR & INSPECT	#7 FINISHING	TUMBLER #7-84

OPR. NO.	TOOL DESCRIPTION	SIZE	SPECIAL TOOL NO.
6	FUNCTIONAL GAUGE	———	INSP. FIXTURE I-5319-2
5	TAP, R.H.	5/8-18 UNF-2B	DRILL JIG T-5319-1
4	DRILL, JOBBER'S LENGTH, STR. SHANK	578 (37/64)	DRILL JIG T-5319-1
3	TAP, R.H.	1-8 UNC-2B	———
2	DRILL, JOBBER'S LENGTH, TAPER SHANK	.875 (7/8)	———

Figure 11–16 Part drawing and production plan for the valve body.

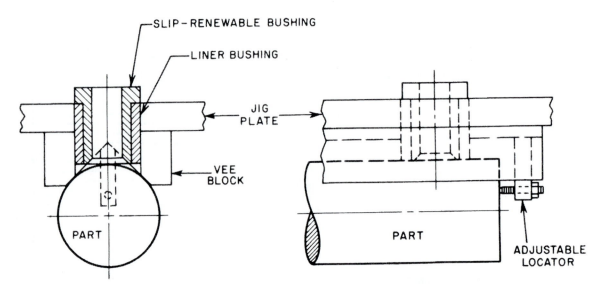

Figure 11-17 Locating the vee-block locator.

Locating the Bushing

Since this jig is intended to drill and tap the part, a slip-renewable bushing should be used. When locating the liner bushing, the primary requirement is center alignment between the vee block and the bushing

the slip-renewable bushing when the jig is turned over for loading and unloading the parts.

hole. The distance between the bushing center and the secondary locator is not quite as critical because an adjustable locator has been specified. The hole should be within ±.010 inch.

Initial Design Sketch

When each area has been thought out and a rough idea of the details of the tool have been decided, the

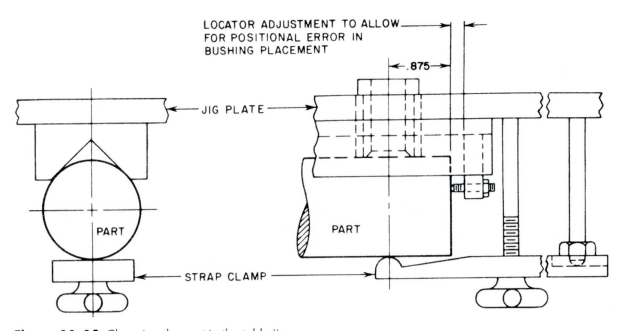

Figure 11-18 Clamping the part in the table jig.

initial sketch should be made (Figure 11–19). Since the main structural member of this table jig is the top plate, it should be made thick enough to withstand the drilling pressure and provide enough bearing surface for the bushings. For these reasons, the plate selected is 1.00 inch thick. The length and width should be 6.00 inches by 7.75 inches to ensure that the part and all its details are within the leg area. This will make the jig more stable and less likely to tip when used. After the sketch is complete and all design problems have been solved, the designer completes the tool drawing (Figure 11–20).

DESIGNING A SANDWICH JIG OR A LEAF JIG

Because the basic construction features of sandwich jigs and leaf jigs are similar, the same rules often apply to both. Analyzing the information found on the part drawing and the production plan for the spacer bracket, as shown in Figure 11–21, discloses the following information:

- The part is flat stock 1.00 inch wide, 5.91 inches long, and .25 inch thick.
- There are two .531-inch holes in one end and one .531-inch hole in the other end.

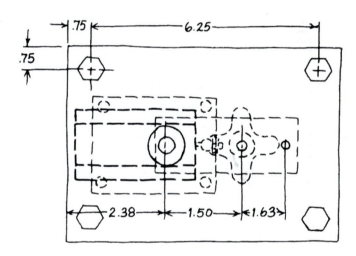

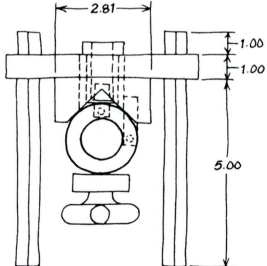

Figure 11–19 Initial design sketch.

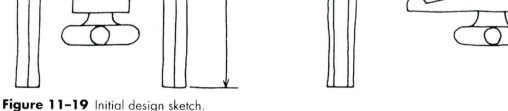

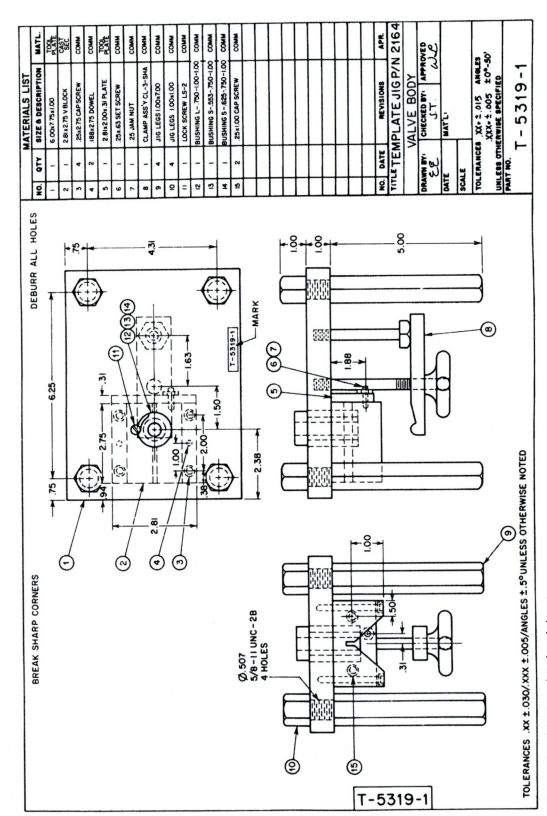

Figure 11-20 Completed tool drawing.

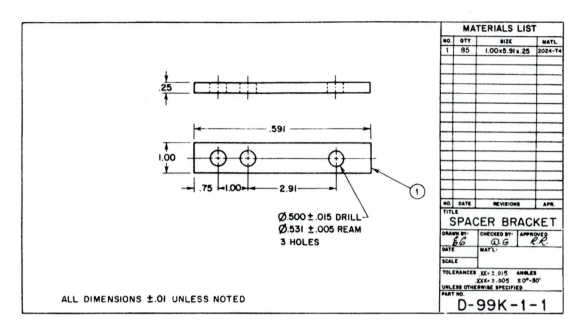

Figure 11-21 Part drawing and production plan for spacer bracket P/N 99K-1.

- The material specified is 2024-T4 aluminum sheet stock, 1.00 inch wide and .25 inch thick.
- The operation required is drilling and reaming three .531 inch holes.
- The production run is eighty-five pieces.
- The blank received is shear-cut on both ends.

Because this part is rather thin, there is a chance it can bend or distort during drilling and reaming.

Therefore, the designer should select a leaf jig or a sandwich jig.

Locating, Supporting, and Clamping the Part

Both long sides on the spacer bracket are finished. One method that could be used to locate the part is a milled slot. A hole is drilled and tapped in the side for

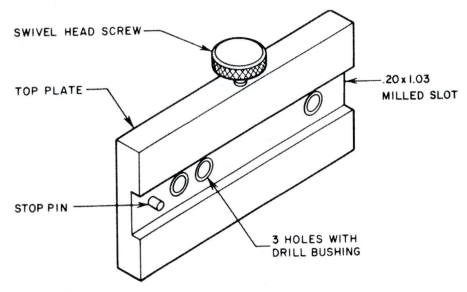

Figure 11-22 Locating and clamping the part.

the .25-inch swivel head screw (Figure 11–22). Locating the top and bottom plates is accomplished by using two bullet-nose dowel and bushing assemblies (Figure 11–23).

Locating the Bushings

When making this jig, the toolmaker should first mill the slot and install the bullet-nose dowel and bushing assemblies. With the jig assembled, the holes can be drilled and reamed in both jig plates. Since the jig must both drill and ream the spacer brackets, the simplest way is to drill from one side, turn the jig over, and ream from the other side. There is only a .031-inch difference between the size of the drill, .500 inch, and the size of the reamer, .531 inch. The outside diameter of both bushings is the same. By drilling and reaming both plates at the same time, the alignment of the plates is perfect. The actual location of the bushing in the plates should be 50 percent of the part tolerance.

After the basic design ideas have been thought out, the designer sketches a rough drawing of the tool (Figure 11–24). The tool drawing is constructed, as in Figure 11–25, once the designer is satisfied that the tool will function as intended.

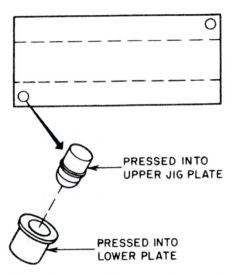

Figure 11-23 Referencing the two plates.

TOOL DESIGN APPLICATION

1. Analyze the part drawing and production plan, as shown in Figure 11–26, to determine all important information.
2. Based on this information, select one specific type of jig and design the appropriate tool.
3. Compare the similarities and differences between the assigned part and the example parts.
4. Prepare an initial sketch to calculate necessary dimensions.
5. Construct the tool drawing, including all necessary data needed to build the tool.

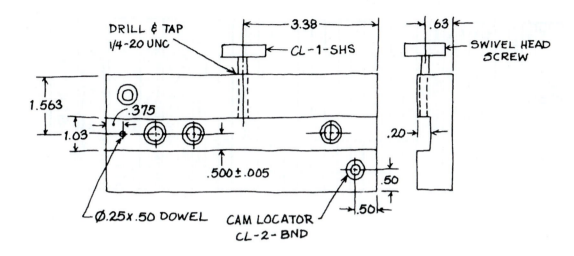

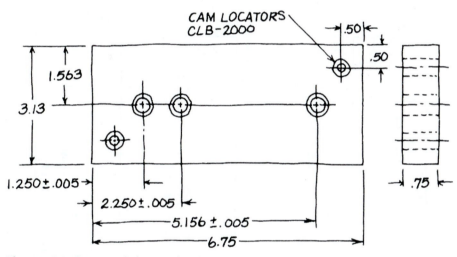

Figure 11-24 Initial design sketch.

Suggestions for Design

- Select locational methods and make calculations by following procedures outlined in this unit and in previous examples.

- Specify all relevant data and sizes on the initial sketch.

- Maintain tolerance values specified on the part drawing.

- Analyze prior operations to determine locational surfaces and work to be performed.

- Make sure the primary design criteria are workability, efficiency, and economy.

SUMMARY

The following important concepts were presented in this unit:

- Plate jigs are very similar to template jigs except that they normally have a built-in clamp. Several variations are commonly used for machining operations. These include:

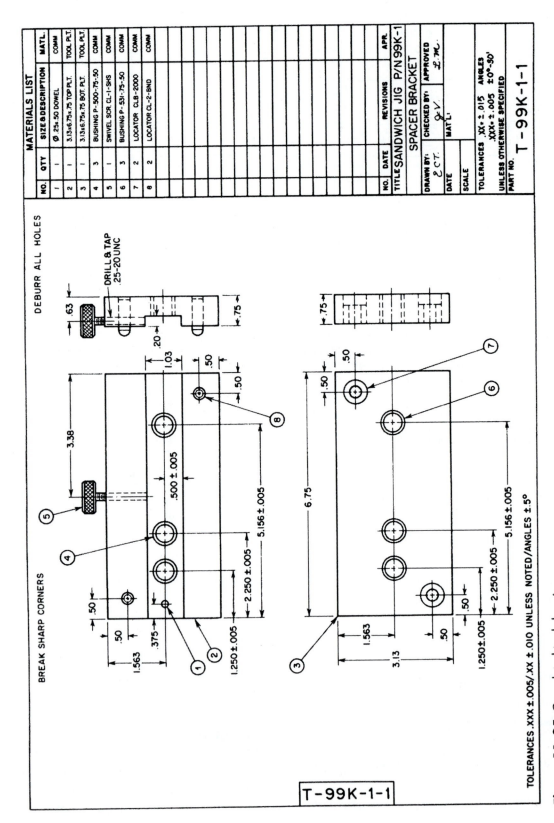

MATERIALS LIST

NO.	QTY	SIZE & DESCRIPTION	MATL.
1	1	Ø.25x.50 DOWEL	COMM
2	1	3.13x6.75x.75 TOP PLT.	TOOL PLT.
3	1	3.13x6.75x.75 BOT. PLT.	TOOL PLT.
4	3	BUSHING P-.500-.75-.50	COMM
5	3	SWIVEL SCR. CL-1-SHS	COMM
6	3	BUSHING P-.53I-.75-.50	COMM
7	2	LOCATOR CLB-2000	COMM
8	2	LOCATOR CL-2-BND	COMM

NO.	DATE	REVISIONS	APR.

TITLE SANDWICH JIG P/N 99K-1

SPACER BRACKET

DRAWN BY: E CT.	CHECKED BY: QV	APPROVED L.M.

DATE

SCALE

MAT'L

TOLERANCES .XX ± .015   ANGLES
.XXX ± .005 ± 0°-50'

UNLESS OTHERWISE SPECIFIED

PART NO. T-99K-1-1

DEBURR ALL HOLES

DRILL & TAP .25-20 UNC

BREAK SHARP CORNERS

T-99K-1-1

TOLERANCES .XXX ±.005/.XX ±.010 UNLESS NOTED/ANGLES ±.5°

Figure 11-25 Completed tool drawing.

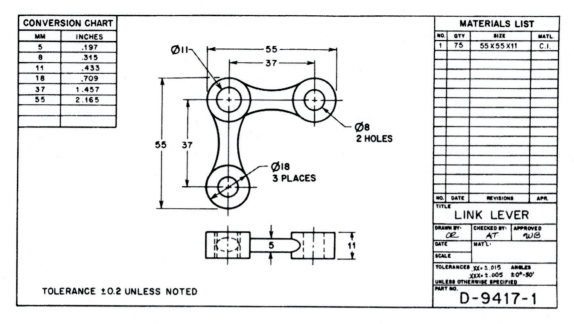

Figure 11-26 Part drawing and production plan.

- Plain plate jigs
- Table jigs
- Sandwich jigs
- Leaf jigs
- The process used to design a plate jig is basically the same as that used for a template jig.
    - Before the design is started, the tool drawing must be studied to obtain all necessary information about the part.

- Careful calculations are made to determine the exact size and location of the locators and bushings.
- Each design should start as a sketch.
    - Sketching helps to formulate the design.
    - Sketches help to reduce problems and show relationships.
    - Once the design is set and the problems solved, the final design drawing is prepared.

REVIEW

1. What is the principal difference between plate jigs and template jigs?
2. What are the common plate jig variations?
3. Which type of plate jig must provide clamps that resist the tool thrust?
4. Which type of plate jig uses two plates?
5. Which type of plate jig can be used as a combination jig?
6. Which type of plate jig is generally smaller than the part?
7. Do plate jigs support the part?
8. How should plate jigs be secured to the machine?
9. When vee-block locators are used, how should the block be positioned with relation to the hole?
10. How can table jigs be kept from tipping over in use?
11. How can the hole alignment be maintained for combination jigs?

UNIT 12

Angle-Plate Jigs and Fixtures

OBJECTIVES

After completing this unit, the student should be able to:

- Analyze part data to determine suitable tool designs.
- Specify locating, supporting, and clamping methods to suit specified sample parts.
- Design one angle-plate jig and one angle-plate fixture to suit specified sample parts.

Angle-plate jigs and fixtures are primarily used to machine parts that cannot be easily machined with plate-type tooling. The size and complexity of these tools vary with the workpiece, but regardless of their size, the basic construction of each is essentially the same.

The main structural member of an angle-plate tool is an angle. The materials commonly used to build these tools are angles, tees, S beams, cast-bracket materials, built-up sections of flat stock, and commercial tool bodies (Figure 12–1).

VARIATIONS AND APPLICATIONS

In the design of any angle-plate tool body, one important point must be kept in mind. The vertical member of the tool must be either the same size or smaller than the horizontal member. This is necessary to maintain the required rigidity in the tool and to ensure accuracy in the part. In the few cases in which this rule cannot be followed, the vertical member must be supported by placing gussets or supports between the two members (Figure 12–2).

Plain Angle-Plate Tooling

Plain angle-plate tools are constructed at a 90-degree angle to the machine table (Figure 12–3). These tools are used to machine part details at angles perpendicular to the machine table.

Modified Angle-Plate Tooling

Modified angle-plate tools are constructed to suit any angle from parallel to perpendicular to the machine table (Figure 12–4). These angle-plate tools are usually used to machine part details that are not perpendicular to the machine table.

Additional Applications

In addition to the primary uses already discussed, angle-plate jigs and fixtures are used as indexing tools (Figure 12–5) or as auxiliary supports (Figure 12–6).

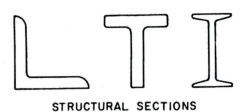

STRUCTURAL SECTIONS

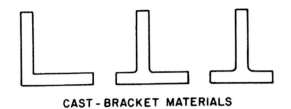

CAST-BRACKET MATERIALS

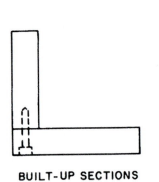

BUILT-UP SECTIONS

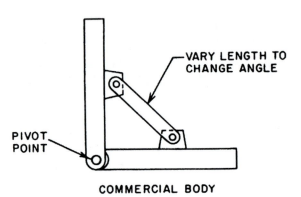

COMMERCIAL BODY

Figure 12-1 Tool body construction.

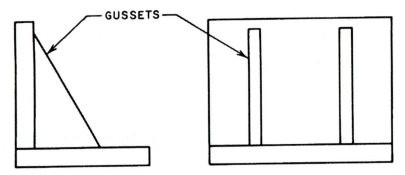

Figure 12-2 Supporting the angle.

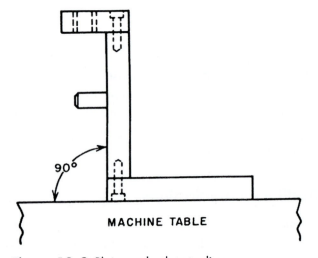

Figure 12-3 Plain angle-plate tooling.

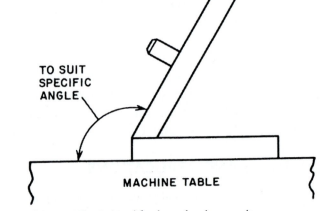

Figure 12-4 Modified angle-plate tooling.

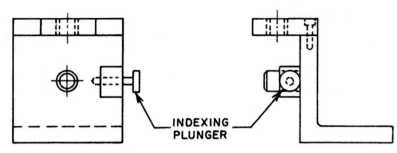

Figure 12–5 Indexing jig.

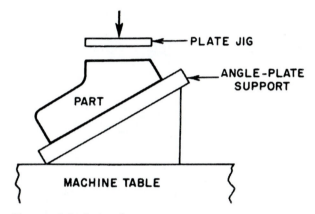

Figure 12–6 Auxiliary support.

DESIGNING AN ANGLE-PLATE JIG

Angle-plate jigs are used to machine a vast assortment of parts. Because of their design, they can accommodate parts of varied shapes. Examples of this adaptability are shown in Figure 12–7.

Once the designer has decided on an angle-plate jig, the evaluating and analyzing begins. Referring to the part drawing and production plan for the coupler, as shown in Figure 12–8, the designer notes the following design data:

- The part is a cylindrical collar, 2.38 inches long, with one 2.38-inch diameter step and one 1.38-inch diameter step. Each step is 1.13 inches wide.
- The part has two bored holes, one 1.188 inches in diameter, the other .594 inch.
- The part has been faced and bored on a lathe.
- The material specified is 1020 steel.
- The production run is 100 pieces.
- The operation to be performed is drilling four .19-inch holes in the large-diameter step and two .19-inch holes in the small-diameter step.
- The blank received for this operation has been faced and bored.

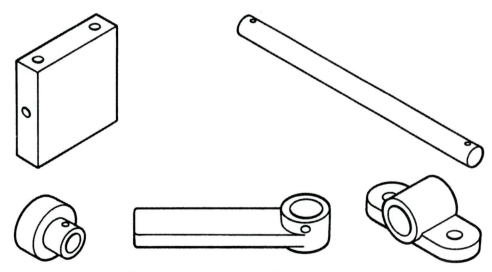

Figure 12–7 Typical parts machined with angle-plate jigs.

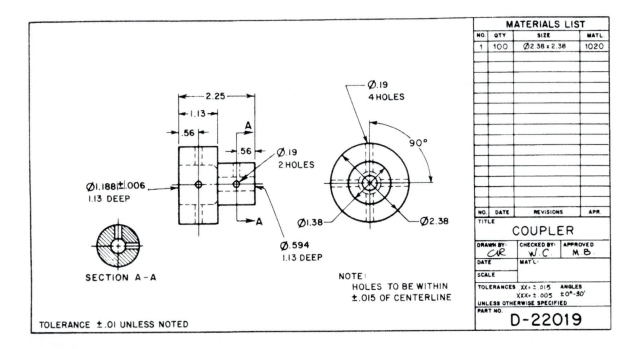

Figure 12-8 Part drawing and production plan.

Locating, Supporting, and Clamping the Part

The simplest, most accurate way to locate the part is by the bored center hole. The step locator is made to fit the large bored hole while passing through the small hole (Figure 12–9). The threaded end of the step locator is used to mount the quick-acting knob that clamps the part in place.

The diameter of the large end of the stepped locator is the most important dimension. This surface is both the support for drilling and the hub around which the part is referenced. Given the information found on the part drawing, the largest allowable diameter of the locating hole is 1.194 inches; the smallest allowable diameter is 1.182 inches. From these sizes, the locator size is determined to be 1.181 ±.0005 inches. This allows .0005-inch clearance between the part and the locator when the smallest part and the largest locator sizes are considered, and .0135 inch when the largest part and the smallest locator sizes are used. The locational tolerance of the hole is ±.015 inch from the centerline; therefore, this locator size is within the tolerance (Figure 12–10). It must be remembered that these sizes are the extremes. In actual practice, the size variance is usually much smaller.

Indexing the Part

Indexing with this tool is performed manually by the operator. The indexing pin used to position the part is a ball plunger, as shown in Figure 12–11, which permits speed and accuracy. In use, the part is first drilled; then

it is rotated until the ball snaps into the drilled hole. Subsequent holes are drilled in the same way. By using a ball plunger, indexing speed is increased because the part only has to be rotated to disengage the ball rather than manually removing an indexing pin. By positioning the ball plunger 90 degrees from the drilling position, four equally spaced holes can be drilled easily.

When the two holes in the small-diameter step are drilled, the larger step holes are again used to reference the part. By positioning the part with the indexing ball in one of the holes, the proper radial location between the hole patterns in both diameters is maintained (Figure 12–12).

Locating the Bushings

The bushings specified for this tool can be either fixed or renewable, depending on the jig body design. Fixed bushings would be used if the top jig were hinged (Figure 12–13). Here the bushings can be swung away to permit easier loading and unloading of parts. This method works well for higher volume runs. For absolute economy, another method is chosen.

By constructing the tool with one fixed bushing and one slip-renewable bushing, the same clearance is obtained but at less cost (Figure 12–14). This type of tool may be slightly slower to load and unload, but for 100 parts the overall increase is minimal.

In either case, the bushings must be located within .0075 inch of the centerline of the locator. Combining the locational tolerances of the bushings with those of the locators could result in the conditions shown in Figure 12–15. The worst possible tolerance condition

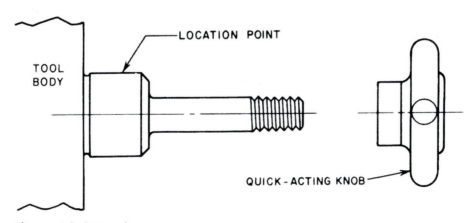

Figure 12–9 Step locator.

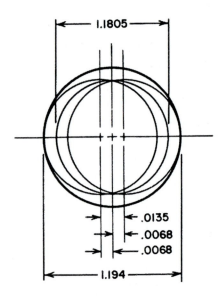

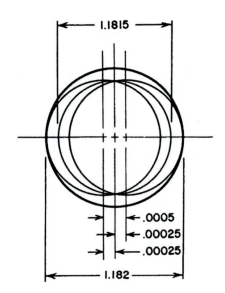

1.194   LARGEST PART SIZE
-1.1805   SMALLEST LOCATOR SIZE
.0135   TOTAL DIFFERENCE

2⟌.0135 (.0068 SHIFT OFF–CENTER
IN EACH DIRECTION

1.182   SMALLEST PART SIZE
-1.1815   LARGEST LOCATOR SIZE
.0005   TOTAL DIFFERENCE

2⟌.0005 (.00025 SHIFT OFF–CENTER
IN EACH DIRECTION

Figure 12–10 Calculating locator size.

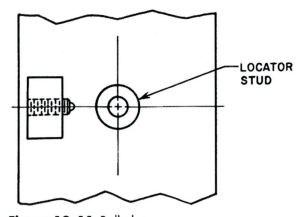

Figure 12–11 Ball plunger.

would place the hole .0118 inch off-center. This would be well within the ±.015-inch part tolerance.

Designing the Tool Body

The body specified for this tool should be an unequal *T*, made from cast-bracket material. This permits adequate rigidity while also providing stability against tipping over. The jig plate could also be cast flat stock. The parts should be doweled and screwed together with the locator fastened (Figure 12–16). Clamping the tool to the machine could be performed either by slots and T-bolts or by strap clamps (Figure 12–17).

Constructing the tool sketch and the final tool drawing are the last steps in designing the angle-plate jig (Figure 12–18).

DESIGNING AN ANGLE-PLATE FIXTURE

Angle-plate fixtures are used extensively throughout manufacturing for a great variety of operations. They are easily adapted to virtually any machine tool and process. Thus far, the fixtures discussed have all been used with the fixture stationary and the tool moving. In this section, the problems and benefits of stationary tooling and rotating fixtures are explored.

Referring to the part drawing and production plan for the hanger bracket, as shown in Figure 12–19, the following facts must be considered when designing the fixture:

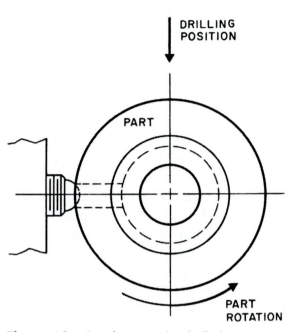

Figure 12–12 Indexing with a ball plunger.

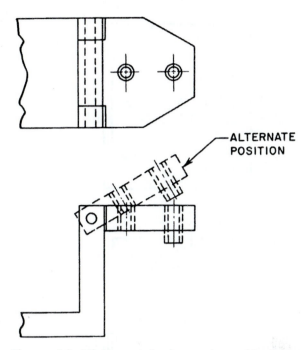

Figure 12–13 Locating bushing in hinged leaf.

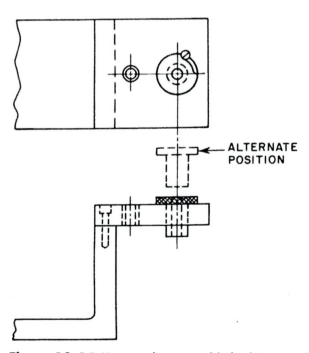

Figure 12–14 Using a slip-renewable bushing.

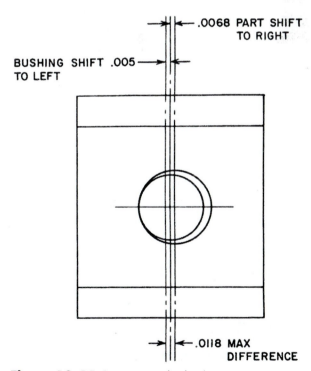

Figure 12–15 Positioning the bushings.

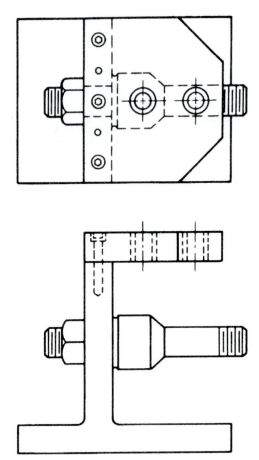

Figure 12-16 Assembled tool body.

- The part is a rough casting approximately 100 millimeters long, 80 millimeters wide, and 80 millimeters high.
- The part has a machined base with four 10-millimeter holes.
- The material specified is cast brass.
- The production run is eighty pieces.
- The operation to be performed is boring a 30-millimeter hole through the cylindrical portion of the casting.
- The part received for this operation has been milled and drilled on the bottom surface.
- The part has a cored hole approximately 20 millimeters in diameter.

Once the design data have been assembled and analyzed, the design process begins.

Locating the Part

The most accurate and efficient method to locate this part is by the holes in its base. Two of the four holes drilled in the base should be selected to locate the part (Figure 12–20). If there is a great deviation in hole diameters, adjustable conical locators can be used to ensure repeated accuracy (Figure 12–21). The bored hole should be machined in relation to these holes.

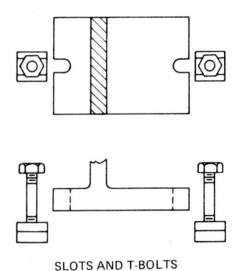

SLOTS AND T-BOLTS

Figure 12-17 Securing the tool to the machine.

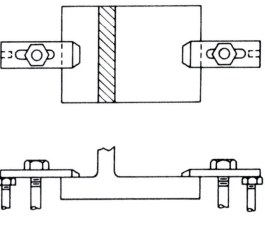

STRAP CLAMP

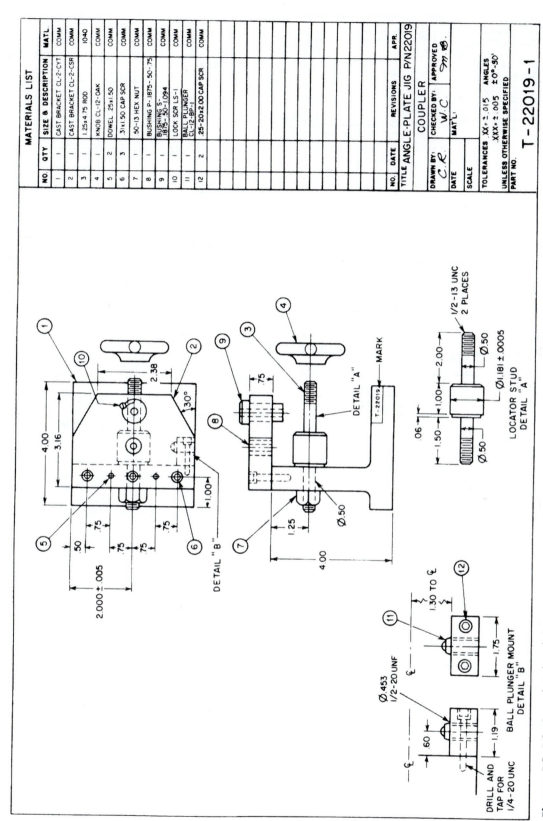

Figure 12-18 Completed tool drawing.

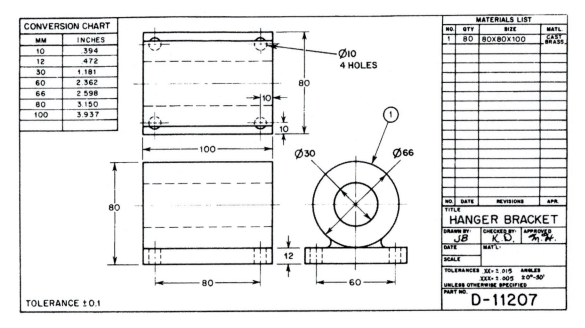

Figure 12-19 Part drawing and production plan.

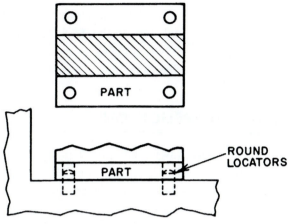

Figure 12–20 Locating the part.

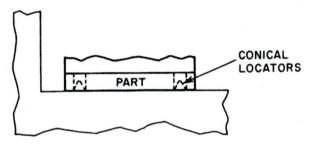

Figure 12–21 Using conical locators.

Assuming that the part received for the boring operation is drilled without excessive variance from the specified size, the locators used should be the solid type. One round pin and one diamond pin will accurately position the part provided the pins are positioned correctly. The proper arrangement of these locators is shown in Figure 12–22.

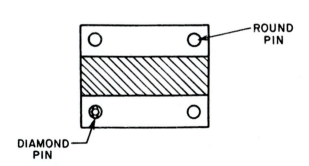

Figure 12–22 Arranging the locators.

Clamping the Part

Several methods could be used to clamp the part. To keep the costs as low as possible, a strap-type clamp was selected for this casting. The strap is the latch type, which pivots on one end and latches on the other end (Figure 12–23). This type was selected for its simplicity and rapid action.

The screw used to secure the clamp should be a socket-head cap screw to reduce the possibility of injury from exposed threads. The strap thickness should be approximately 15 millimeters to resist bending. The screws should be approximately M10.

Designing the Tool Body

The material specified for this tool should be cast-bracket material in an angle form. Locating the tool body to the faceplate it is to be mounted on can be accomplished in many ways. The simplest and easiest method is to provide a boss on the back side of the angle plate, which will fit the center hole of the faceplate. Once centered,

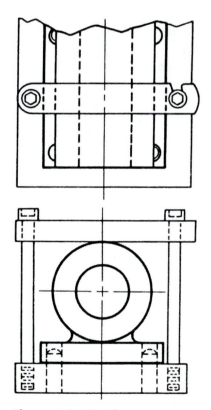

Figure 12–23 Clamping the part.

the fixture is bolted to the faceplate through holes that match the slots in the faceplate. These bolts should be countersunk and designed so that the nuts are behind the faceplate (Figure 12–24).

Occasionally with larger tools, it may be necessary to provide a counterbalance to reduce vibration and smooth the rotation of the faceplate. In these cases, the designer must specify a weight equal in size to the amount of mass offset by the tool. The counterbalance should be positioned directly opposite the heaviest part of the fixture (Figure 12–25).

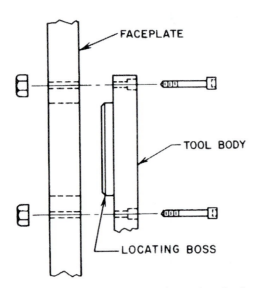

Figure 12–24 Mounting the tool to the faceplate.

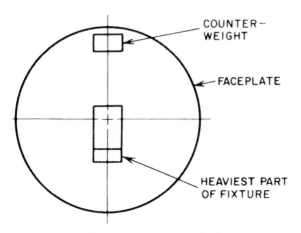

Figure 12–25 Mounting counterbalance.

Making the design sketches and completing the tool drawing are the final steps in the design process (Figure 12–26). Remember to check every detail to ensure its safety and value to the tool.

TOOL DESIGN APPLICATION

1. Analyze the part drawing and production plan, as shown in Figure 12–27, to determine all relevant information necessary to design an angle-plate jig and an angle-plate fixture.
2. Compare similarities and differences between the assigned part and previous examples.
3. Prepare design sketches to calculate required sizes and dimensions.
4. Include all necessary data in the complete tool drawing.

Suggestions for Design

- Note the shape of the part—select the most appropriate locating device.
- Check the possibility of using toggle-action clamps for both tools.
- Check the possibility of straddle-milling the stepped end.
- Check prior operations to determine the accuracy and features of the part.
- Use holes to locate work whenever possible.

SUMMARY

The following important concepts were presented in this unit:

- Angle-plate jigs and fixtures are variations of plate jigs and fixtures that are used to mount workpieces at angles to their mounting features. The two basic types are plain and modified.
  - Plain angle-plate jigs and fixtures mount the workpieces at a 90-degree angle.
  - Modified angle-plate jigs and fixtures mount the workpieces at an angle other than 90 degrees.
- Before the design is started, the tool drawing must be studied to obtain all necessary information about the part.

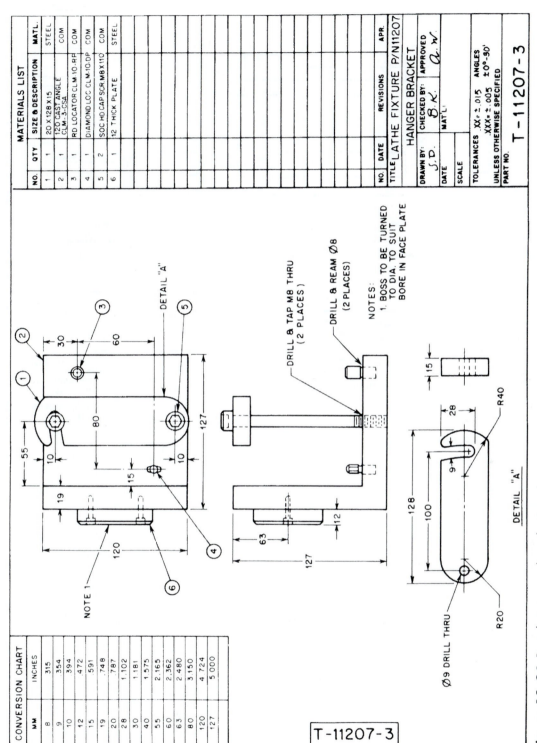

MATERIALS LIST

NO.	QTY	SIZE & DESCRIPTION	MATL.
1	1	20 X 128 X 15	STEEL
2	1	120 CAST ANGLE CLM-3-CSA	COM
3	1	RD LOCATOR CLM-10-RP	COM
4	1	DIAMOND LOC CLM-10-DP	COM
5	2	SOC HD CAP SCR M8 X 110	COM
6	1	12 THICK PLATE	STEEL

NO.	DATE	REVISIONS	APR.

TITLE LATHE FIXTURE P/N 11207
HANGER BRACKET

DRAWN BY: J.D.	CHECKED BY: B.K.	APPROVED a.w
DATE	MAT'L:	
SCALE		

TOLERANCES .XX=±.015 ANGLES
.XXX=±.005 ±0°-30'
UNLESS OTHERWISE SPECIFIED

PART NO. **T-11207-3**

NOTES:
1. BOSS TO BE TURNED
TO DIA. TO SUIT
BORE IN FACE PLATE

DRILL & TAP M8 THRU
(2 PLACES)

DRILL & REAM Ø8
(2 PLACES)

DETAIL "A"

NOTE 1

Ø9 DRILL THRU

R20

R40

DETAIL "A"

CONVERSION CHART	
MM	INCHES
8	.315
9	.354
10	.394
12	.472
15	.591
19	.748
20	.787
28	1.102
30	1.181
40	1.575
55	2.165
60	2.362
63	2.480
80	3.150
120	4.724
127	5.000

T-11207-3

Figure 12-26 Part drawing and production plan.

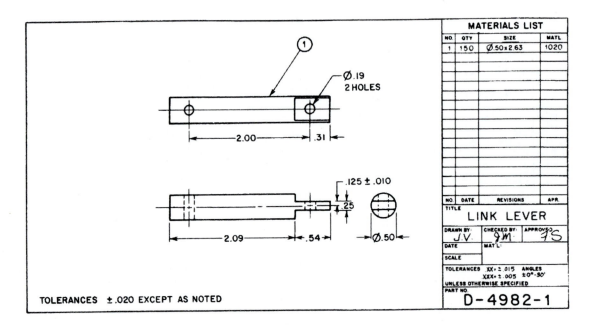

Figure 12-27 Part drawing and production plan.

- Careful calculations must be made to determine the exact size and location of the locators used to position the part.
- Cutter location and referencing are accomplished using a set block.
- Commercially available components should be used where possible.

- Each design should start as a sketch.
  - Sketching helps to formulate the design.
  - Sketches help to reduce problems and show relationships.
- Once the design is set and the problems solved, the final design drawing is prepared.

REVIEW

1. What are the two primary types of angle-plate jigs and fixtures?
2. Which type is used for what application?
3. What must the size relation of height to width be for angle plates?
4. What device can be used to increase indexing speed?
5. How should jig plates be installed for solid-type angle-plate jigs?
6. How are angle-plate tools normally held on machine tables?
7. Must all fixtures remain stationary while the tool moves?
8. What type of screw should be used, in the interest of safety, for rotating tools?
9. What method can be used to reference the tool to a machine faceplate?
10. What should be used on the faceplate to prevent vibration with large tools?

UNIT 13

Channel and Box Jigs

OBJECTIVES

After completing this unit, the student should be able to:

- Analyze part data to determine suitable tool designs.
- Specify locating, supporting, and clamping methods, and details to suit sample parts.
- Determine the specific type of jig required, and design a tool, to suit a sample part.
- Construct the required tool drawings.

CHANNEL JIGS

Channel and box jigs are the most expensive and detailed types of jigs in common use. However, despite this complexity and added cost, when properly designed and used, these jigs can save countless hours of machining time and parts handling. Once a part is loaded into one of these tools, it is seldom removed until all machining is complete. Channel and box jigs are designed to machine part details on more than one surface without repositioning the work in the tool.

The channel jig is the simplest and most basic form of closed jig (Figure 13–1). As its name implies,

the main structural member of this tool is a channel. In use, the channel jig is capable of machining parts on three surfaces.

In addition to their multidirectional machining capability, channel jigs provide better stability and support for thin parts than do open jigs. Work can be mounted in channel jigs in almost any way imaginable. Clamping is also easily performed in this jig. Generally, the work is mounted against the top and one side of the jig while it is clamped from the other side (Figure 13–2).

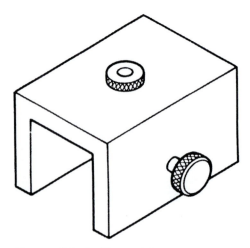

Figure 13–1 Channel jig.

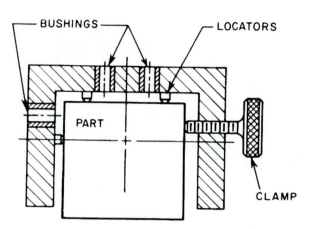

Figure 13-2 Locating and holding work in a channel jig.

DESIGNING A CHANNEL JIG

Once the designer selects a channel jig to perform work on a specific part, the part analysis begins. Referring to the part drawing and production plan for the spacer ring, as shown in Figure 13–3, the following design data are extracted.

- The part is a cylindrical ring 120 millimeters in diameter and 150 millimeters thick.
- The part has an 80-millimeter bored hole with two internal grooves.
- The material specified is cast bronze.
- The production run is 200 pieces.
- The operation required is drilling six 16-millimeter clearance holes through the sides, and drilling and tapping six M12 × 1.75-millimeter holes in one end.
- The blank received has been bored, faced, and grooved.

With these data, the design process begins.

Locating the Part

Since the part has a machined bore and is faced on both ends, these are the surfaces that will be used to locate the part. The locator selected for this jig must be relieved as shown to prevent binding (Figure 13–4). Relieving the locator also provides a place for the drill to run out. In

cases where drill runout cannot be provided by relieving the locator, small indentations or a groove slightly larger than the drill size should be used (Figure 13–5).

The size tolerance of the locator must again permit the part to be machined within its specified limits. Considering the extreme permissible dimensions in part size, the largest bore would be 80.1 millimeters; the smallest would be 79.9 millimeters. Based on the smallest size, the locator size should be 79.88 ± 0.01 millimeters. Again considering the extreme permissible dimensions, the maximum clearance between the locator and part would be 0.115 millimeter off-center (Figure 13–6). This is well within the off-center allowance of 0.25 millimeter specified.

Clamping the Part

In this jig, clamping is a secondary function of the locator. By drilling and tapping a hole, as shown in Figure 13–7, a strap clamp can be used to hold the part in place. When the locator is used to clamp the part, it should be 5 to 20 millimeters shorter than the workpiece. This permits positive clamping and prevents the clamp from "bottoming out" against the locator. The strap used to clamp the part should be approximately 10 millimeters thick to prevent its bending when clamped.

Constructing the Tool Body

Construction of the tool body begins by specifying the size of channel to be used. If stock channel is unavailable, or if the part is too large for standard channel sizes, a fabricated channel may be substituted (Figure 13–8). In the case of this tool, a piece of stock cast channel 203 millimeters by 203 millimeters by 150 millimeters long can be specified.

The locator stud is centrally mounted inside the channel by means of two M10 screws. The screws used to secure the locator are socket-head cap screws. Their mounting holes are counterbored so that the cap screws will be slightly below the surface of the tool to prevent any possible interference with the drill-press spindle.

Since the basic shape of this tool does not readily afford a suitable clamping surface, a secondary or

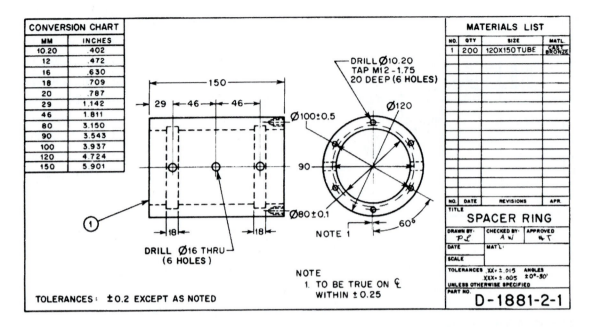

CONVERSION CHART	
MM	INCHES
10.20	.402
12	.472
16	.630
18	.709
20	.787
29	1.142
46	1.811
80	3.150
90	3.543
100	3.937
120	4.724
150	5.901

DRILL Ø10.20
TAP M12-1.75
20 DEEP (6 HOLES)

Ø100±0.5

Ø80±0.1

DRILL Ø16 THRU
(6 HOLES)

Ø120

NOTE 1

60°

NOTE
1. TO BE TRUE ON ℄
   WITHIN ±0.25

TOLERANCES: ±0.2 EXCEPT AS NOTED

MATERIALS LIST			
NO.	QTY	SIZE	MATL.
1	200	120X150 TUBE	CAST BRONZE

NO.	DATE	REVISIONS	APR.

TITLE
SPACER RING

DRAWN BY: P.L.	CHECKED BY: A W	APPROVED 4 T
DATE	MAT'L:	
SCALE		

TOLERANCES XX= ±.015 ANGLES
XXX= ±.005 ±0°-30'
UNLESS OTHERWISE SPECIFIED

PART NO.
D-1881-2-1

PRODUCTION PLAN						
P/N **1881-2**	PART NAME **SPACER RING**			QUANTITY **200**		ORDER NO **79-9215**
DRAWING NO. **D-1881-2-1**	PROCESS PLANNER **T.L. PRICE**		REVISION NO	DATE	PAGE **1**	OF **1**
OPR. NO.	DESCRIPTION			DEPT.	MACH. TOOL	
1	BORE Ø80 THRU & GROOVE 18x90 DIA.			#27 LATHE	TURRET LATHE #27-63	
2	FACE BOTH ENDS TO 150 LENGTH			#27 LATHE	TURRET LATHE #27-63	
3	DRILL & TAP 6-12X1.75 HOLES IN END			#66 DRILLING	RADIAL DRILL #66-1	
4	DRILL 3-Ø16 HOLES IN EACH SIDE			#66 DRILLING	RADIAL DRILL #66-1	
5	DEBURR & INSPECT			#7 FINISHING	TUMBLER #7-1215	
5	FUNCTIONAL GAUGE			———	INSP. FIXTURE T-1881-2-1	
4	DRILL JOBBER'S LENGTH TAPER SHANK			16mm	DRILL JIG T-1881-2-1	
3	TAP, 2 FLUTE DRILL JOBBER'S LENGTH TAPER SHANK			M12X1.75 10.20mm	DRILL JIG T-1881-2-1	
OPR NO.	TOOL DESCRIPTION			SIZE	SPECIAL TOOL NO.	

Figure 13–3 Part drawing and production plan.

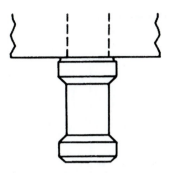

Figure 13-4 Cylindrical locator.

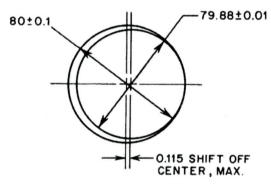

Figure 13-5 Providing drill runout.

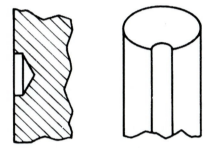

80±0.1   79.88±0.01

0.115 SHIFT OFF
CENTER, MAX.

Figure 13-6 Dimensional differences.

auxiliary clamp must be used. The clamp shown in Figure 13–9 is one style that could be used to hold the tool on the drill-press table. The advantages of this clamp are speed, versatility, and range.

Locating the Bushings

Since the primary locator in this tool is a cylinder, the circular hole pattern for the six M12 × 1.75-millimeter holes must be concentric to this locator. To ensure this, the designer should specify a close tolerance between the positions of the hole circle and the locator. A locational tolerance of ±0.03 millimeter should provide an adequate allowance for the toolmaker and a sufficiently accurate tool to meet the demands of the part drawing.

Adding this tolerance variable to the size tolerance of the locator means that the extreme permissible dimensions would place the holes off-center in either direction by a maximum of 0.145 millimeter (Figure 13–10). This is well within the specified allowable variance of 0.25 millimeter on the part drawing.

By specifying the method described to position the locator, the designer can make the toolmaker's task much easier. When using this method, the designer should specify the shoulder length to be slightly greater than the thickness of the channel. This permits accurate referencing of the locator, as shown in Figure 13–11, and makes the jig borer work much easier and less time-consuming. The shoulder can also aid the toolmaker in locating the hole positions in the sides of the tool. When all the holes are properly positioned and machined to size, the shoulder can be removed by grinding it flush with the top of the tool.

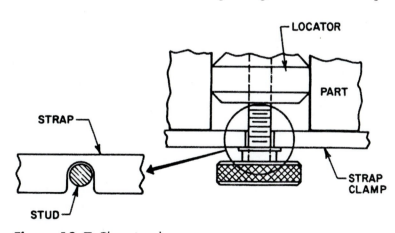

Figure 13-7 Clamping the part.

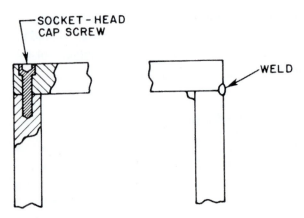

Figure 13-8 Fabricated channel.

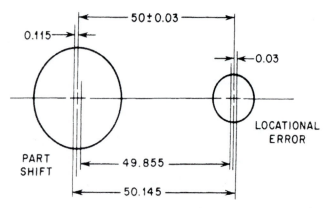

Figure 13-10 Dimensional differences between the part and the bushings.

Locating the bushings for the six holes drilled in the sides is done in much the same manner. Here again, the main requirement is the proper relation to the locator center. The tolerance specified for these holes should also be ±0.03 millimeter.

The final steps in designing this channel jig are completing the rough design sketch and the tool drawings (Figure 13–12).

BOX JIGS

Box, or tumble, jigs, are normally made in the form of a box or framework around a part (Figure 13–13). While the capability to machine parts on all surfaces exists with these tools, some parts mounted in box jigs are only machined on two or three sides. This is because box jigs also offer good support to frail workpieces and excellent tolerance control.

DESIGNING A BOX JIG

Once a box jig has been selected as the most logical tool to use for a machining operation, the tool designer must gather and evaluate all the data concerning the part. Referring to the part drawing and production plan for the guide block, as shown in Figure 13–14, the following information and design data are extracted:

- The part is 70 millimeters long, 50 millimeters wide, and 30 millimeters thick.
- There is one step 27 millimeters wide and one 15-degree angle machined on the part.

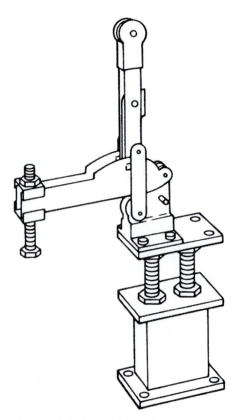

Figure 13-9 Auxiliary clamp.

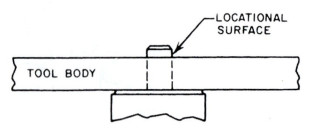

Figure 13-11 Referencing the work.

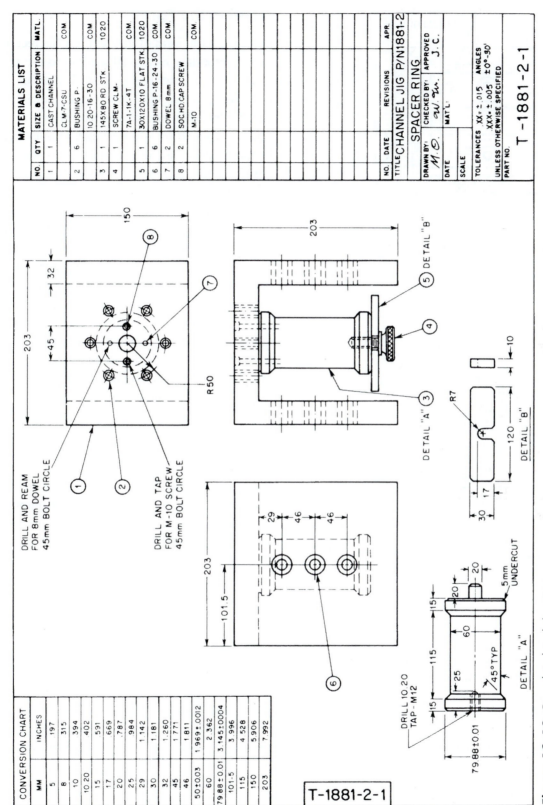

Figure 13-12 Completed tool drawing.

MATERIALS LIST

NO.	QTY	SIZE & DESCRIPTION	MATL.
1	1	CAST CHANNEL	
		CLM-7-CSU	COM
2	6	BUSHING P.	COM
		10 20-16-30	1020
3	1	145X80 RD STK	1020
4	1	SCREW CLM-	
		7A-1K-4T	COM
5	1	30X120X10 FLAT STK.	1020
6	6	BUSHING P-16-24-30	COM
7	2	DOWEL 8mm	COM
8	2	SOC. HD. CAP SCREW	
		M-10	COM

NO.	DATE	REVISIONS	APR.

TITLE CHANNEL JIG P/N1881-2

SPACER RING

DRAWN BY: M.O.	CHECKED BY: W.M.	APPROVED J.C.
DATE	MAT'L:	
SCALE		

TOLERANCES .XX= ±.015 ANGLES ±0°-30'
.XXX= ±.005
UNLESS OTHERWISE SPECIFIED
PART NO.

T-1881-2-1

CONVERSION CHART	
MM	INCHES
5	.197
8	.315
10	.394
10.20	.402
15	.591
17	.669
20	.787
25	.984
29	1.142
30	1.181
32	1.260
45	1.771
46	1.811
50±.003	1.969±.0012
60	2.362
79.88±.01	3.145±.0004
101.5	3.996
115	4.528
150	5.906
203	7.992

T-1881-2-1

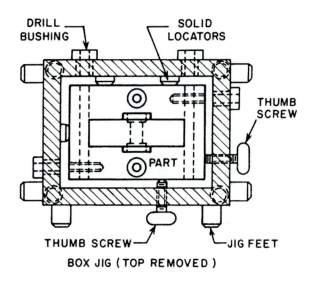

BOX JIG (TOP REMOVED)

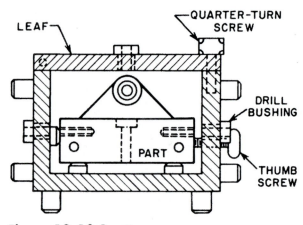

Figure 13-13 Box jig.

- The material specified is 1040 steel.
- The production run is 150 pieces.
- The part has three M14 tapped holes and two 8-millimeter holes drilled 10 millimeters deep.
- The five holes in the part are located on four surfaces.
- The part received for drilling has been milled and deburred.

Using these data, the designer begins formulating the required design ideas and alternatives. From these initial ideas and impressions, the design of the box jig is developed.

Locating the Part

Since the basic form of this part is a flat rectangle, the most appropriate locational method to use is the six-point method described in Unit 3. Starting with the part positioned on three locators, as shown in Figure 13–15, the solid side of the tool is first considered. This surface will act as the primary location point. Two button locators should be positioned as shown in Figure 13–16. The channel used to construct the tool body should be approximately 15 to 20 millimeters thick to resist the tool thrust and maintain the required rigidity. The tertiary locator consists of a single button that contacts the part on its short side (Figure 13–17).

As a precaution against improper loading, the designer must select a method to make the tool foolproof. The simplest device to use for this is a pin positioned as shown in Figure 13–18. This pin will allow the part to be installed only in the correct position. Once the part is located properly, the designer must select a method and device to hold this position during the machining cycle.

Clamping the Part

Clamping is performed by two screw clamps positioned as shown in Figure 13–19. With these clamps, adequate clamping force is obtained without interference with the machining of the part. By specifying swivel-head screw clamps rather than plain screws, the chance of marring the workpiece is reduced.

The final part of the tool body the designer must consider is the leaf. With this tool, the leaf does not contain any bushings. In other, more complex jigs, bushings could be positioned in the leaf. When the leaf is used to support bushings, a positive means must be provided to ensure the leaf's correct location. Whenever possible, the leaf should be the last place selected to position bushings, because there is always the possibility of error from wear, misalignment, or movement of the leaf. To secure the leaf, a quarter-turn screw is used.

Locating the Bushings

One important point must be remembered. The bushings are to be positioned relative to the locators that control the dimensional position of the holes. For

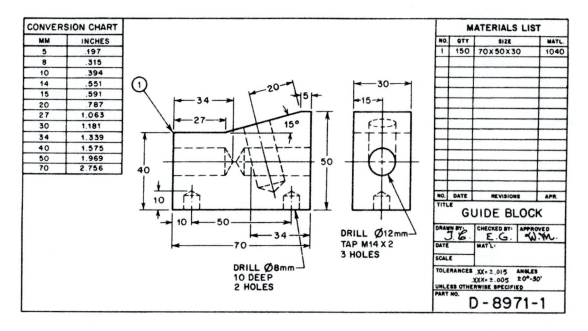

CONVERSION CHART	
MM	INCHES
5	.197
8	.315
10	.394
14	.551
15	.591
20	.787
27	1.063
30	1.181
34	1.339
40	1.575
50	1.969
70	2.756

MATERIALS LIST

NO.	QTY	SIZE	MATL.
1	150	70X50X30	1040

NO.	DATE	REVISIONS	APR.

TITLE
GUIDE BLOCK

DRAWN BY: J.G. CHECKED BY: E.G. APPROVED W.M.

DATE   MAT'L:

SCALE

TOLERANCES .XX=±.015 ANGLES
.XXX=±.005 ±0°-30'
UNLESS OTHERWISE SPECIFIED

PART NO.
D-8971-1

DRILL Ø12mm
TAP M14 X 2
3 HOLES

DRILL Ø8mm
10 DEEP
2 HOLES

PRODUCTION PLAN

P/N 8971	PART NAME GUIDE BLOCK		QUANTITY 150	ORDER NO. 77-14992
DRAWING NO. D-8971-1	PROCESS PLANNER T. WONG	REVISION NO.	DATE	PAGE 1 OF 1

OPR. NO.	DESCRIPTION	DEPT.	MACH. TOOL
1	CUTOFF 30X50mm STOCK TO 70mm LONG	#68 CUTOFF RM	ABRASIVE CUTOFF #68-1
2	MILL FLAT & ANGLE	#37 MILLING	HORZ. MILL #37-10
3	DEBURR	#7 FINISHING	TUMBLER #7-4
4	DRILL 5 HOLES	#66 DRILLING	RADIAL DRILL #66-19
5	TAP 3 HOLES	#66 DRILLING	RADIAL DRILL #66-19
6	DEBURR & INSPECT	#7 FINISHING	TUMBLER #7-4

OPR. NO.	TOOL DESCRIPTION	SIZE	SPECIAL TOOL NO.
6	FUNCTIONAL GAUGE	———	INSP. FIXTURE T-8971-2
5	TAP, 4 FLUTE	M14X2	———
4	DRILLS, JOBBER'S LENGTH, TAPER SHANK	8mm 12mm	BOX JIG T-8971-2
2	MILLING CUTTERS, INTERLOCK	100X30X25 100X40X25(15°) 90X10X25	VISE JAW FIXTURE T-8971-1

Figure 13-14 Part drawing and production plan.

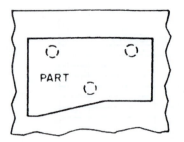

Figure 13-15 Primary locator.

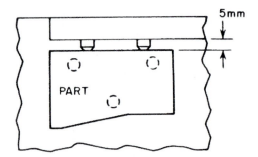

Figure 13-16 Secondary locator.

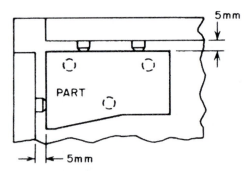

Figure 13-17 Tertiary locator.

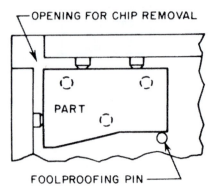

Figure 13-18 Foolproofing the tool.

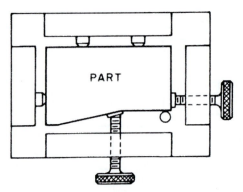

Figure 13-19 Clamping the part.

example, the bushings for the larger hole are positioned relative to locators A and B (Figure 13–20). The bushings for the small hole are relative to locators A and C.

The position of the bushings relative to horizontal and vertical location is normally controlled by the outside form of the box jig. In the case of this tool, one hole must be machined at an angle. To accurately position the bushing to drill this hole, the side opposite the bushing must be machined to the same angle (Figure 13–21). The locational tolerance of the bushings in the tool must again be between 30 and 50 percent of the part tolerance.

Once all the data have been assembled and evaluated, and a design has been planned, the designer constructs the rough sketches and tool drawings (Figure 13–22).

TOOL DESIGN APPLICATION

1. Analyze the part drawing and production plan, as shown in Figure 13–23, to determine all relevant information necessary to design a channel or box jig.
2. Compare similarities and differences between the assigned part and previous examples.
3. Prepare design sketches to calculate sizes and dimensions.
4. Include all necessary data in the completed tool drawing.

Suggestions for Design

• Select either type of tool. Both tools will work for this part.
• Base tool selection on economy and efficiency.

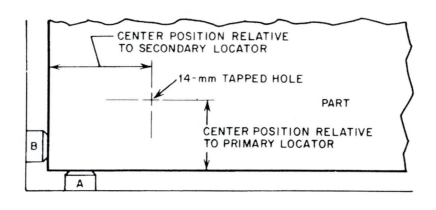

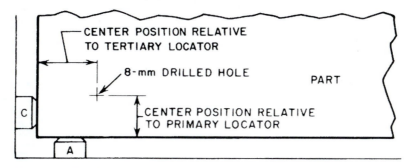

Figure 13-20 Positioning the bushings.

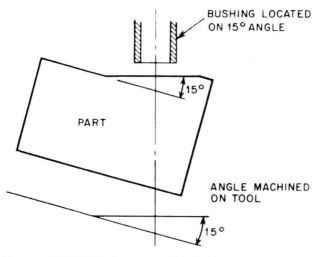

Figure 13-21 Referencing the bushings to the tool contour.

SUMMARY

The following important concepts were presented in this unit:

- The channel jig is the simplest form of closed jig. It is capable of machining parts on as many as three different surfaces.

- The box, or tumble jig, is the most detailed and complex form of jig. It is capable of machining parts on as many as six different surfaces.
- The process used to design either a channel jig or a box jig is the same as that used for a template jig.
  - Before the design is started, the tool drawing must be studied to obtain all necessary information about the part.
  - Careful calculations are made to determine the exact size and location of the locators and bushings.
- Commercially available components should be used when possible.
- Each design should start as a sketch.
  - Sketching helps to formulate the design.
  - Sketches help to reduce problems and show relationships.
  - Once the design is set and the problems solved, the final design drawing is prepared.

REVIEW

1. What are channel and box jigs used for?
2. Which is the most basic form of closed jig?
3. Why must locators be relieved?

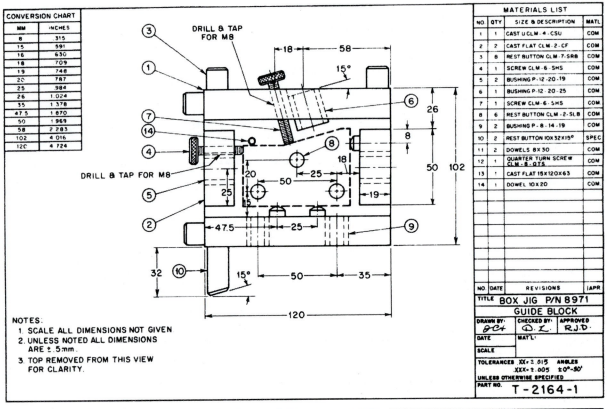

CONVERSION CHART	
MM	INCHES
8	.315
15	.591
16	.630
18	.709
19	.748
20	.787
25	.984
26	1.024
35	1.378
47.5	1.870
50	1.969
58	2.283
102	4.016
120	4.724

NOTES:
1. SCALE ALL DIMENSIONS NOT GIVEN
2. UNLESS NOTED ALL DIMENSIONS ARE ± .5 mm.
3. TOP REMOVED FROM THIS VIEW FOR CLARITY.

MATERIALS LIST			
NO.	QTY	SIZE & DESCRIPTION	MATL
1	1	CAST U CLM - 4 - CSU	COM
2	2	CAST FLAT CLM - 2 - CF	COM
3	8	REST BUTTON CLM - 7 - SRB	COM
4	1	SCREW CLM - 6 - SHS	COM
5	2	BUSHING P - 12 - 20 - 19	COM
6	1	BUSHING P - 12 - 20 - 25	COM
7	1	SCREW CLM - 6 - SHS	COM.
8	6	REST BUTTON CLM - 2 - SLB	COM
9	2	BUSHING P - 8 - 14 - 19	COM
10	2	REST BUTTON 10 X 32 X 15°	SPEC
11	2	DOWELS 8 X 30	COM
12	1	QUARTER TURN SCREW CLM - 8 - QTS	COM
13	1	CAST FLAT 15 X 120 X 63	COM
14	1	DOWEL 10 X 20	COM

NO.	DATE	REVISIONS	APR.

TITLE **BOX JIG P/N 8971**

GUIDE BLOCK

DRAWN BY:	CHECKED BY:	APPROVED
JC+	D.L.	R.J.D.
DATE	MAT'L:	
SCALE		

TOLERANCES .XX = ±.015   ANGLES
.XXX = ±.005   ±0°-30'
UNLESS OTHERWISE SPECIFIED

PART NO.   **T - 2164 - 1**

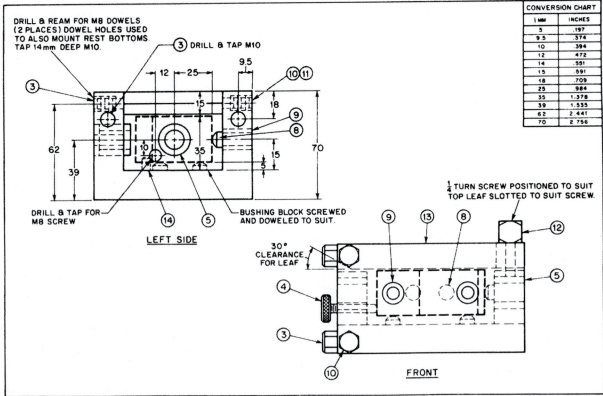

DRILL & REAM FOR M8 DOWELS (2 PLACES) DOWEL HOLES USED TO ALSO MOUNT REST BOTTOMS. TAP 14mm DEEP M10.

DRILL & TAP M10

DRILL & TAP FOR M8 SCREW

BUSHING BLOCK SCREWED AND DOWELED TO SUIT.

LEFT SIDE

¼ TURN SCREW POSITIONED TO SUIT TOP LEAF SLOTTED TO SUIT SCREW.

30° CLEARANCE FOR LEAF

FRONT

CONVERSION CHART	
MM	INCHES
5	.197
9.5	.374
10	.394
12	.472
14	.551
15	.591
18	.709
25	.984
35	1.378
39	1.535
62	2.441
70	2.756

Figure 13–22 Completed tool drawing.

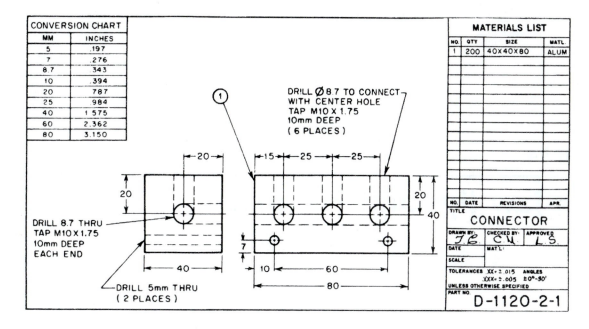

CONVERSION CHART	
MM	INCHES
5	.197
7	.276
8.7	.343
10	.394
20	.787
25	.984
40	1.575
60	2.362
80	3.150

DRILL Ø8.7 TO CONNECT WITH CENTER HOLE TAP M10 X 1.75 10mm DEEP (6 PLACES)

DRILL 8.7 THRU TAP M10 X 1.75 10mm DEEP EACH END

DRILL 5mm THRU (2 PLACES)

MATERIALS LIST

NO.	QTY	SIZE	MATL.
1	200	40 X 40 X 80	ALUM

NO.	DATE	REVISIONS	APR.

TITLE   **CONNECTOR**

DRAWN BY:	CHECKED BY:	APPROVED
J.C.	C.M.	L.S.
DATE	MATL:	
SCALE		

TOLERANCES XX = ±.015   ANGLES
.XXX = ±.005   ±0°-30'
UNLESS OTHERWISE SPECIFIED

PART NO.   **D-1120-2-1**

Figure 13-23 Part drawing and production plan.

PRODUCTION PLAN

P/N 1120-2	PART NAME CONNECTOR		QUANTITY 200	ORDER NO. 551491-78
DRAWING NO. D-1120-2-1	PROCESS PLANNER J. FLYNN	REVISION NO.	DATE	PAGE 1 OF 1

OPR. NO.	DESCRIPTION	DEPT.	MACH. TOOL
1	CUTOFF 40X40 STOCK TO 80mm LONG	#68 CUTOFF RM.	ABRASIVE CUTOFF #68-1
2	DRILL 7-Ø8.7 HOLES	#66 DRILLING	DRILL PRESS #66-4
3	TAP 7-M10X1.75 HOLES	#66 DRILLING	DRILL PRESS #66-4
4	DEBURR & INSPECT	#7 FINISHING	TUMBLER #7-1215
4	FUNCTIONAL GAUGE	——	INSP. FIXTURE T-1120-2-2
3	TAP-2 FLUTE	M10-1.75	DRILL JIG T-1120-2-1
2	DRILL JOBBER'S LENGTH TAPER SHANK	8.7mm	DRILL JIG T-1120-2-1

OPR. NO.	TOOL DESCRIPTION	SIZE	SPECIAL TOOL NO.

4. How can a clamp be prevented from bottoming out when the locator is used to clamp?

5. Why must screws be counterbored below the surface of the tool?

6. Why should the designer specify that the locator shoulder be longer than the jig plate thickness?

7. How many surfaces can be machined without moving the part in a box jig?

8. What method should be used to locate a flat rectangular part?

9. What is the simplest foolproofing device?

10. Why should the leaf be the last place to put bushings?

11. How should the outside of the jig be designed for angular holes?

UNIT 14

Vise-Jaw Jigs and Fixtures

OBJECTIVES

After completing this unit, the student should be able to:

- Analyze part data to determine suitable tool designs.
- Specify locating and supporting methods, and details to suit sample parts.
- Design one vise-jaw jig and one vise-jaw fixture to suit specified sample parts.

THE MACHINE VISE

Vise-jaw jigs and fixtures are the simplest and most basic form of workholding device. The adaptability, versatility, and overall economy of these tools make them one of the most efficient and cost-effective workholders for modern production.

The basic component used with these jigs and fixtures is the machine vise (Figure 14–1). Vise-jaw jigs and fixtures are designed to replace the standard jaws. This allows the machine vise to serve as a special workholder for any number of different parts,

Figure 14–1 Standard swivel-base machine vise.

simply by using a special set of jaws for each part. Not only does this save time in making special tools, but also tool costs can be greatly reduced. Several variations of the machine vise are available to further increase its usefulness (Figure 14–2).

LOCATING WORK IN VISE-JAW WORKHOLDERS

Locating the workpiece in a vise-jaw tool is performed in much the same way as in other types of jigs and fixtures. Almost every locational device and

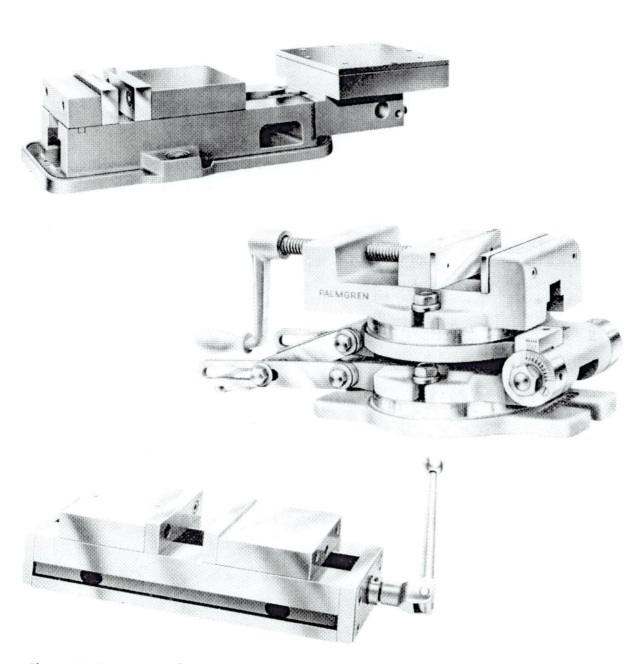

Figure 14–2 Variations of the standard vise.

method discussed in this text can be used with vise-jaw tooling.

The only restrictions to using this type of workholder are the size of the part and the capacity of the vise. Usually, if the part will fit into the vise, a vise-jaw jig or fixture can be used.

Locating with Pins and Blocks

The most common method of locating workpieces is with pins or blocks. With this method, the part is located either by its outside edges or by holes in the part (Figure 14–3). To maintain the required repeatability, it is a good practice to place the locating elements against the fixed jaw to prevent any misalignment resulting from the movement of the movable jaw.

Locating with Nests

Parts that do not have holes or machined edges to locate from can sometimes be positioned by using a nest. Nest-type locators are made by either the machining method or the casting method.

Machined nests are normally used for symmetrical parts that do not have any intricate details. This style is used where the part profile can easily be produced with standard tooling (Figure 14–4). Since making nests is very time-consuming and costly, machined nests are normally restricted to tools that are used in very long production runs.

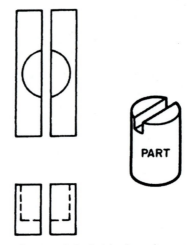

Figure 14–4 Machined nest.

Cast nests are primarily used to hold unsymmetrical parts or those having intricate details (Figure 14–5). Plastic or epoxy resins or low-melt alloy metals are used to make these nests.

Nests made from plastic or epoxy resins are normally used for short-run production work or to machine unusually shaped one-of-a-kind parts. These materials are mixed cold and poured into a shell. The part, which is coated with a releasing agent, is pressed between two shells and left until the plastic compound hardens (Figure 14–6). Ground glass, metal filings, or ground walnut shells are sometimes added to the mixture to act as a filler. These materials also increase the wear resistance of the plastic.

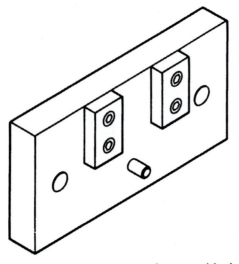

Figure 14–3 Locating with pins or blocks.

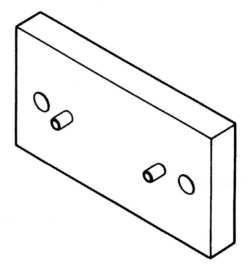

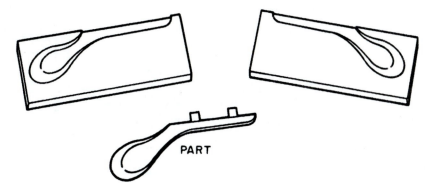

Figure 14–5 Cast nest.

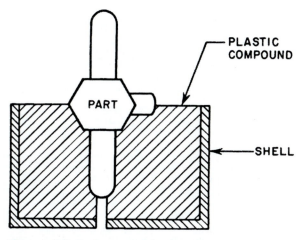

PLASTIC COMPOUND

PART

SHELL

Figure 14–6 Casting with plastic compounds.

Nests made from low-melt alloys, such as bismuth, require a slightly different process. In the making of a nest of this type, the part is suspended inside the mold (Figure 14–7). To increase wear resistance, the mold cavity can be filled with hardened ball bearings before the molten metal is poured. The melting point of these alloys is between 147 and 227 degrees Celsius.

The disadvantage of low-melt alloys for tooling is the cost of the equipment needed to melt the metal. Their advantage over other forms of casting materials is their ability to be reclaimed by melting when no longer needed.

DESIGNING A VISE-JAW JIG

Designing a vise-jaw jig should require little time. All that is required is a suitable device to locate the part and support it against the tool thrust. Even locating the drill bushings for this type of tool presents few problems.

Referring to the part drawing and production plan for the strap support as shown in Figure 14–8, the designer extracts the following information:

- The part is a flat bar .75 inch wide, .38 inch thick, and 17.75 inches long.
- The material specified is 1020 steel.
- The machining operation required is drilling one .25–.26-inch hole in one end and two .31–.32-inch holes in the other end.

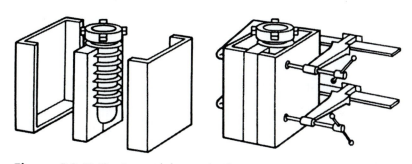

Figure 14–7 Casting with low-melt alloys.

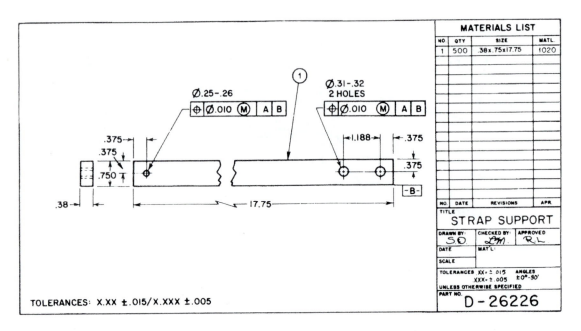

Figure 14-8 Part drawing and production plan.

- The blank received is rolled bar-stock milled on each end.
- The size of the production run is 500 pieces.

Using this information, the designer begins the design.

Locating the Part

The simplest method of locating this part is a milled shelf on the stationary jaw (Figure 14–9). To reference the lengthwise movement of the part, an adjustable locator is used. The width of the shelf should be about

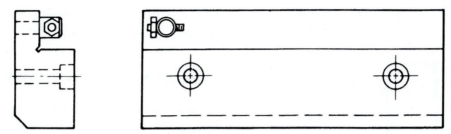

Figure 14-9 Locating with a milled shelf.

75 percent of the part width to permit the needed clamping room and keep the primary locator on the solid jaw.

The movable jaw can be either solid or milled (Figure 14–10). If the jaw is milled, enough clearance must be left between the jaws to allow the vise to clamp without bottoming out. For a part of this size, a clearance of .050 to .060 inch is sufficient.

Locating the Bushings

Since this jig is intended to drill the holes at both ends of the part, one of two possible designs can be selected. With the first, two separate jig plates are designed, one for each end of the part (Figure 14–11). The second method uses only one jig plate that is double-ended. This type is first used in one position, then rotated to the second position (Figure 14–12). With either jig plate, to ensure the necessary accuracy, the jig plate and jaw should be notched (Figure 14–13).

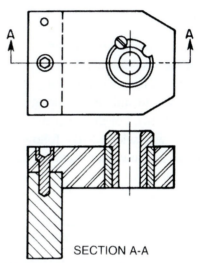

SECTION A-A

Figure 14-11 Single jig plates.

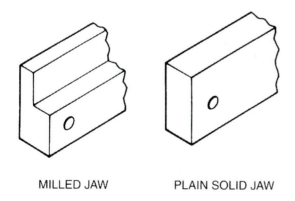

MILLED JAW   PLAIN SOLID JAW

Figure 14-10 Variations for the movable jaw.

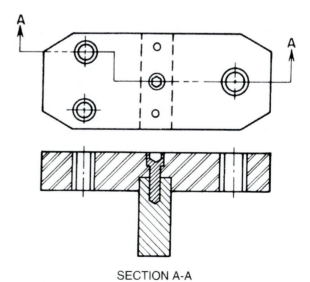

SECTION A-A

Figure 14-12 Double-ended jig plate.

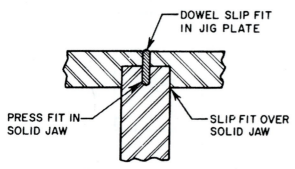

Figure 14-13 Locating the jig plate on the jaw.

Securing the Jaws

Vise-jaw jigs and fixtures are held in place by the mount holes found in the vise casting. A good practice for a tool designer is to have a chart listing standard vise-jaw dimensions to aid in the design of vise-jaw tools (Figure 14–14). For the strap support, an 8-inch machine vise was selected. By following the size data for the vise as shown in the chart, many calculations can be avoided.

The final step in the design of this jig is preparing the tool drawing (Figure 14–15).

DESIGNING A VISE-JAW FIXTURE

Vise-jaw fixtures are basically the same as vise-jaw jigs (Figure 14–16). The main differences between them are the absence of drill bushings and the addition of a set block in the fixture. The design process for both tools is identical.

Referring to the part drawing and production plan for the sight bracket (Figure 14–17), the following data are extracted and used to design the vise-jaw fixture:

- The part is a flat rectangle 85 millimeters long, 40 millimeters wide, and 12 millimeters thick.
- The machining operation required is gang-milling three slots: two 10 millimeters wide, one 3 millimeters wide, and all 15 millimeters deep.
- The blank received is saw-cut on both ends and has two 8-millimeter drilled holes.

STANDARD VISE - JAW SIZES					
VISE	A	B	C	D	E
KURT-D30/PD30	$3\frac{1}{16}$/77.8	1/25.4	$1\frac{3}{4}$/44.5	$\frac{9}{16}$/14.3	$\frac{5}{16}$/7.9
D40	$4\frac{1}{16}$/104.7	$1\frac{1}{4}$/31.8	$2\frac{1}{2}$/63.5	$\frac{11}{16}$/17.5	$\frac{3}{8}$/9.5
D50	$5\frac{1}{8}$/130.2	$1\frac{1}{2}$/38.1	$3\frac{1}{8}$/79.4	$\frac{13}{16}$/20.6	$\frac{7}{16}$/11.1
D60/PD60/SCD60	$6\frac{1}{8}$/155.6	$1\frac{3}{4}$/44.5	$3\frac{7}{8}$/98.4	$\frac{15}{16}$/23.8	$\frac{1}{2}$/12.7
D80	$8\frac{1}{8}$/206.4	$2\frac{1}{4}$/57.2	$4\frac{3}{4}$/120.7	$1\frac{7}{32}$/30.9	$\frac{5}{8}$/15.9
D100/SCD100	$10\frac{1}{8}$/257.2	3/76.2	$5\frac{3}{4}$/146.1	$1\frac{5}{8}$/41.3	$\frac{1}{4}$/19.1
PALMGREN 80B	4/101.6	$1\frac{1}{2}$/38.1	$2\frac{1}{4}$/57.2	$\frac{3}{4}$/19.1	$\frac{1}{4}$/6.4
45 V	$4\frac{1}{2}$/114.3	$1\frac{1}{4}$/31.8	3/76.2	$\frac{5}{8}$/15.9	$\frac{1}{4}$/6.4
60B	6/152.4	2/50.8	4/101.6	1/25.4	$\frac{5}{16}$/7.9
80B	8/203.2	$2\frac{1}{4}$/57.2	6/152.4	$1\frac{1}{8}$/28.6	$\frac{3}{8}$/9.5

Figure 14-14 Standard vise-jaw dimensions.

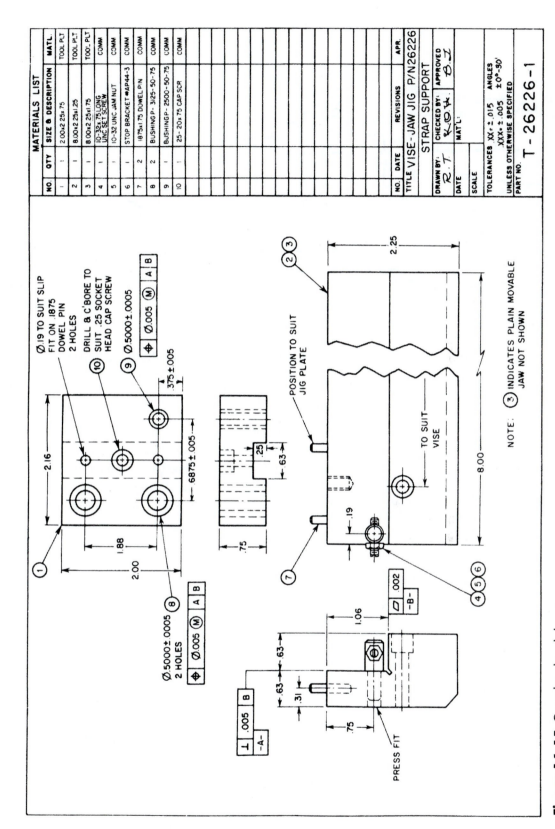

Figure 14-15 Completed tool drawing.

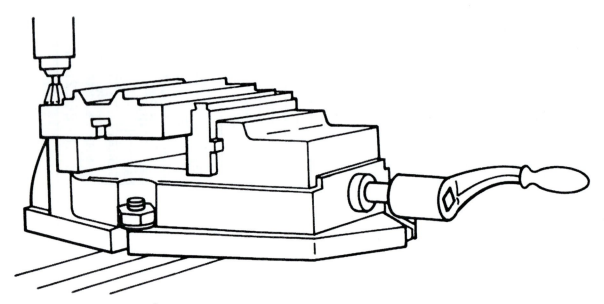

Figure 14–16 Vise-jaw fixture.

• The production run is 150 pieces.
• The material specified is 7075-T6 aluminum.

Starting with this information, the designer begins designing the vise-jaw fixture.

Locating the Part

The best locational surfaces to use in positioning this part are the two drilled holes. As shown in Figure 14–18, the locating pins are positioned within ±0.10 millimeter on 55-millimeter centers. This tolerance permits slight variations in the hole locations. With the locators 0.05 millimeter smaller than the holes, the parts will still keep the required accuracy.

The locators selected for this part are one solid pin and one diamond pin. The diamond pin must be positioned to resist the rotation around the solid pin (Figure 14–19). The length of these locators is 6 millimeters—half the part thickness. This also ensures the fast, efficient loading and unloading of the parts. Since the part is totally located against the solid jaw, the movable jaw should be made flat.

Positioning the Set Block

Since the position of this fixture is stable and not likely to change from part to part, a set block can be used to reference the cutters. This set block, as shown in

Figure 14–20, is made to reference two surfaces at the same time—the side of one outside cutter and the depth of the slot.

In this fixture, the solid jaw should be selected to mount the set block (Figure 14–21). The block can be either machined into the jaw or mounted separately. In either case, it should be positioned so that the same feeler gauge can be used to reference both cutter planes.

The final step is completing the sketches and the tool drawing (Figure 14–22). In the design of tools for gang milling or straddle milling, the size of the collars that separate the cutters is important. To maintain the required spacing, these collars must be made to a specific size. While standard collars can be used in some cases, for this fixture the size is nonstandard. Special collars must be made. In any case, it is a good practice to store these collars with the fixture so that they can be found easily if needed for repeat production of the part.

TOOL DESIGN APPLICATION

1. Analyze the part drawings and production plan for the connector rod (Figure 14–23). Determine all relevant information necessary to design a vise-held jig and a vise-held fixture.

2. Compare similarities and differences between the assigned part and the examples.

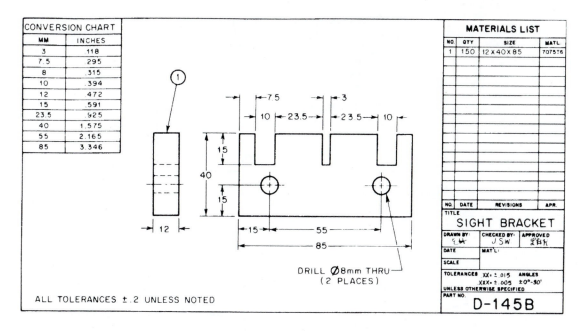

CONVERSION CHART	
MM	**INCHES**
3	.118
7.5	.295
8	.315
10	.394
12	.472
15	.591
23.5	.925
40	1.575
55	2.165
85	3.346

ALL TOLERANCES ±.2 UNLESS NOTED

DRILL ⌀8mm THRU
(2 PLACES)

MATERIALS LIST

NO.	QTY	SIZE	MATL
1	150	12 x 40 x 85	7075T6

NO.	DATE	REVISIONS	APR.

TITLE
SIGHT BRACKET

DRAWN BY: ΨH   CHECKED BY: J SW   APPROVED LBK
DATE   MAT'L:
SCALE

TOLERANCES XX= ±.015   ANGLES
XXX= ±.005   ±0°-30'
UNLESS OTHERWISE SPECIFIED

PART NO.
D-145B

PRODUCTION PLAN

P/N 145B	PART NAME SIGHT BRACKET		QUANTITY 150	ORDER NO. 77-10109
DRAWING NO. D-145B	PROCESS PLANNER MARK HARPER	REVISION NO	DATE	PAGE 1 OF 1

OPR. NO.	DESCRIPTION	DEPT.	MACH. TOOL
1	CUTOFF 40X12 STOCK TO 85mm	#68 CUTOFF RM.	HORZ. BANDSAW #68-20
2	DRILL 2 - 8mm HOLES	#66 CUTOFF RM.	DRILL PRESS #68-20
3	DEBURR	#7 FINISHING	TUMBLER #7-1207
4	MILL 2-10mm SLOTS 1 - 3mm SLOT	#37 MILLING	HORZ. MILL #37-10
5	DEBURR & INSPECT	#7 FINISHING	TUMBLER #7-1207
5	FUNCTIONAL GAUGE	———	INSP. FIXTURE I-145B-3
4	CUTTERS, PLAIN MILLING	2-10X25X100 1-3X25X100	MILL FIXTURE T-145B-2
2	DRILL JOBBER'S LENGTH, STRAIGHT SHANK	8mm	DRILL JIG T-145B-1
OPR. NO.	TOOL DESCRIPTION	SIZE	SPECIAL TOOL NO.

Figure 14-17 Part drawing and production plan.

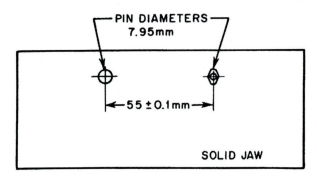

Figure 14-18 Locating with pin-type locators.

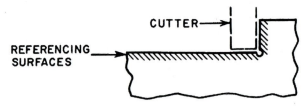

Figure 14-19 Positioning the diamond pin.

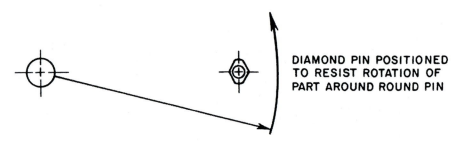

Figure 14-20 Relation of set block and cutters.

3. Prepare an initial sketch to calculate necessary dimensions.
4. Complete the tool drawing, including all necessary data needed to build the tool.

Suggestions

- Note the shape of the part and select the most appropriate locating device.
- Decide which part features can best be used to locate the part.

SUMMARY

The following important concepts were presented in this unit:

- Vise-jaw jigs and fixtures are some of the simplest and least expensive jigs and fixtures.
  - When a vise-jaw tool is used to mount workpieces, the designer is actually only designing a locating device rather than building a complete workholder.
  - The vise, which holds the jaws, is the major part of these workholders.
- Before the design is started, the tool drawing must be studied to obtain all necessary information about the part.
- Careful calculations must be made to determine the exact size and location of the locators used to position the part.
- Bushings are used to locate and support the drills used with vise-jaw jigs.

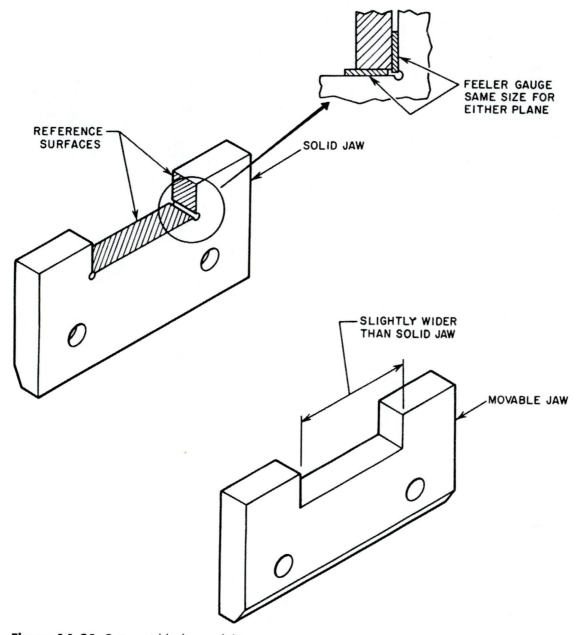

Figure 14-21 Cutter set block on solid jaw.

- Cutter location and referencing with vise-jaw fixtures is accomplished using a set block.
- Commercially available components should be used when possible.
- Each design should start as a sketch.
  - Sketching helps to formulate the design.
  - Sketches help to reduce problems and show relationships.
  - Once the design is set and the problems have

been solved, the final design drawing is prepared.

REVIEW

1. What are the main features of vise-jaw jigs and fixtures that make them the most efficient and cost-effective workholders?

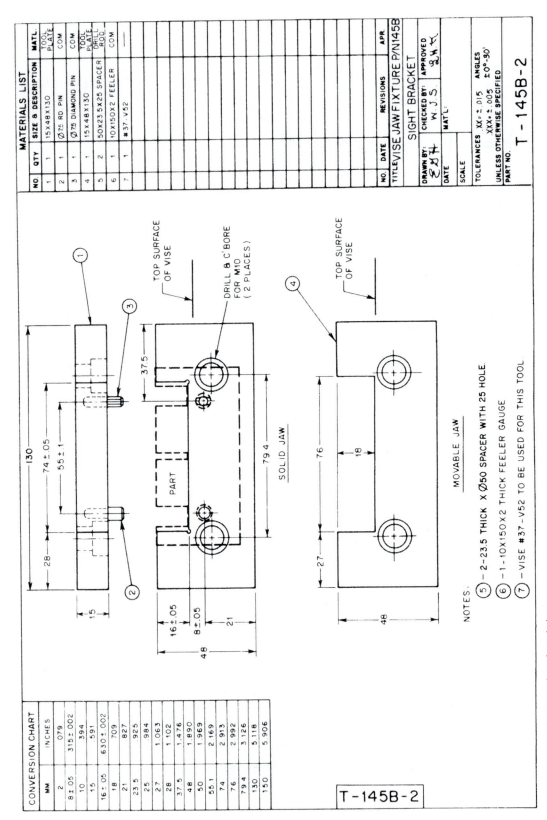

Figure 14-22 Completed tool drawing.

MATERIALS LIST

NO.	QTY	SIZE & DESCRIPTION	MATL.
1	1	15x48x130	TOOL PLATE
2	1	Ø.75 RD PIN	COM
3	1	Ø.75 DIAMOND PIN	COM
4	1	15x48x130	TOOL PLATE
5	2	50x23.5x25 SPACER	DRILL ROD
6	1	10X150X2 FEELER	COM
7	1	#37-V52	—

NO.	DATE	REVISIONS	APR.

TITLE VISE JAW FIXTURE P/N145B

SIGHT BRACKET

DRAWN BY: ЕДН   CHECKED BY: W.J.S.   APPROVED: ЗНТ

DATE   MAT'L

SCALE

TOLERANCES .XX=±.015   .XXX=±.005   ANGLES ±0°-30'

UNLESS OTHERWISE SPECIFIED

PART NO. **T-145B-2**

CONVERSION CHART

MM	INCHES
2	.079
8±.05	.315±.002
10	.394
15	.591
16±.05	.630±.002
18	.709
21	.827
23.5	.925
25	.984
27	1.063
28	1.102
37.5	1.476
48	1.890
50	1.969
55.1	2.169
74	2.913
76	2.992
79.4	3.126
130	5.118
150	5.906

TOP SURFACE OF VISE

DRILL & C'BORE FOR M10 (2 PLACES)

SOLID JAW

PART

TOP SURFACE OF VISE

MOVABLE JAW

NOTES:

⑤ — 2-23.5 THICK X Ø50 SPACER WITH 25 HOLE.

⑥ — 1-10X150X2 THICK FEELER GAUGE

⑦ — VISE #37-V52 TO BE USED FOR THIS TOOL.

T-145B-2

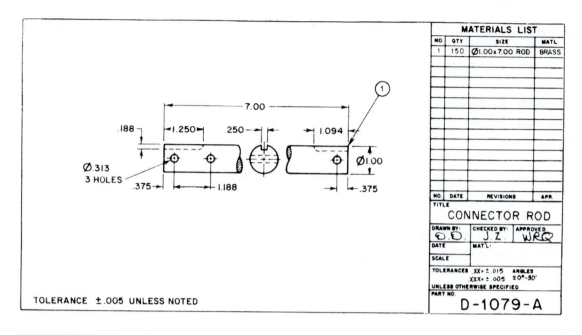

NO.	QTY	SIZE	MATL.
1	150	Ø1.00x7.00 ROD	BRASS

MATERIALS LIST

NO.	DATE	REVISIONS	APR.

TITLE
CONNECTOR ROD

DRAWN BY:	CHECKED BY:	APPROVED
Q.Q.	J.Z.	WRQ
DATE	MAT'L:	
SCALE		

TOLERANCES .XX= ±.015   ANGLES
.XXX= ±.005   ±0°-30'
UNLESS OTHERWISE SPECIFIED

PART NO.
D-1079-A

TOLERANCE ±.005 UNLESS NOTED

PRODUCTION PLAN

P/N 1079-A	PART NAME CONNECTOR ROD		QUANTITY 150	ORDER NO 78-459
DRAWING NO. D-1079-A	PROCESS PLANNER K. OTT	REVISION NO	DATE	PAGE 1 OF 1

OPR. NO.	DESCRIPTION	DEPT.	MACH. TOOL
1	CUTOFF Ø1.00 BAR 7.00 LONG	#68 CUTOFF RM	CUTOFF MACH. #68-214
2	DRILL 3-Ø.313 HOLES	#66 DRILLING	DRILL PRESS #66-14
3	DEBURR	#7 FINISHING	
4	MILL 2-.188-.25 WIDE SLOTS	#37 MILLING	HORZ. MILL #37-642
5	DEBURR & INSPECT	#7 FINISHING	TUMBLER #7-1204

5	FUNCTIONAL GAUGE	———	INSP. FIXTURE I-1079-A-3
4	MILLING CUTTER	6.00x.250x1.00	MILL FIXTURE
2	DRILL, JOBBER'S LENGTH STRAIGHT SHANK	.3125	DRILL JIG T-1079-A-1

OPR. NO.	TOOL DESCRIPTION	SIZE	SPECIAL TOOL NO

Figure 14–23 Part drawing and production plan.

2. What is the basic component used with vise-jaw jigs and fixtures?

3. What are the restrictions to using vise-jaw jigs and fixtures?

4. What are the most common methods used for locating with vise-jaw tools?

5. What two methods are commonly used to make nests?

6. Which nest is primarily used for symmetrical parts?

7. Which nest is more practical for nonsymmetrical parts?

8. What materials are commonly used to make nests for nonsymmetrical parts?

9. What two methods could be used for jigs that must drill multiple-hole patterns?

10. How can the relation of the jig plate and vise jaw be accurately controlled?

11. Which jaw should be used to locate the part?

12. What should be done with the spacing collars when the tool is not in use?

13. What provision must be made with regard to the feeler gauge when calculating the set block size?

14. How can jaws made from bismuth alloys be made more wear-resistant?

Specialized Workholding Topics

UNIT 15

Power Workholding

OBJECTIVES

After completing this unit, the student should be able to:

- Describe the primary types of power-workholding systems.
- Identify the major elements of a power-workholding system.
- Define the basic actions of a power-workholding system.
- Specify the primary functions of each power component.
- Describe the primary advantages of power workholding.

Power workholding was discussed briefly earlier in the text, but now we will take a closer look at its various forms and applications. Power clamps are generally used for high-volume production where the added cost of these clamping devices can be offset by the higher output of parts. However, with the newer power-clamping systems and simpler installations, this form of clamping is also very useful for almost any clamping application.

In most cases today, power-clamping systems are being installed directly on a machine tool and are used in conjunction with whatever type of fixture is mounted on the machine. Figure 15–1 shows a typical setup of a power-clamping system applied in this manner. Since the entire power supply is self-contained on the machine tool, the only power-clamping devices required on the jig or fixture are the clamps. When the workholder is installed on the machine tool, the necessary hoses are then connected to the power supply and the system is ready to use. Once finished, the hoses are disconnected, the workholder is changed, and the fixture-mounting process is repeated.

The most popular type of power-clamping system in use today is the air-assisted, self-contained hydraulic system, which is shown in Figure 15–2.

TYPES OF POWER-WORKHOLDING SYSTEMS

Power-workholding systems have been part of manufacturing for quite a long time. However, power workholding was not as widely used until the air-assisted hydraulic system was developed. Older hydraulic systems required the hydraulic fluid to be piped throughout the shop at very high pressures. This was not only messy but also dangerous. Today the two primary forms of power clamp are the air-assisted hydraulic system and the straight pneumatic (air) clamping system.

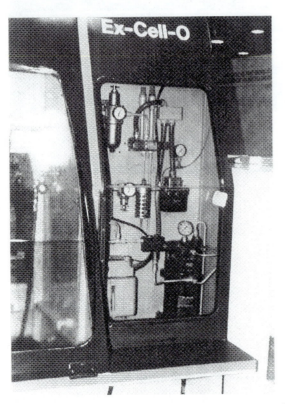

Figure 15–1 Power-clamping system mounted to a machine tool (*Courtesy of Jergens, Inc.*).

Figure 15–2 Hydraulic power-clamping system (*Courtesy of Jergens, Inc.*).

Air-Assisted Hydraulic Workholding Systems

The air-assisted hydraulic systems most commonly used for workholding operations are divided into three groups of components. As shown in Figure 15–3, the typical system starts with the first group of components, the *shop air* system, and ends with the workholding devices.

The term *shop air* refers to the compressed air system found in machine shops. This system provides power, in the form of pneumatic pressure, to drive a variety of different pieces of shop equipment. Everything from air tools and coolant systems to power-workholding systems relies on this power source. The most common shop air system uses compressed air at a pressure of 90 to 100 psi (pounds per square inch). The shop air system consists of the air inlet, the filter/regulator/lubricator device, the safety valve, and a clamp/release valve.

The second group of components is the hydraulic booster system, which consists of the booster, the

check valve (or other type of valve), and the manifold. The final group is the clamping system, which generally includes the variety of different clamping elements used to hold, position, and support the workpiece.

In operation, the system simply uses shop air to activate the booster, which in turn hydraulically intensifies the pressure to the clamps and other devices. While there are many different pressure ratios, the most common is 30 to 1, which yields 3000 psi of hydraulic pressure from 100 psi of air pressure. As shown in Figure 15–3, each of these components is interconnected with hoses.

In addition to the air-operated booster system described previously, there are other power sources used for these workholding systems. The two primary types that fall into this group are the electric booster and the hydraulic pump. The electric booster units are used in applications where compressed air is not available or where electrical power is preferred. They operate in essentially the same way as the air-type units. Hydraulic pump systems are used for applications where many devices will operate from the same power source. They can deliver a higher rate and volume of fluid to the components than a booster system can. Booster systems, both air and electric, normally have a self-contained reservoir with a limited amount of fluid. Although more than adequate for most workholding

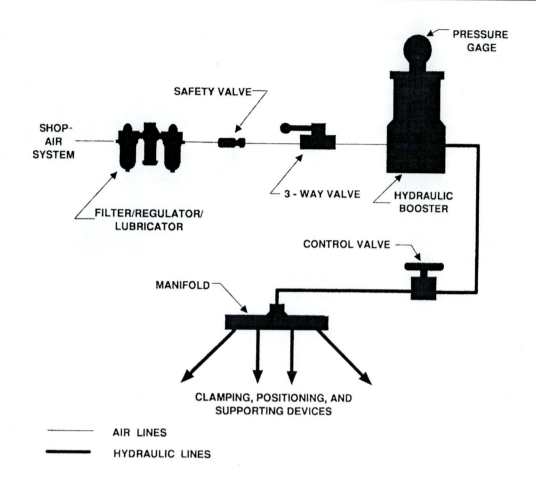

Figure 15-3 The layout and components of a typical air-assisted hydraulic workholding system.

applications, the boosters do have limitations. Therefore, for larger applications with many workholding devices, a hydraulic pump and a larger reservoir should be used.

The one drawback to using any of these systems is the basic law of "no hydraulic pressure–no clamping pressure." If any of these elements leaks or is disconnected, the pressure is lost and the clamps loosen. Several devices are available to reduce this problem. Rotary valves may be used for applications where the workholder must be rotated. Likewise, for fixtures that require the hoses to be disconnected, *accumulators* can be used to maintain the hydraulic pressure. The accumulator is a device that is installed in the line between the clamps and the power source. The pallet-decoupler unit, shown in Figure 15–4, contains

Figure 15-4 A pallet-decoupler unit uses an accumulator charged with either fluid or gas to maintain hydraulic pressure when the power source is removed (*Courtesy of Carr Lane Manufacturing Company*).

an accumulator. Accumulators are charged with either fluid or gas and maintain the necessary pressure in the system when the power is disconnected.

Another group of components that uses a mechanical locking principle instead of a pressurized system is shown in Figure 15–5. These clamps are built using a mechanical lock that is activated with hydraulic pressure. After locking, the hydraulic pressure may be removed and the clamps stay locked firmly against the part. The basic operation of this clamp is shown in Figure 15–6, which illustrates the basic wedge-lock principle behind this concept.

The locking action of this clamping system is much the same as that of a wedge-action clamp. Working on the principle of the inclined plane, this locking device works in much the same way as a tapered shank drill works. In use, the piston is driven to one side of the clamp body by the hydraulic force. The locking angles on the piston and locking pin engage and lock together. Since this lock is mechanical, once the clamp is activated, the hydraulic connections can be removed and the clamp will not loosen. The clamp is loosened by reconnecting the hydraulic pressure to the release port. This drives the piston back, off the locking pin.

Figure 15–5 These hydraulic clamps are made with an internal locking device that mechanically locks the clamps against the workpiece when the power source is removed (*Courtesy of Jergens, Inc.*).

Pneumatic Workholding Systems

Pneumatic, or air-operated, workholding systems are another form of power-workholding system gaining wider acceptance today. Unlike the air-assisted hydraulic system, these clamps and acces-

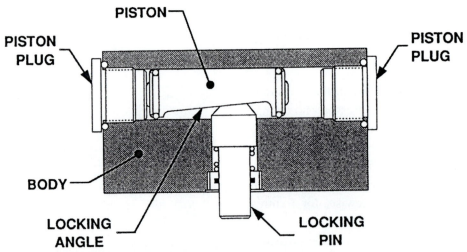

Figure 15–6 The basic principle behind these mechanically locked hydraulic clamps is the locking angle machined on the piston and locking pin (*Courtesy of Jergens, Inc.*).

sories operate completely on air pressure. One application of pneumatic workholding is shown in Figure 15–7. Here the clamps are used to hold two parts for the machining operation.

Air-clamping systems are well suited for applications where hydraulic fluid could cause problems. Typically, air clamps are used for light machining, inspection, assembly, and similar operations. Although these clamping systems are gaining wider acceptance,

the primary drawback to their widespread application in heavy machining is the size of the ram needed to generate the holding forces. To generate the same force, a typical pneumatic cylinder must be approximately five times larger than a hydraulic cylinder. Another problem with air-type clamps is the air itself. Air will compress; hydraulic fluid will not. Therefore, for applications where the compression of the air could cause workholding problems, hydraulic systems are generally preferred.

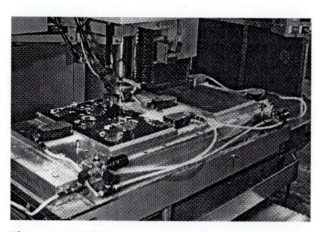

Figure 15–7 Typical pneumatic clamping system (*Courtesy of DE-STA-CO Division, Dover Corporation*).

BASIC OPERATION OF POWER-WORKHOLDING SYSTEMS

Typically, power-workholding systems use a variety of different devices to hold, or locate, the workpiece. The three general functions of any power-workholding system are to position, support, and clamp a workpiece (Figure 15–8). Positioning devices are used to hold the workpiece against a locator in a plane other than that used for clamping. Supporting elements are used below the part and are intended to provide the necessary support to resist any cutting forces during the machining cycle. The clamping devices are designed to lock the part in the fixture and prevent movement

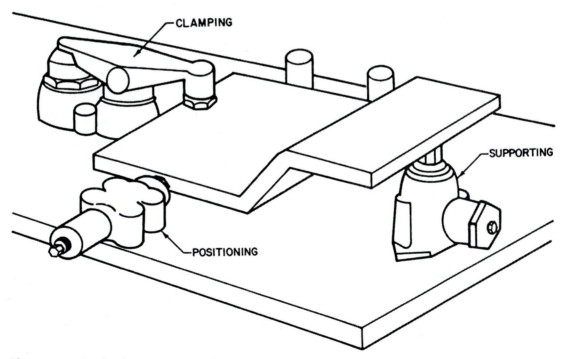

Figure 15–8 The basic actions of a power-clamping system are positioning, supporting, and clamping.

during the machining cycle. The specific type and application of the power-workholding system for any particular part is dependent on the part itself. As with other jigs and fixtures, the part is the primary factor in any design. Figure 15–9 shows several applications of clamping devices used for power-workholding operations.

BENEFITS OF POWER WORKHOLDING

Increased production speed and productivity are not the only reasons for using a power-workholding system. Such systems offer several other advantages over plain manual clamps, including consistent and repeatable operation, controlled forces for either light or heavy clamping, automatically adjusting work supports, remote clamp operation, and automatic sequencing.

Consistent and Repeatable Operation

One of the major benefits of a power-workholding system is the ability to supply consistent and repeatable clamping forces. Manually operated clamps depend on the strength of the operator, but power clamps are controlled by a power source, so operator fatigue has no bearing on the clamping pressure. This results in consistent clamping pressure, part after part, which safely allows much higher machining feed rates. And with retracting clamps, swing clamps, or similar self-positioning power clamps, the position of the clamp on the workpiece is controlled by the clamp. This eliminates many problems caused by operator fatigue or inexperience. Once properly set and positioned, a power clamp performs the same throughout the complete production run.

Controlled Clamping Forces

In cases where either light or heavy clamping forces are needed to hold a workpiece, the clamping force of a power system can be adjusted to suit the workpiece's requirements. Normally, this feature of a power-workholding system is used for parts that have varying thicknesses, brittle materials, or similar characteristics. Thus, if the workpiece is delicate, as with some die-cast parts, this feature alone may justify the use of power clamps. On the other hand, if the workpiece needs extreme pressure to provide enough holding force, power clamps can also be adjusted to give the additional pressure.

Automatically Adjusting Work Supports

For operations where the workpiece must be supported to prevent distortion or deflection during the machining cycle, automatically adjusting work supports are well suited. As shown in Figure 15–10, these supports are positioned under the workpiece and adjust automatically to suit almost any part. The workpiece shown in this example has a step that requires the supports to make contact with the surface at two different heights. Since these work supports are automatically adjusted, this difference, as well as any other difference in the supporting surface, is easily fixed.

Remote Clamp Operation

In many fixturing operations, more than one clamp is needed to hold the workpiece. So, quite often, one or more of the clamps may be hard to reach. For example, a large workpiece might need six or more clamps to com-

Figure 15–9 Typical applications of power clamping (*Courtesy of Vlier Engineering*).

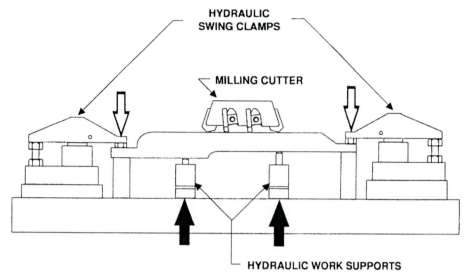

Figure 15–10 Hydraulic work supports are used to prevent part distortion during machining.

pletely hold the part against its locators. Placing these clamps around a large workpiece will often present a safety hazard to the operator. Since power clamps normally operate together, from a single point away from the cutters, the hazard of the operator reaching over the part to tighten a clamp, close to a cutter, is removed. Also, the time saved per year with a power-workholding system over several individual manual clamps can offset the cost of the system. The single remote operating point greatly reduces the time and expense of manually positioning and clamping each clamp.

Automatic Sequencing

The ability of a power-workholding system to operate the clamps and other devices in a specific order is very important. Such systems can provide this sequencing function with a sequencing valve installed in the hydraulic circuit, which activates the clamps and other devices at the proper time.

For example, it may be necessary to clamp the part in a specific order to reduce the chance of deforming the part when clamped. Here supports are often used to reduce the chance of bending the part. When supports are used, the first operation is positioning them under the part. Once the supports are in position, contacting the workpiece, the clamps are brought in contact with the part.

One other case where the sequence of operation is important is when a clamp must be moved out of the way during a machining operation. Here the sequencing valve can be used to move the clamp out of the way as the cutter passes. When the cutter is clear, the valve then reclamps the part. In almost every case, a power-sequencing arrangement is much faster and more reliable than manually operated clamps.

SUMMARY

The following important concepts were presented in this unit:

- The two primary power-workholding systems in use today are the air-assisted hydraulic system and the straight pneumatic (air) clamping system.
- The most common type of system is the air-assisted hydraulic system. Its major parts are:
  - The shop air system, which consists of the air inlet, the filter/regulator/lubricator device, the safety valve, and the clamp/release valve.
  - The hydraulic booster system, which consists of the booster, check valve, and manifold.
  - The clamping system, which includes the various clamps and supports used to actually hold, position, and support the workpiece.

- The benefits of using a power-workholding system include:
  - Consistent and repeatable operation
  - Controlled clamping force
  - Automatically adjusting work supports
  - Remote clamp operation
  - Automatic sequencing

REVIEW

1. List the three groups of components commonly found in air-assisted hydraulic workholding systems.
2. Which of the components converts the air pressure into hydraulically intensified pressure?
3. In addition to the air-operated booster system, what two other types of power source may be used for a hydraulic workholding system?

4. What is the primary purpose of an accumulator in a power-workholding system?
5. What principle is used to provide a positive lock with the mechanical-locking hydraulic clamps?
6. What is the ratio between the shop-air pressure and the hydraulic pressure?
7. What is the most common pressure, in pounds per square inch, used for air-assisted hydraulic workholding?
8. What is the primary drawback to using pneumatic clamping systems?
9. What are the three general functions of a power-workholding system?
10. List four advantages to using a power-workholding system.

UNIT 16

Modular Workholding

OBJECTIVES

After completing this unit, the student should be able to:

- Describe the primary purpose of modular workholding systems.
- Identify the major types of modular workholding systems.
- Define the primary advantages of modular workholding.
- Describe the process of building a modular workholder.
- Describe the process of documenting modular workholders.

Throughout the history of manufacturing, jigs and fixtures have played an important role in increasing productivity. The two major categories of workholding tool that have developed over the years are general-purpose workholders and special-purpose workholders. Although these two forms of fixturing cover just about every tooling requirement, neither is economically suited for every product. For this reason, a third form of fixturing, modular-component workholding, has evolved to fill the gap between general-purpose and special-purpose fixturing.

Modular fixturing is a workholding system that uses a series of reusable standard components to build a wide variety of special-purpose workholding devices. Modular fixtures are assembled with a variety of standard, off-the-shelf tooling plates, supports, locating elements, clamping devices, and similar units (Figure 16–1). These elements are attached with a variety of common hardware items and are assembled in different combinations to build an almost unlimited number of jigs and fixtures.

In the broadest sense, any standard or commercial jig and fixture component may be considered to be modular. Likewise, the term *modular-component workholding system* can be applied when describing any workholder that uses a series of reusable standard components. However, in manufacturing today, this term is normally applied only to fixturing systems that are composed of a variety of matched and coordinated interchangeable elements.

The process of building modular workholders is quite easy. The various parts are designed to work together, so building a workholder is merely a process of assembling the necessary elements. Once the basic assembly methods are mastered with simple jigs and fixtures, the assembler can use imagination and experience to build more detailed workholders. Modular tooling systems can provide fixturing devices for just about any type of part.

Figure 16–1 Modular fixtures are assembled for a variety of standard tooling components (*Courtesy of Carr Lane Manufacturing Company*).

MODULAR FIXTURING SYSTEMS

Modular-component workholding systems are available in three basic styles or types: subplate, "T"-slot, and dowel pin. In addition, within each of these categories are some variations as well as slightly different accessories made by individual manufacturers. The specific type, or style, of modular-component workholding system used will normally depend on the designer's requirements and the tasks the tooling will perform. The following is a brief description of the types of systems available and their individual strengths and weaknesses.

Subplate Systems

The most elementary and basic type of modular fixturing system is the subplate system, which uses a series of flat grid plates, angle plates, multisided tooling blocks, and similar components as major structural elements (Figure 16–2). These components may be

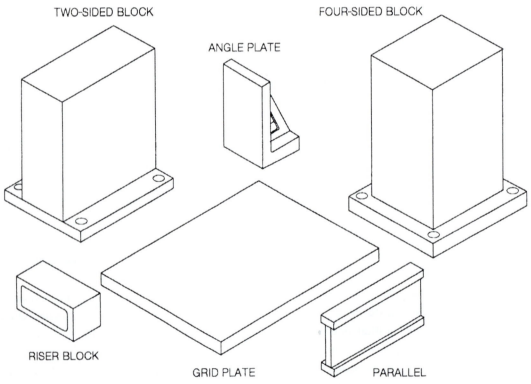

Figure 16–2 Subplate systems use a series of flat grid plates, angle plates, multisided tooling blocks, and similar components as major structural elements.

used individually or combined to assemble different workholders (Figure 16–3).

Workpieces are mounted to these devices with standard commercial jig and fixtures. As shown in Figure 16–4, subplate systems may have either dowel-pin or "T"-slot arrangements for mounting the additional attachments and accessories. Here the workpieces are mounted on a collection of standard locators and are held in place with commercial strap clamps.

In addition to being used for some relatively simple workholders, subplates are used for mounting other workholders to a machine tool. Subplate modular workholders can be used for almost any type of fixturing operation. But many times they are used for mounting very large workpieces (Figure 16–5).

Subplate-type modular workholding systems offer the tool designer many price advantages over the more complete "T"-slot or dowel-pin systems. But for all their cost benefits, they do not have the versatility of the more complete systems. Despite their limited number of components, however, there are ways to use the lower cost subplate elements to construct virtually any workholder, no matter how complex. One method is by using a *master plate* in the fixturing arrangement.

With the master plate tooling method, a dedicated tooling plate acts as an adapter between the workpiece and the modular workholding elements. Rather than being mounted directly to the modular elements, the workpiece is first attached to a dedicated tooling plate, which is then mounted to the modular ele-

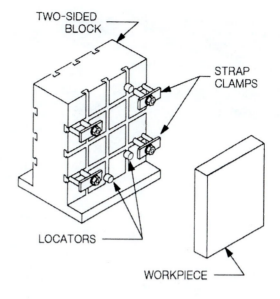

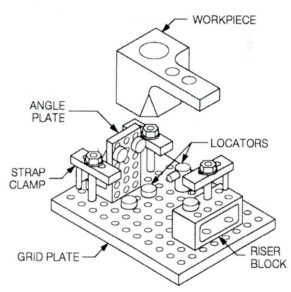

Figure 16–4 Subplate systems may have either dowel-pin or "T"-slot arrangements for mounting the attachments and accessories.

ments. The end result is a combination of modular and dedicated tooling elements, used together, to fixture the workpiece.

The cast housing, shown in Figure 16–6A, is a type of workpiece that could benefit from this type of fixturing arrangement. As shown, the workpiece must be bored from both sides, and the small end has six drilled and tapped holes. There is also a milled flat on the side of the housing that must be machined as well.

The workholder shown in Figure 16–6B is an example of a master plate that could be used for this

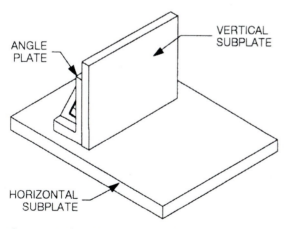

Figure 16–3 These structural elements may be used individually or combined to assemble different workhold-

Figure 16–5 Subplate modular workholders are many times used for mounting very large workpieces (*Courtesy of MidState Machine Company*).

workpiece. Here the workpiece is located on the plate against the dowel pins. The four clamps securely hold it to the master plate, which is attached to the modular elements with the four mounting holes in the corners. The two diagonal holes are used to locate the master plate on the modular elements with two dowel pins.

Figure 16–7 shows how the complete arrangement appears if the workpiece and master plate are mounted on a four-sided tooling block. This setup could be used to mill the flat and perform all the operations needed at the small end of the workpiece. The master plate and workpiece could then be mounted on a set of parallels to perform the operations needed on the large end of the workpiece.

With this master plate arrangement, the workpiece will normally require a variety of different operations and setups. In these situations, the master plate acts

as a pallet arrangement. A workpiece can be fixtured in a lathe for a turning operation, on a horizontal machining center for milling operations, and finally, on a vertical mill for drilling and tapping. All these operations can be performed while the workpiece remains attached to the master plate. The workpiece is removed from the master plate when all the required operations are completed. The master plate is then recycled and used for the next workpiece.

"T"-Slot Systems

The "T"-slot system uses a series of precisely machined base plates, mounting blocks, and other elements having machined and ground "T"-slots. These are used to mount and attach the additional accessories. Regardless of the shape of the base

A

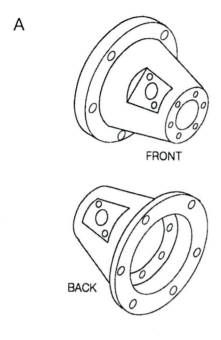

FRONT

BACK

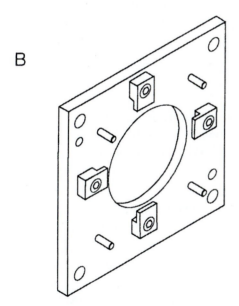

Figure 16–6 The master plate tooling method uses a dedicated tooling plate between the workpiece and the modular workholding elements.

B

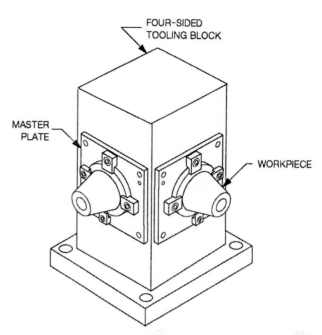

FOUR-SIDED TOOLING BLOCK

MASTER PLATE

WORKPIECE

Figure 16–7 Mounting the workpiece and master plate on a four-sided tooling block.

plates or the type of component, these "T"-slots are machined exactly perpendicular and parallel to each other (Figure 16–8).

The principal advantages of the "T"-slot system are its adaptability, strength, and ease of positioning the components. The "T"-slot design does not require

the components to be located at fixed points and so permits more movement of the components on the base plates. Therefore, the parts are generally easier to fixture when the workholding elements are located in "T"-slots. Furthermore, these "T"-slots, by design, are stronger than dowel pins.

The principal disadvantage of the "T"-slot system is in its repeatability from one tool to the next. When a tool is made the second or subsequent time, more care must be taken to make sure the components and elements are properly positioned on the base. With a dowel-pin system, each locating hole in any component is a fixed reference point. However, the only fixed reference points on a "T"-slot system are at the intersections of the "T"-slots. By design, the typical "T"-slot system does not have as many fixed reference points as a dowel-pin system has. This being the case, repositioning the elements in a "T"-slot system sometimes requires precise measurement. Figure 16–9 shows a workholder built with a "T"-slot system.

Dowel-Pin Systems

The dowel-pin system is very similar in its basic design to "T"-slot-type tools. Both the overall size

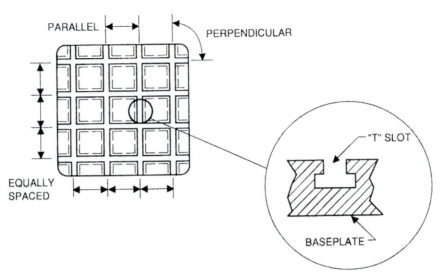

Figure 16-8 Configuration of a "T"-slot baseplate.

capabilities and range of components are very similar; the major difference is in the grid pattern of holes used to locate and mount the other accessories (Figure 16–10). The major advantage to using a dowel-pin system is in the automatic positioning of the components from one tool to the next—if a tool must be built more than once, the components are faster and easier to locate with the dowel pins.

The principal problem with this type of system is in clamping. Because of the spacing between the holes, the components and arrangements made for

Figure 16-9 "T"-slot modular fixture (*Courtesy of Halder Technik/Flexible Fixturing Systems, Inc.*).

clamping are somewhat more cumbersome than those found with "T"-slot systems. The fixed locating points in the base plates provided by the dowel pins, while desirable for repeatability, do not permit movement of the components. To overcome this problem, many elements used with the dowel-pin modular systems use slots in the individual components to achieve adjustability.

Dowel-pin systems are made with two different hole styles. As shown in Figure 16–11A, one style uses alternating tapped holes separated with dowel-pin holes. This arrangement permits both dowel pins and screws to be used to locate and mount the components. The second style, shown in Figure 16–11B, combines both the locating functions and the mounting functions in the same hole by mounting a locating bushing on top of a tapped hole. This setup typically requires a special-purpose locating screw, as shown, which both locates and holds the components. Figure 16–12 shows a workholder built with a dowel-pin system.

Standard Modular Elements

The specific elements and components in each modular fixturing system are decided by its manufacturer. These elements, while slightly different in design from one system to the next, fall into several basic categories. The principal parts of any modular component tooling system include *mounting bases* and *plates, locating* and *supporting elements,* and *clamping elements.*

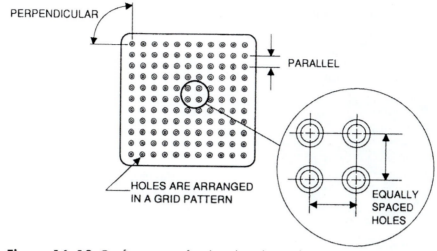

PERPENDICULAR

PARALLEL

HOLES ARE ARRANGED
IN A GRID PATTERN

EQUALLY
SPACED
HOLES

Figure 16–10 Configuration of a dowel-pin base plate.

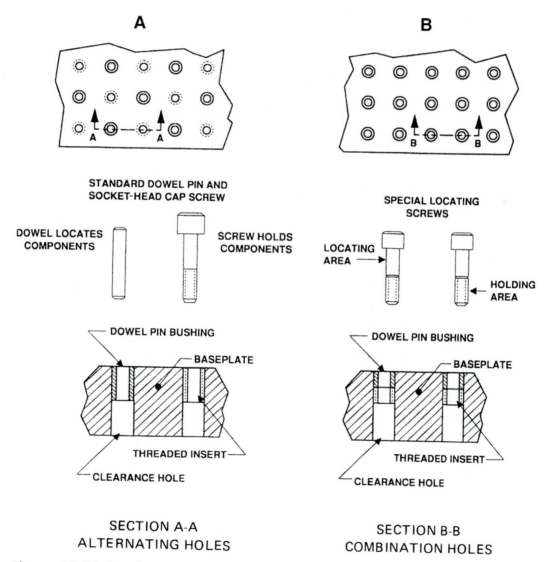

A

B

STANDARD DOWEL PIN AND
SOCKET-HEAD CAP SCREW

DOWEL LOCATES
COMPONENTS

SCREW HOLDS
COMPONENTS

SPECIAL LOCATING
SCREWS

LOCATING
AREA

HOLDING
AREA

DOWEL PIN BUSHING

BASEPLATE

THREADED INSERT

CLEARANCE HOLE

DOWEL PIN BUSHING

BASEPLATE

THREADED INSERT

CLEARANCE HOLE

SECTION A-A
ALTERNATING HOLES

SECTION B-B
COMBINATION HOLES

Figure 16–11 Dowel-pin systems are made with either alternating mounting holes or combined mounting holes.

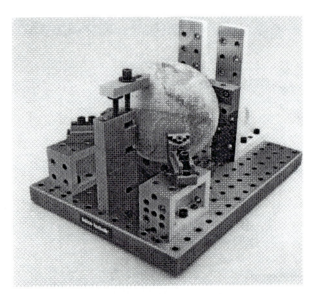

Figure 16-12 Dowel-pin modular fixture (*Courtesy of Bluco Corporation*).

A

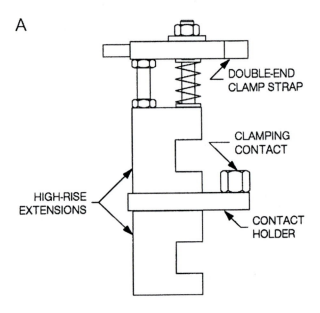

B

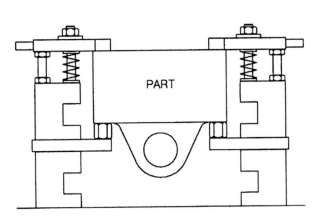

Figure 16-13 High-rise clamp (*Courtesy of Carr Lane Manufacturing Company*).

The mounting bases and plates are available in several styles and sizes. The most common types of mounting plates are rectangular, square, and round. In addition to these flat grid plates, most systems also have a variety of angle-plate bases, multisided tooling blocks, sine tables, and rotary table bases.

Locating and supporting elements are available in several different styles and types, but they are similar in most tooling systems. Mounting blocks are another form of locating element primarily used in locating other elements. These components position locators or clamping devices at specific heights off the mounting base. They may be used individually or combined to achieve any desired height.

The clamping elements commonly used with modular component tooling are mechanical clamps. Most resemble standard jig and fixture clamping devices; however, some special-purpose clamping devices are used almost exclusively with modular fixturing systems. A few examples of these clamps are covered next. In addition to the standard mechanical clamps, each modular fixturing system can be adapted with some type of power-assisted clamping device if desired.

High-Rise Clamps. A variation of the standard strap clamp used with many modular fixturing systems is the high-rise clamp (Figure 16–13A). Rather than a single clamp, the high-rise clamp is composed of a series of individual parts assembled to suit the requirements of a specific workpiece. As shown, the major elements in this clamp are the clamping strap, contact elements, and riser blocks. These elements may be set to suit any specific workpiece requirements. Both the clamp strap and the support elements are tapped to accommodate either standard contacts or swivel contact bolts. Extension units are also available for these contacts to precisely position both the clamp and the support as required by the setup. The high-rise clamp design permits almost any size workpiece to be held. By the simple addition of riser elements, as shown in

Figure 16–13B, the workpiece may be clamped directly to the base plate or elevated to clear any obstructions.

Up-Thrust Clamps.

The up-thrust clamp, as shown in Figure 16–14A is a unique design that holds a workpiece with pressure applied from below. This design is well suited for holding workpieces that must be elevated off the machine table and securely held (Figure 16–14B). It permits the workpiece to be clamped against its top surface, which is especially useful for parts that have irregularities and would otherwise be difficult to locate. Instead of trying to locate the workpiece on an irregular opposite side, the up-thrust clamps allow the workpiece to be both located and machined on the same surface.

The clamp is operated by pushing the cam handle down. This moves the clamping element upward against the workpiece. The jaw element rotates and has two clamp openings, one for thinner parts and a second for thicker parts. The smaller opening will hold parts up to 1.375 inches thick, while the larger opening can handle workpieces up to 2.750 inches thick. When mounted, the clamping surface of the fixed (top) jaw is exactly 4.000 inches off the mounting surface. Precise adjustments may be made by rotating the cam screw to move the bottom jaw up or down. The underside of the top jaw element has a precision-ground locating surface and acts as a precision locator for the clamped workpiece.

Screw Edge Clamps.

Screw edge clamps combine the action of a screw clamp and a locator in a single unit. As shown in Figure 16–15A, they are available in both a fixed and an adjustable configuration. The clamping screw is made with a 4- to 6-degree slant to help keep the workpiece against the locating surface. These clamps are often used with either a single-edge or double-edge support, as shown in Figure 16–15B, to locate the opposite side of the part.

The locating surface of the fixed clamp is at a set height off the mounting surface and corresponds to the height of most of the other locating devices. The adjustable screw edge clamp is similar to the fixed type, but has an adjustable screw locator. The adjustable locating surface permits the clamp to be used for applications where the height of the workpiece, in relation to the clamp screw, must be varied. A series of special accessory contacts is also available with some of these clamps to suit specific part shapes (Figure 16–15C). Should any additional height be required for clamping, riser blocks, as shown in Figure 16–15D, may also be used between the clamp and the tooling plate.

Pivoting Edge Clamps.

Another style of edge clamp is the pivoting edge clamp, as shown in Figure 16–16, which uses a pivoting lever arrangement to apply the holding force. These clamps are often used with a matching backstop unit to hold the workpiece between two jaw elements. The pivoting lever arrangement applies the holding forces in both forward and downward directions simultaneously, as shown.

These clamps are available in several styles. The type shown at view A has a serrated gripping surface

A

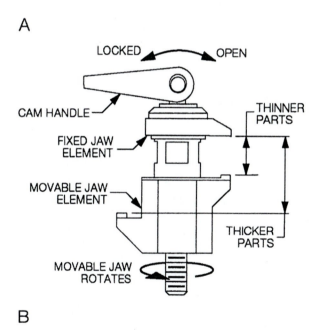

B

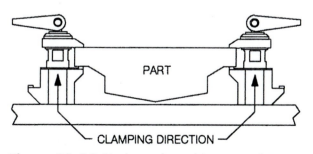

Figure 16–14 Up-thrust clamp (*Courtesy of Carr Lane Manufacturing Company*).

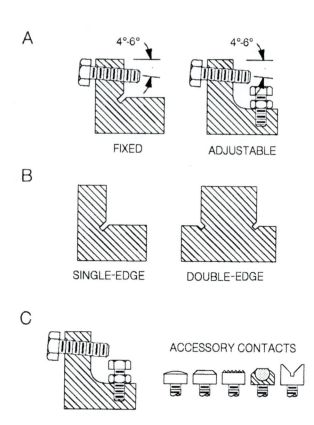

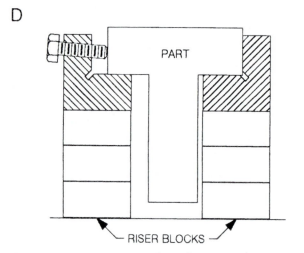

Figure 16–15 Screw edge clamps and accessories (*Courtesy of Carr Lane Manufacturing Company*).

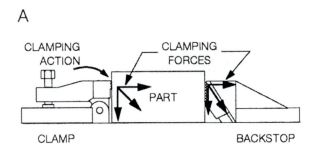

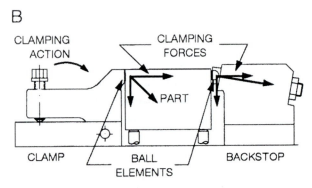

Figure 16–16 Pivoting edge clamps (*Courtesy of Carr Lane Manufacturing Company and SafeTech Corporation*).

and is often used to hold parts where surface marring is not a problem. The clamp style shown at view B uses a ball element in each member to equalize the clamping pressure and to accommodate slight workpiece surface irregularities.

Cam Edge Clamps. The cam edge clamp, as shown in Figure 16–17A, is a slightly different approach to edge clamping workpieces. It uses a flat spiral cam design to generate the holding force against the workpiece. The cam has a 180-degree throw and a 5/32-inch clamping range. The nose element of the clamp is mounted on a pivot and is designed to apply the clamping force both forward and down against the workpiece simultaneously. This action applies 1500 pounds of horizontal force against the workpiece while exerting 100 pounds of downward force to hold the workpiece securely. These clamps may be mounted directly to the base plate or on an adjustable work support/positioner, as shown at view B in Figure 16–17. For applications where an adjustable support is required, the cam edge clamp may also be mounted on a spring-loaded work support, as shown at view C.

Ball Element Clamps. In addition to the standard solid clamping devices already discussed, there is a series of clamping components made with an internal ball element. Though many systems offer a few ball element clamps, the S.A.F.E.® (Self-Adapting Fixture

A

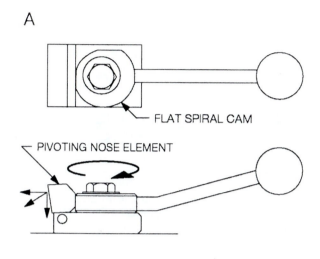

FLAT SPIRAL CAM

PIVOTING NOSE ELEMENT

B

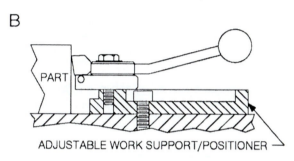

ADJUSTABLE WORK SUPPORT/POSITIONER

C

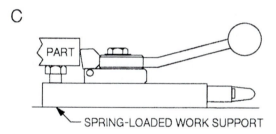

PART

SPRING-LOADED WORK SUPPORT

Figure 16–17 Cam edge clamp (*Courtesy of Carr Lane Manufacturing Company*).

Element) Modular Fixturing System is built exclusively around the ball element design.

The design of the S.A.F.E. ball elements permits the contact surface of the locators, supports, and clamps to conform to the workpiece shape, thus eliminating almost all chance of part distortion during clamping. Figure 16–18 shows the basic action of these ball elements. In this example, the same workpiece is fixtured with both conventional clamps and the S.A.F.E. clamping arrangement. As shown in view A, when the part is placed in the conventional clamps, the curved bottom rests only on the edges of the support.

Then, when the clamps are tightened, the clamping action actually deforms the basic shape of the part. After machining, when the part is released, it returns to its preclamped shape, resulting in a machined surface that is concave rather than flat.

When the same workpiece is clamped with the S.A.F.E. clamps, as shown at view B of Figure 16–18, the ball elements rotate within their sockets to conform to the shape of the part. So, when the workpiece is clamped, the ball elements adapt to the workpiece form and the part can be securely clamped with no distortion. When the clamps are released and the part is removed, the workpiece retains its flat machined surface. In addition to its nondeforming characteristics, the design of these fixturing elements also permits more accurate part location and a firmer, more secure clamping arrangement.

Another method frequently used to hold odd-shaped or irregular parts is tabs cast or forged into the part. These tabs make locating and clamping much easier. The tabs are usually drilled to suit a threaded fastener. The fasteners used for these applications vary to suit the requirements of the part, but hex-head and socket-head cap screws are two common ones. Though many workpieces can be mounted directly to a base plate or custom-made riser element, if there are irregularities in the part, distortion can result if the tabs are not properly supported.

The S.A.F.E. ball element fixture units are unique clamp variations intended to hold the workpiece with a threaded fastener rather than a clamping bar. The ball element in these units has a precision clearance hole through the ball to allow the screw to pass through the ball and attach to the clamp body. These fixture units may be attached with a surface-mounted arrangement, or where height is important the screw can be installed in a counterbored hole, Figure 16–19A. This arrangement provides ample clamping force, and the ball elements minimize any distortion when clamping. A specialty riser mount, shown at view B in the figure, may also be used where additional locating height is necessary.

Additional accessories that are available with these systems include machine vises, magnetic chucks, indexing units, and other similar fixturing devices. In addition to the standard elements and devices included

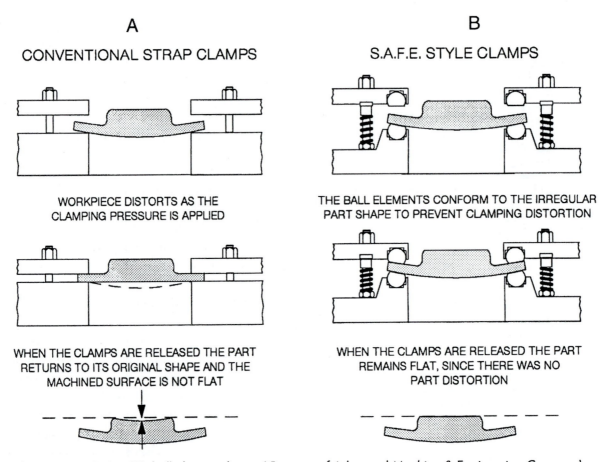

A

CONVENTIONAL STRAP CLAMPS

WORKPIECE DISTORTS AS THE
CLAMPING PRESSURE IS APPLIED

WHEN THE CLAMPS ARE RELEASED THE PART
RETURNS TO ITS ORIGINAL SHAPE AND THE
MACHINED SURFACE IS NOT FLAT

B

S.A.F.E. STYLE CLAMPS

THE BALL ELEMENTS CONFORM TO THE IRREGULAR
PART SHAPE TO PREVENT CLAMPING DISTORTION

WHEN THE CLAMPS ARE RELEASED THE PART
REMAINS FLAT, SINCE THERE WAS NO
PART DISTORTION

Figure 16–18 S.A.F.E. ball element clamps (*Courtesy of Advanced Machine & Engineering Company*).

with each tooling system, parts such as special locators, supports, or other elements may be custom-made and fitted to suit any special requirement.

MODULAR FIXTURING APPLICATIONS

No single type of fixturing is the complete answer to every workholding requirement. The best way to approach workholding is to use a combination of special-purpose, general-purpose, and modular workholders. Using all three fixturing forms allows each to complement the other. As shown in Figure 16–20, for applications where large numbers of parts are run on a regular basis, special-purpose, or dedicated, workholders should be used. Where a moderate to small number of parts are run on an irregular basis, modular workholders may be the better choice. Likewise, for small runs or one-of-a-kind simple parts, general-purpose or temporary workholders, such as vises, collets, or chucks, may be the most economical.

Special-purpose fixtures should be used for parts that are in current production, while modular fixturing is best suited for parts produced on an as-needed basis. Although the special-purpose workholders handle the bulk of everyday production, modular workholders can be used on an as-ordered basis to make small numbers of special parts. In addition to short-run production, modular workholders are also well suited for prototype or experimental tools and as replacements for special-purpose workholders that must be removed from production for maintenance.

Modular elements may also be used to build special-purpose workholders. Depending on the requirements of the fixture and the workpiece, both special-purpose and modular components can be used together to build a workholder. Although special-purpose tools are generally made with specialized parts, where possible, modular elements can replace some specialized elements. Likewise, with more detailed fixturing, special components may

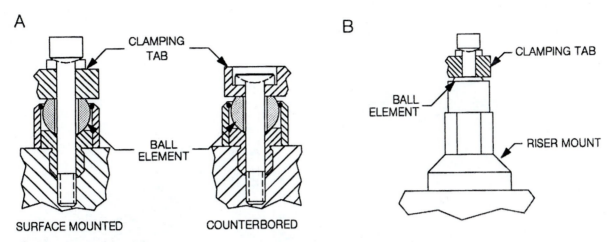

Figure 16-19 Clamping with cast or forged tabs and ball element clamps (*Courtesy of Advanced Machine & Engineering Company*).

sometimes be necessary for a modular workholder as well. Although modular workholders are normally disassembled after each use, they can be left together. When building any fixture, a combined approach, using all three forms of fixtures, will usually be the most effective, efficient, and economical.

Most of the modular workholders discussed here are typically for machining operations. However, machining is not the only application these tools can

be used to perform. The system shown in Figure 16–21 is specifically designed for welding operations. Though a single-table setup is shown here, for very large work either larger tables may be used or several tables may be combined to suit the weldment. This type of modular fixturing system differs only in its elements and application; the basic principles behind the other forms of modular fixturing are also applied when assembling these workholders.

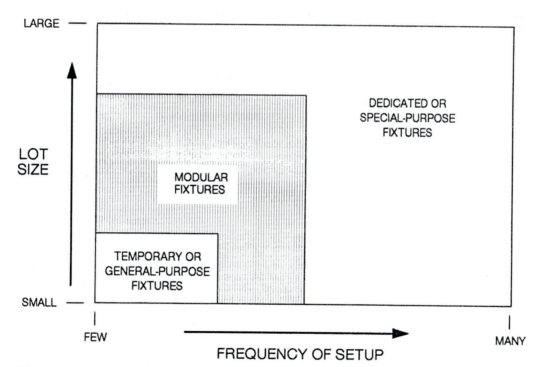

Figure 16-20 The tooling form selected is usually based on the lot size and the frequency of setup.

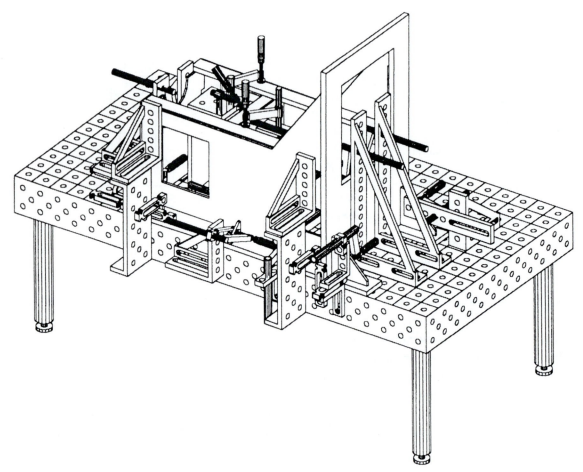

Figure 16–21 Modular fixturing system for welding (*Courtesy of Bluco Corporation*).

Constructing Modular Tooling

Modular tooling sets consist of a variety of standard and special components such as base plates, bushings, locator straps, spacers, clamps, bolts, and screws (Figure 16–22). These interchangeable components are made to tolerances as close as ± .0002 – .0004 inch (± .0.005 – .01 millimeter). They are easily assembled into variable forms to suit almost any workpiece (Figure 16–23).

When building a modular tool, the first step is choosing a base plate large enough to suit the part and all the modular components. Next, the assembler selects the components that locate and support the part and assembles them on the base plate. The final details are then added to specialize the tool. Finally, these components, such as bushings or clamps, are aligned and set to suit the part and then clamped into place. Figure 16–24 shows a modular fixture assembled for use. Figure 16–25 shows the same fixture holding the workpiece.

Several methods, or techniques, can be used to construct the modular components into a desired tool:

- Build the tool around an actual part.
- Build the tool around a mock-up or model of the part.
- Build the tool to specific dimensions without any part.

Of the methods listed, the most accurate is building the tool around a part. However, if the job is a first run or a prototype where no parts yet exist, a model of wood or other material can be used. If neither of these is available, the tool can be built by setting each component by the dimensions on the part drawing.

The range of modular tooling sets is almost unlimited. However, some jobs require a locator or

Figure 16–22 Partial display of the elements in a typical modular fixturing system (*Courtesy of Carr Lane Manufacturing Company*).

Figure 16–23 Constructing a modular workholder.

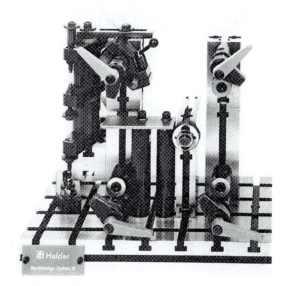

Figure 16–24 Assembled modular fixture without a workpiece (*Courtesy of Halder Technik/Flexible Fixturing Systems, Inc.*).

Figure 16–25 Assembled modular fixture with a workpiece in place (*Courtesy of Halder Technik/Flexible Fixturing Systems, Inc.*).

other detail that is not in a set. When this happens, a part can be made to suit the particular requirement and installed on the tool. This is unusual, however, and the making of special details should be avoided.

Once the tool is built and the parts produced conform to the desired specifications, the tool is released to production. When the job is complete, the tool is returned to the toolroom, where the assembler disassembles it and returns the parts to the storage rack so they can be used to make other tools.

Recording the tool data for each modular tool is very important. After the tool is built, but before it goes to production, the components used to construct it should be listed. The tool is also photographed from several angles or is videotaped. The data, when filed with the part drawing, as shown in Figure 16–26, provide a permanent record of the tooling, greatly reducing the time to build a tool if additional production runs are needed in the future.

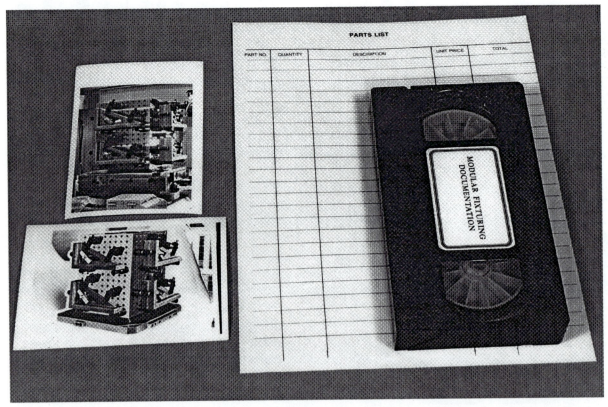

Figure 16–26 Once assembled, the fixture should be photographed or videotaped and the components noted on a parts list to keep a record of the workholder (*Courtesy of Carr Lane Manufacturing Company*).

Figure 16–27 Multipart workholders are as easy to assemble as single-part fixtures (*Courtesy of Stevens Engineering, Inc.*).

Advantages of Modular Tooling

Modular tooling components offer a wide range of features and components that greatly reduce the costs of building special tooling. Despite its rather high initial cost, modular tooling can and does save money and time.

Reduced lead time is the chief advantage of modular tooling systems. Savings of 80 percent or more over conventional tooling are common. As a general rule, each hour of lead time required for a modular tool equals 6 hours of lead time for a conventional jig or fixture.

Reusability is another advantage of modular fixturing. Once a tool has been built and used, it can be disassembled and its elements used for other fixtures. Although a tool may be stored rather than disassembled, the more common method is to return the components to active use for other tools. Since building time is minimal, the time required to rebuild a fixture is often less than that required to find the existing workholder.

Versatility is also an advantage of modular fixturing. Almost any tool can be built with these sets. Even multipart workholders as shown in Figure 16–27, are quite easy to assemble. Modular tools also serve as excellent alternatives to conventional tooling for very short runs to work out problems before hard tooling is built. Once a production job begins, modular component fixturing systems can provide a backup if needed when a production jig or fixture must be repaired or modified.

SUMMARY

The following important concepts were presented in this unit:

- The three primary types of modular workholding systems used today are the subplate system, the "T"-slot system, and the dowel-pin system.

– The subplate system is the most basic type and uses flat grid plates, angle plates, or multisided tooling blocks and similar components as major structural elements.

– The "T"-slot system uses components with a variety of precisely machined "T"-slots to mount and align each element.

– Dowel-pin systems are available with either alternating tapped holes and dowel-pin holes or a combination of the two.

• Modular workholders may be constructed in many different ways. The most common methods are:

– Building the workholder around the part.

– Building the workholder around a mock-up or model of the part.

– Building the workholder to specific dimensions without the part.

• The primary advantages of using a modular workholding system are:

– Reduced lead time in building workholders

– Reusability of the various components

– Versatility of the modular construction

REVIEW

1. What are the three major categories of workholding devices?

2. Which of the three major categories of workholder uses a series of reusable standard components to build a variety of fixtures?

3. What devices are mainly used to assemble modular elements?

4. What are the three most common modular component fixturing systems?

5. Which system is stronger and does not require fixed locating points to mount the components?

6. Which system is easier to reassemble and has automatic repositioning of the elements?

7. Which system is the most basic and elementary form of modular fixturing?

8. What is the adapter between the workpiece and the modular elements called?

9. What two hole arrangements are used with dowel-pin systems?

10. Which hole arrangement uses special fasteners?

11. Which hole arrangement uses standard fasteners?

12. How are the slots in a "T"-slot system positioned?

13. What type of pattern is used with dowel-pin systems?

14. List three benefits gained from using modular workholders.

15. Which type of clamp can locate a workpiece from its top surface?

16. What are the two common forms of screw edge clamps?

17. Which style of modular fixturing clamp uses a flat spiral cam?

18. What two components are used for a pivoting edge clamp arrangement?

19. Which form of clamp eliminates workpiece distortion by conforming to the contours of the part?

20. What type of detail can be added to an odd-shaped or irregular workpiece to make locating and clamping easier?

21. In addition to machining, what other application is well suited for modular workholding?

22. What are the three methods used to assemble a modular fixture?

23. Which type of fixture should be used for large runs of the same part?

24. Which type of fixture should be used for "as-ordered" parts?

25. How should a modular fixture be documented?

Welding and Inspection Tooling

OBJECTIVES

After completing this unit, the student should be able to:

• Identify the basic design objectives for welding fixtures.

• Identify the major types of workholders used for welding.

• Describe the basic types of gauging fixtures.

• Describe the primary functions of gauging fixtures.

Welding and inspection are everyday operations in manufacturing. Like many other areas, these operations can be simplified and improved through the use of appropriate jigs and fixtures. Although welding is specified in the examples, the methods and techniques listed will apply equally well to other assembly operations such as brazing, soldering, riveting, and stapling.

TOOLING FOR WELDING OPERATIONS

Welding is one of the most efficient and economical methods used to join metals. For this reason, it is the primary assembly method used by industry. However, since large amounts of heat are used in welding, distortion is a problem. Jigs and fixtures play an impor-

tant role in reducing or eliminating this problem in production welding operations.

The terms *jig* and *fixture*, when applied to welding operations, do not carry the same meaning they do when applied to machining. A jig used for welding is normally a fixed-position tool; a fixture is a tool made to rotate around either a horizontal or vertical axis.

Welding jigs and fixtures are designed to be constructed from arc-welded sections of standard structural shapes. Using standard plate, angle, channel, pipe, and other sections greatly reduces the cost of the tool and allows greater flexibility in design. Arc-welding the tool permits any needed changes in either design or alignment to be made quickly and easily. Machined elements used for positioning or aligning parts should be kept to a minimum, since these add greatly to the cost of the tool.

Types of Jigs and Fixtures

The jigs and fixtures used for welding can generally be limited to three basic types: tacking, welding, and holding.

Tacking jigs and fixtures are used to hold the parts of an assembly in their proper position so they can be tack-welded together. These tools are generally used for assemblies that must be held together in several places to prevent warping or distortion when welding is complete. Parts assembled in a tacking jig or fixture are removed after tacking and either finished without special tools or transferred to a holding jig or fixture.

Welding jigs or fixtures are used to hold the parts of an assembly in position for welding. The difference between welding and tacking is the amount of welding performed. The tacking tool is used only when the part is to be tack-welded. When the part is to be completely welded together, a welding jig or fixture is used. Welding jigs and fixtures are normally built heavier than tacking tools to resist the added forces caused by the heat within the part.

Holding jigs and fixtures are used to finish tack-welded assemblies. Like welding tools, holding jigs and fixtures must be made rigid enough to prevent distortion and warping.

Locating and Holding Workpieces

The basic principles of locating and holding that apply to machining tools can also be applied to welding tools. While the degree of precision is generally

less with welded assemblies, the parts must still be held and clamped to maintain their proper relation. Figure 17–1 and Figure 17–2 illustrate two typical welding fixtures and how each is located and clamped.

The most convenient and adaptable clamp to use for welding operations is the toggle-style clamp, Figure 17–3. Variations of this type of holding device are shown in Figure 17–4.

Basic Design Considerations

In addition to locating and holding the part, the designer must also consider several other factors before a welding jig or fixture can be designed.

Heat dissipation is an important consideration with any welding tool. Several methods can be used to ensure that proper heat is maintained in the weld area. The primary factor that determines the amount of heat required is the metal being joined.

When metals such as steel and other poor heat conductors are joined, the excess heat should be carried off to prevent overheating the weld. To do this,

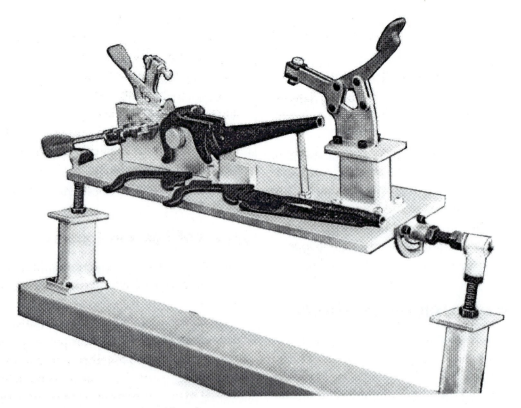

Figure 17–1 Typical welding fixture.

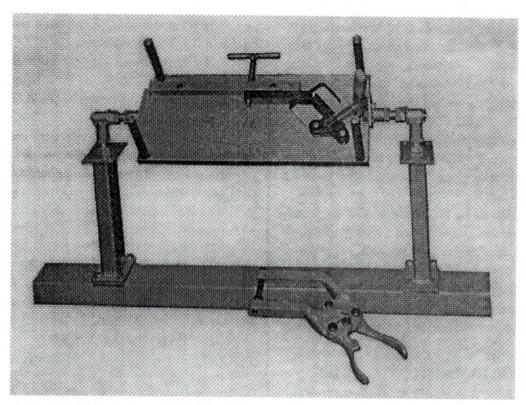

Figure 17-2 Turnover-type welding fixture.

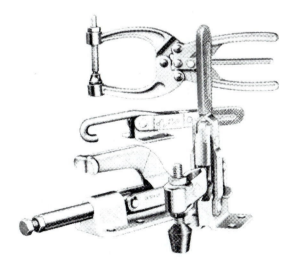

Figure 17-3 Standard toggle clamp.

Figure 17-4 Variations of the standard toggle clamp.

backup bars of copper, titanium, or beryllium can be used. For metals that are good conductors of heat, such as copper or aluminum, too rapid cooling becomes the problem. To prevent this, the fixture or jig must be made to contact the part in as small an area as possible (Figure 17–5).

Clamping supports must be provided to prevent distorting the work while it is in a heated condition. Whenever possible, place clamps directly over the supporting elements.

Locators should be positioned so that the distortion will cause the part to loosen rather than tighten against the locators. If this is not possible, either power or manual ejectors should be built into the tool.

Foolproofing is one feature that is necessary for any type of welding jig or fixture. Each tool must be designed so the part will only fit into its proper position.

MODULAR FIXTURING FOR WELDING

Welding fixtures are typically the most common devices used to align and retain the various pieces for welding. Nearly all welding fixtures are designed and built to suit the specific requirements of a single assembly. For this reason, most welding fixtures are quite expensive and are often justified only when fabricating many units.

When a limited number of fabricated units are needed for very short production runs or prototype work, other methods of positioning and holding the parts are usually employed. Here a tack fixture may be used to position and manually hold the pieces with a bewildering arrangement of "C"-clamps, bar clamps, framing squares on cast iron plates, or other custom elements. When time and money permit, temporary fixtures may also be built up on tack tables to properly fixture these pieces. Regardless of the quantities involved, each of these fabricated units must still be properly fixtured to ensure that they are constructed correctly.

However, another fixturing alternative is available for one-of-a-kind or short-run welding tasks. This fixturing alternative is the **Demmeler modular fixturing system** (Figure 17–6). This system is an ideal alternative to either building expensive permanent welding fixtures or assembling makeshift temporary fixtures. Modular fixturing offers all the advantages of permanent fixturing at a unit cost usually associated with temporary tools. While the concept of modular workholding is usually applied with machining fixtures, the idea is also good for welding tools. Here specialized workholders can be constructed quickly and easily from a series of standardized elements.

As shown, the basic system consists of a welded three-dimensional (3-D) worktable that provides a sturdy platform for mounting a variety of standard angles, blocks, and other accessories. This table is

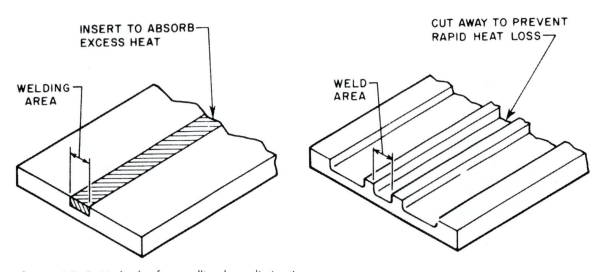

Figure 17–5 Methods of controlling heat dissipation.

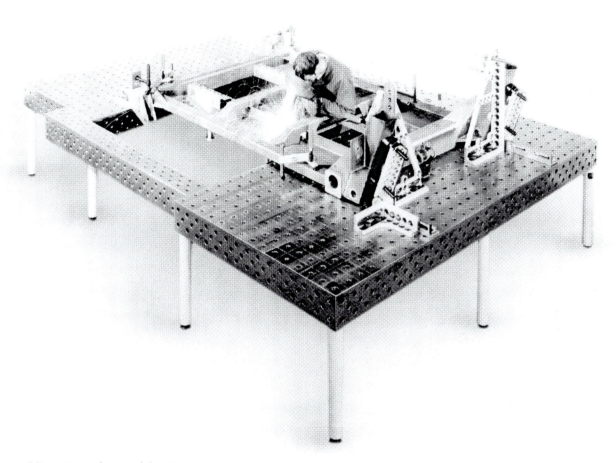

Figure 17–6 Demmler Modular Fixturing System.

constructed of high tensile strength steel and is heavily ribbed to ensure stability and flatness to within .0004"/foot. The 3-D worktables furnished with the system may be used individually or several may be joined together, depending on the size and complexity of the fabricated assembly. Several 3-D worktable sizes are available to suit a wide array of different size weldments. The standard 3-D worktable sizes range from 1000 mm × 1000 mm × 200 mm to 3000 mm × 1500 mm × 200 mm (39.4" × 39.4" × 7.9" to 118" × 59.1" × 7.9"). These tables are also furnished with either four legs or six legs, depending on the table size. These legs are all 850 mm (33.5") long and have 40 mm (1.6") of adjustment to level the complete table.

Each 3-D worktable is made with a grid pattern of mounting holes to attach the various fixturing elements. These holes are 28 mm (1.1") in diameter and are spaced 100 mm (3.9") apart across the face of the 3-D worktable. These holes and the grid pattern are accurate to within .0015"/foot and are included in all the structural elements furnished with the system. To accommodate precise positioning between holes, several of the structural elements are also made with 28 mm (1.1") wide slots. These allow the fixturing elements to be positioned exactly where they are required.

All these fixturing elements and workpiece positioners are attached to the 3-D worktable with a unique positioning and clamping bolt. This bolt, as shown in Figure 17–7, is designed to both precisely position and securely clamp the individual elements. In use, the bolt is inserted through the fixturing element into the 3-D worktable. Depending on the setup, these bolts may also be used to align and attach individual fixturing elements. Once installed, an integral

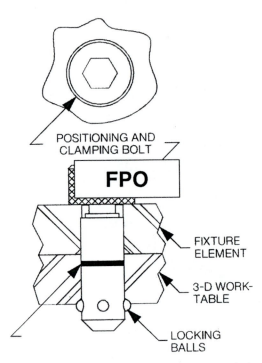

Figure 17–7 Positioning and clamping bolt.

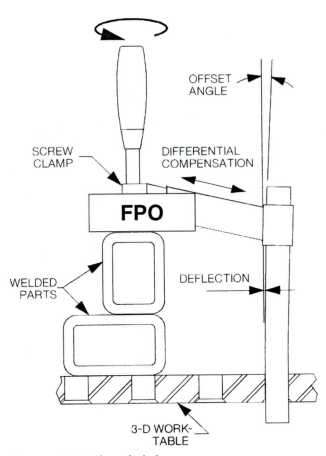

Figure 17–8 Threaded clamp.

"O"-ring prevents the bolt from turning when tightened. The "O"-rings also clean dirt from the mounting hole as they are installed.

These bolts are hardened and provide 3 tons of clamping force while withstanding up to 25 tons of shear force. Once installed, they are snugged up by turning the knurled end. This action extends the locking balls, centering the bolt shank and engaging the elements in a chamfered recess in the bottom of the mounting holes. When everything is properly positioned, the bolts are securely tightened with a hex wrench.

The system also contains a variety of fast-acting clamps to securely hold the pieces being welded. These clamps are available in two general styles: toggle clamps and threaded clamps. Each clamp style is also available in a wide array of different styles to suit virtually any clamping requirement. While the toggle action clamps have a relatively standard design, the threaded clamps are quite unique. These threaded clamps are furnished with a compensating mechanism to ensure the clamping force is applied perpendicular to the thread. As shown in Figure 17–8, as the clamp is tightened, the clamping tube deflects slightly. The compensating mechanism then neutralizes the effects of this deflection.

The clamping tubes are available in several lengths and in two general styles. One design has the clamping arm welded to the tube, whereas the other permits the clamping arm to move up or down the clamping tube. The height of the welded style clamp is set by simply inserting or retracting the clamping tube in the mounting hole. For those situations where a clamp is positioned between holes, the sliding clamp design allows the clamp to be positioned by raising or lowering the clamp arm on the clamping tube.

Objectives in Design

The following is a list of the general objectives of jig and fixture design for welding operations:

- The primary function of a jig or fixture is to establish and hold the proper position of the parts throughout the welding cycle.
- A jig or fixture must repeatedly produce parts within the specified tolerances with little or no distortion.
- Proper heat control in the weld area must be maintained.
- Whenever possible, all welding should be done in a flat, horizontal plane.
- Jigs and fixtures must be foolproof and easy to operate.
- Only essential dimensions should be located and clamped rigidly.
- All welded areas must be easily accessible.
- Any heavy tools must be supported by mechanical devices. The operator should not have to do heavy lifting.
- As many operations as possible should be performed before moving the part in the tool.

INSPECTION FIXTURES

Every part made must meet a standard for size and shape if it is to perform its design function. While it is quite possible to measure each dimension separately, this is not the most cost-effective means to ensure part quality and conformity to dimension. To satisfy the requirements of speed and accuracy, gauging or inspection fixtures are used.

The main requirement of an inspection fixture is accuracy. Each inspection fixture should contain only those elements needed to check the specified sizes of forms. Individual gauges that check only one size are preferred over complicated tooling if the dimension being gauged is independent of other part features. An example of this is the size of the threads in a hole. While the location of the hole is important to the part, the size of the thread is independent of the location.

There are two general types of inspection fixtures: gauging and measuring.

Gauging Fixtures

Gauging fixtures are used to check a part against a preset standard size. For example, the fixture shown

in Figure 17–9 is used to check the inside and outside diameters of a ring. If both are within the prescribed tolerance, the ring will drop into the fixture. However, if the outside diameter is too large or the inside is too small, the ring will not fit. While useful, this gauge is limited because it cannot determine whether the outside diameter is too small or the inside is too big.

One alternative to using this tool would be to use separate gauges to check each diameter (Figure 17–10). However, the problem now is the location of the hole. A part such as the one shown in Figure 17–11 could easily pass inspection. Its usefulness in an assembly is doubtful, though, because the hole is not in the center of the part.

To satisfy the majority of requirements, another form of gauge has been developed, whose principal function is determining whether the part will perform as intended. The gauge represents the mating part, and the workpiece is checked against the gauge.

Measuring Fixtures

Measuring fixtures can indicate exactly where and by how much a part is out of tolerance. As shown in Figure 17–12, the part is located by its center hole and

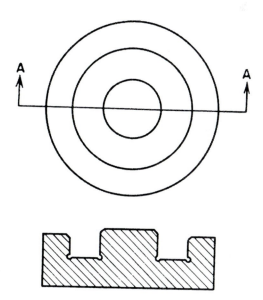

SECTION A-A

Figure 17–9 Receiver gauge.

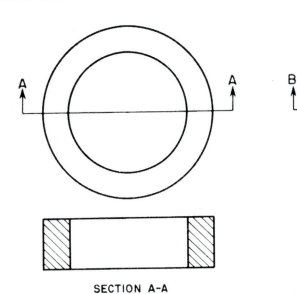

SECTION A-A   SECTION B-B

Figure 17-10 Separate gauges.

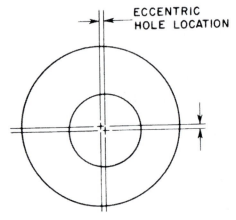

Figure 17-11 Possible condition of "good" part checked with separate gauges.

rotated past a dial indicator to check the runout of the outside diameter.

Measuring fixtures are useful, but they are often too fragile for shop use and must remain in the inspection room. Gauging fixtures, however, are built together and can normally be used in the shop.

Measuring fixtures, to accurately inspect conformance of the part to the required specifications, must inspect the part with relation to the appropriate datums. As geometric dimensioning and tolerancing become more popular for tool drawings, this requirement will become easier to meet because of the datum specifications on the engineering drawing. For example, if the

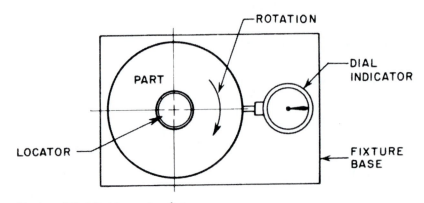

Figure 17-12 Measuring fixture.

part shown in Figure 17–13 were to be inspected, the first requirement would be to ensure the proper relationship of the referenced datums to the measuring fixture and the dial indicator used to inspect the part. The requirement specified for the part is that the center section must be aligned with the two ends within .010 inch. The runout specification shown here specifies circular runout, which means that the indicator must be positioned to inspect the runout of the center section along a singular circular path around the part. If total runout were specified, the indicator would also require lateral movement across the face of the center section.

The datum surfaces of this part are its two cylindrical ends. The easiest and simplest method that could be used to locate these datums is Vee blocks. Figure 17–14 shows one type of inspection fixture that could be used to inspect the circular runout of the part. Since runout is specified as a total reading, the part when inspected must not show any more than .010 inch of total movement of the indicator in one revolution. This requirement may be shown on some prints by the abbreviation FIM or TIR (full indicator movement or total indicator reading).

Supplement Gauges

Several other gauges are used to check parts. They can be used either as part of a fixture or separately as needed. The most common types are flush-pin, fixed-limit, and template gauges.

Flush-pin gauges are used mostly as depth indicators. A pin or rod is used to check the relationship of a surface to the top of the gauge. Figure 17–15 shows the top edge of the part resting against the underside of the gauge and the pin checking the depth of the slot. If the slot is too shallow, the pin will not

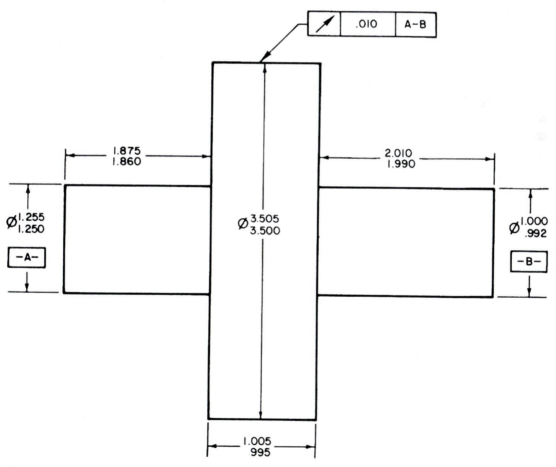

Figure 17–13 Geometrically dimensioned part.

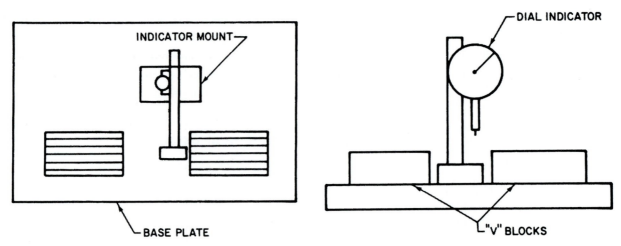

Figure 17–14 Typical inspection fixture.

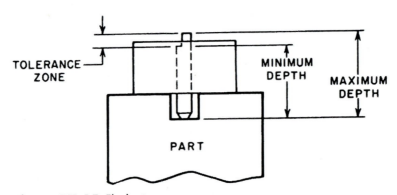

Figure 17–15 Flush-pin gauge.

drop in far enough. If the slot is too deep, it will drop in too far. Figure 17–16 shows how this gauge could be used in an inspection fixture.

Fixed-limit gauges are generally used to check the upper and lower limits of size, as shown in Figure 17–17, and are made to suit the specified maximum and minimum dimensional sizes of the part. Only one gauging element should fit a feature if it is within the specified tolerance. The use of this gauge is shown in Figure 17–18. The smaller end (go) is inserted into the hole. If it will not go in, the hole is too small and the piece is rejected. If the larger end (no-go) passes through a workpiece, the hole is too big and this part is also rejected. The ideal part will accept the small end of the gauge, but reject the large end. For this reason, the gauge has received the name "go–no-go."

Common go–no-go gauges are the snap gauge, ring gauge, and plug gauge. Within each of these are

further divisions for testing specific forms such as threads, tapers, and straight features.

Template gauges are primarily used to check contour forms such as radii, angles, or threads (Figure 17–19). They compare a surface against a master of the exact shape. When used, they can easily detect any variation from the desired form.

SUMMARY

The following important concepts were presented in this unit:

- In welding, the terms *jig* and *fixture* have different meanings than they do when referring to machining operations.
  - The term *jig* describes a workholding device that is mounted in a fixed position.

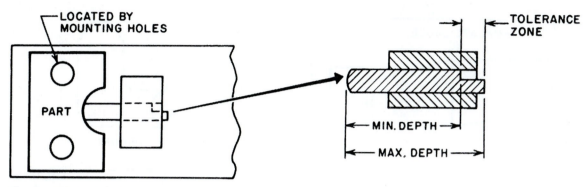

Figure 17-16 Flush-pin gauge used in an inspection fixture.

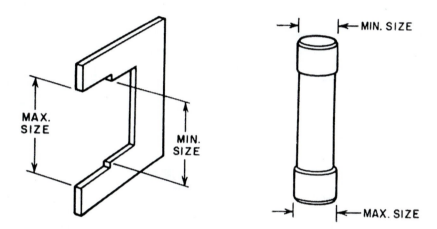

Figure 17-17 Fixed-limit gauges.

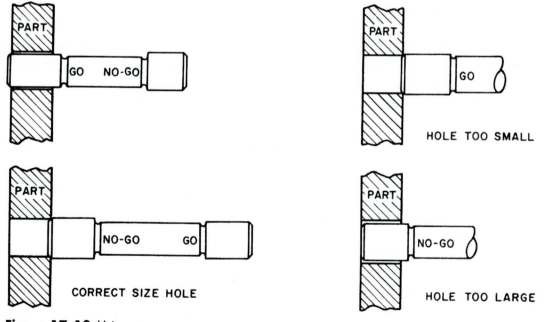

Figure 17-18 Using go–no-go gauges.

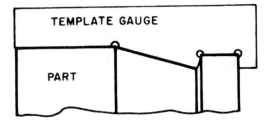

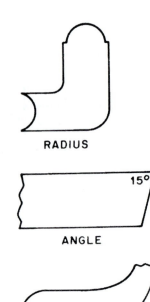

RADIUS

ANGLE

15°

THREAD

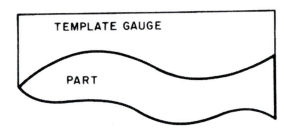

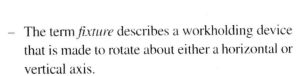

Figure 17-19 Template gauges.

- The term *fixture* describes a workholding device that is made to rotate about either a horizontal or vertical axis.
- The three common types of welding jigs and fixtures are tacking, welding, and holding.
- Toggle clamps are the most common type of clamp used for welding jigs and fixtures.
- In welding, heat control is an important consideration. The two most commonly used methods of controlling heat in a jig or fixture are:
  - Using cutaway portions to prevent rapid heat loss through the jig or fixture.
  - Installing an insert to absorb excess heat.
- The primary requirement in an inspection workholder is accuracy.
- Individual gauges are preferred over complicated arrangements whenever possible.
- The two primary forms of inspection fixture are gauging and measuring.
  - Gauging fixtures check the part against a standard of known size and can only determine if a part is in or out of tolerance.
  - Measuring fixtures actually measure a part and can indicate exactly where and by how much a part is out of tolerance.

REVIEW

1. When discussing welding tools, what do the terms *jig* and *fixture* mean?
2. When constructing welding tools, what type of material should be used to help keep the cost down and add design flexibility?
3. List the three basic types of jigs used for welding.
4. What type of clamp is the most adaptable and convenient for use with welding workholders?
5. What is the purpose of a backup bar in a welding jig or fixture?
6. How should the supports be positioned in a welding workholder?
7. What is the difference between a gauging fixture and a measuring fixture?
8. List three common supplemental gauges.
9. What do the abbreviations FIM and TIR mean?
10. What terms are used to refer to the upper and lower limits of a fixed gauge?

UNIT 18

Low-Cost Jigs and Fixtures

OBJECTIVES

After completing this unit, the student should be able to:

- Identify the basic low-cost workholding devices.
- Describe the methods used to add low-cost fixturing elements to new workholders.
- Specify the advantages of using low-cost fixturing components.

As the demand for more efficient production increases, so too does the requirement for low-cost jigs and fixtures. Tooling was previously thought to be a necessary part of the overhead for the production of any part. However, today tool engineers and designers must produce these jigs and fixtures at the lowest possible cost while maintaining the accuracy and quality of the part. To accomplish these goals, new and innovative ideas are being used to lower tooling costs. The principal types of low-cost jigs and fixtures being produced today use commercially available tooling components in conjunction with the locators and supports specifically designed for the workpiece.

CHUCKS AND CHUCKING ACCESSORIES

Chucks are available in many styles and types. From very small drill chucks to massive lathe chucks, there is an almost endless variety for the designer to choose from. The plain 6- to 12-inch chuck is the most common for most workholding applications. These chucks have long been useful workholding devices for a variety of cylindrical parts. However, with the development of several novel accessories and chuck variations, the basic lathe chuck is even more versatile. These accessories make the standard chuck an extremely useful workholder for a wide range of cylindrical and noncylindrical parts.

Chuck Jaws

Replaceable chuck jaws, or soft jaws, are some of the more common chuck accessories. As shown in Figure 18–1, these jaws are available in several variations. The two general styles are the individual jaws, shown at view A, and the full grip jaws, shown at view B.

Another style of chuck jaw that is very useful for many different parts is the Inserta Jaw®. As shown in Figure 18–2, these jaws are machined into the desired form and mounted on either the end or the top of a

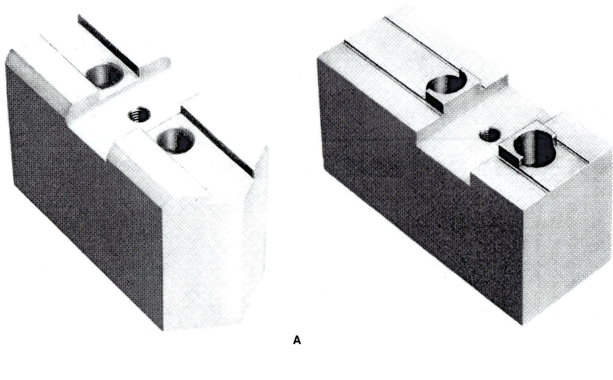

A

B

Figure 18-1 Replaceable chuck jaws, or soft jaws, are available in two general styles, individual and full-grip (*Courtesy of Huron Machine Products, Inc.*).

Figure 18–2 Inserta Jaw® chuck jaws (*Courtesy of Starwood Enterprises, Inc.*).

permanently mounted chuck jaw. This not only speeds the changeover process but also reduces the amount of storage space required for each set of jaws. The top-mounted jaws also have the capability of being machined on both ends, so a single set of jaws can be used for two different parts.

Soft jaws are often mounted to the lathe chuck and turned to suit the workpiece. The workpiece is then loaded for the machining operations. Though this method is well suited for many applications, it does tie up a machine tool. Another method for turning soft jaws is with a turning fixture (Figure 18–3). This fixture is mounted on a nonproduction machine and used to machine all the soft jaws for all the machines. In this way, the production machines can keep working and only be stopped to change the jaws.

The expanding mandrel, as shown in Figure 18–4, is another device frequently used for low-cost workholders. This tool only needs to be turned down to the required diameter for the part to be mounted and machined. The mandrel shown also has the advantage of being manually operated, which further reduces the cost.

Specialty Chucks

Though the standard lathe chucks are some of the most common chucks found in the shop, they are not the only type used for workholding. Another form of

workholder well suited for odd-shaped parts that must be turned is the precision lathe fixture, shown in Figure 18–5. This fixture is mounted to a lathe faceplate and adjusted to suit the particular configuration of the part. The large plates shown on top of the fixture are the counterweights used to maintain the balance of the fixture. The part may be mounted directly to the fixture stage or in another fixture mounted to the fixture as shown.

Another workholder variation that uses a chuck is shown in Figure 18–6. This workholder uses a standard rotary table to which a heavy-duty drilling fixture is attached. This attachment permits a drill bushing to be positioned at the proper height and center distance relative to the workpiece. The large handle on the unit is used to raise the jig plate up to allow the part to be removed, and is lowered to the stop ring on the vertical column to set the proper bushing spacing. The radial spacing of the holes is controlled by the movement of the rotary table. The double-ended jig plate can be fitted with two different liner bushings to further extend the versatility of this fixture.

Castings, forgings, and other irregularly shaped workpieces can present a variety of workholding problems. This is especially true when the workpiece must be positioned relative to an axis or where multiple features and uneven surfaces must be machined. The Clamp Chuck System®, as shown in Figure 18–7, is a

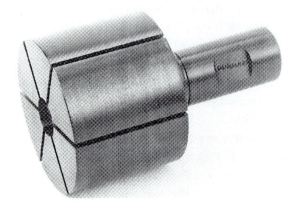

Figure 18-3 A turning fixture is often used to machine all soft jaws to prevent production interruptions (*Courtesy of Huron Machine Products, Inc.*).

Figure 18-4 Expanding mandrel (*Courtesy of Dunham Tool Company*).

Figure 18-5 Precision lathe fixture (*Courtesy of Universal Vise and Tool Company*).

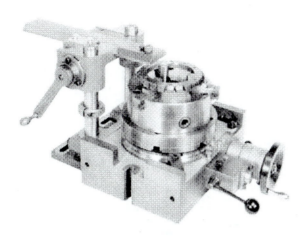

Figure 18–6 Heavy-duty drilling fixture attached to a rotary table (*Courtesy of Universal Vise and Tool Company*).

universal workholding system that uses a draw-down clamping action to hold many different workpieces.

The Clamp Chuck System® may be used for a wide range of turning, milling, boring, and grinding operations. The unit may be spindle-mounted or fitted with an optional base unit for table-mounted operations. When it is spindle-mounted, a draw tube is attached to the machine draw bar to provide the clamping actions. For machines not equipped with a draw bar, an air cylinder may be attached to the clamp chuck to provide the clamping action.

When table-mounted, the Clamp Chuck unit may be mounted either in a plain mill base or in a hydraulic mill base. When the plain mill base is used, the Clamp Chuck is fitted with the air cylinder unit. The hydraulic mill base unit has a self-contained hydraulic clamping cylinder that furnishes the clamping force.

Clamp Chuck units are available in sizes ranging from 4.25 to 12.00 inches in diameter. As shown in Figure 18–8, the basic Clamp Chuck consists of a body, a tooling plate, a clamping plate, and the draw tube. The tooling plate is typically used as the locating element and is fitted with a variety of standard and custom locating devices. A simple adapter may

Figure 18–7 The Clamp Chuck System® (*Courtesy of Clamp Chuck Systems*).

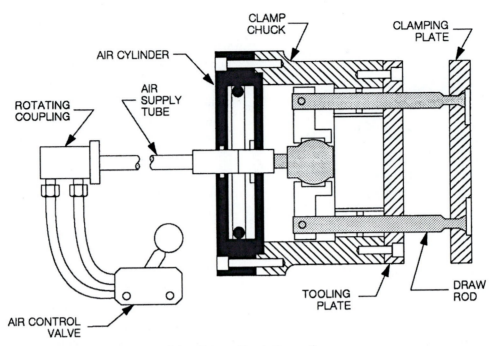

Figure 18-8 Components of the Clamp Chuck System®.

also be attached to permit the tooling plate to mount standard collet-type tooling.

The clamping plate is attached to the draw tube with the draw rods and clamps the workpiece as the draw bar is retracted. The clamping plate may be replaced with finger clamps or with special-purpose clamps on some workpieces. Both the tooling plates and the clamping plates are intended to be customized to suit the workpiece. When needed, the air cylinder and accessories are mounted to the body as shown.

Figure 18-9 shows how this system may be configured for a variety of applications. With the arrangement shown at view A, the workpiece is mounted on a solid locator attached to the tooling plate, and a pair of finger clamps are used in place of the clamping plate to hold the workpiece. Even special shapes may be securely and accurately held, as shown at view B. Here the workpiece is clamped on the projected shaft with a pair of special draw fingers. These fingers both clamp the part and draw the spherical locating surface into the locating recess.

Another unique workholding system is the Prohold Workholding Fixture®, as shown in Figure 18-10, which consists of a series of chucks mounted inside a variety of multisided tooling blocks. These units are

made in three general styles: two-sided, four-sided, and flat pallet. Their design permits up to twelve workstations on a single four-sided tooling block, or up to eight workstations on a two-sided tooling block. The internal chucking components are hardened, ground, and sealed within the tooling block to eliminate any problems caused by dirt and debris.

As shown in Figure 18-11, the typical four-sided unit has three chucks on each face. A 2.015-inch center hole permits a workpiece up to 2.00 inches to pass completely through the chuck. The holes on each face are aligned so that, when necessary, a part could pass completely through the fixture. The hardened master jaws have a travel of .250 inch and are furnished with standard .026 × 90-degree serrations for jaw mounting. These serrations suit standard hard jaws as well as soft jaws and provide full size adjustment within the range of each chuck.

For odd-shaped parts or those needing more support, full grip soft jaws may also be mounted on the fixture and machined in place or machined in a secondary operation on another machine tool. Each chuck is operated independently, and an unlimited variety of setups may be made on a fixture.

A

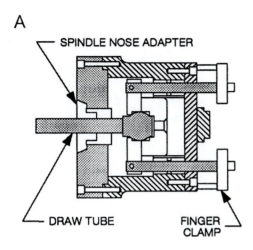

B

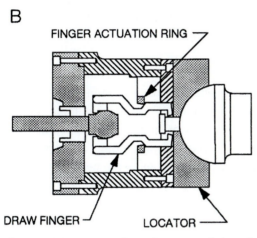

Figure 18–9 Typical applications of the Clamp Chuck System®.

Figure 18–10 The Prohold Workholding Fixture® (*Courtesy of Enterprise Prohold*).

Each of the tooling blocks is available in either vertically or horizontally mounted variations. The permanent chucking center lines establish fixed locational coordinates and permit consistent accuracy and repeatability to within .0007 inch. The distance between the chucking center lines is held to within ±.0005 inch. Each chucking face is flat, square, and aligned to the base to within ±.001 inch.

COLLETS AND COLLET ACCESSORIES

Collets are often used for a variety of workholding setups. They are available in several types and styles, but the 5C collet, as shown in Figure 18–12, is the most common in most shops. These collets are available in many standard and special sizes and styles.

An advantage to using 5C collets for workholding is their standardized mounting approach. Regardless of the size or style, all 5C collets are mounted with their threaded end in a draw bar, draw tube, or nut arrangement, which allows them to be used interchangeably on different machines.

Standard 5C collets may be used for many different workholding tasks. However, the machinable 5C collets, as shown in Figure 18–13, are a variation that further extends the usefulness of the basic design. As shown, these machinable collets are available as *5C emergency collets*, *step chucks* and *closers*, and *expanding collets*. Each style is made with a hardened and ground standard collet mount. The gripping end of the collet is soft and is machined to suit the workpiece. Once machined, the collet may be hardened or left soft.

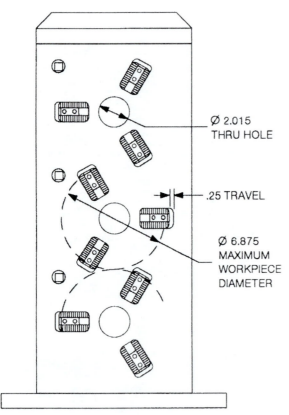

Ø 2.015
THRU HOLE

.25 TRAVEL

Ø 6.875
MAXIMUM
WORKPIECE
DIAMETER

Figure 18–11 The typical four-sided Prohold unit has three chucks on each face.

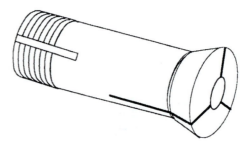

Figure 18–12 5C collet.

Specialty Collet Devices

Specialty collet devices further extend the versatility of the various collet forms. These devices provide alternate mounting methods for either the collet or the workpiece, or both.

Collet Blocks. The machined 5C emergency collets are generally mounted in machine spindles. A variety of other mounting devices are also available for off-spindle mounting. The 5C indexing head is one of the more common off-spindle mounting devices, but a simpler variation of the 5C indexing head is the collet block, shown in Figure 18–14, which is commercially

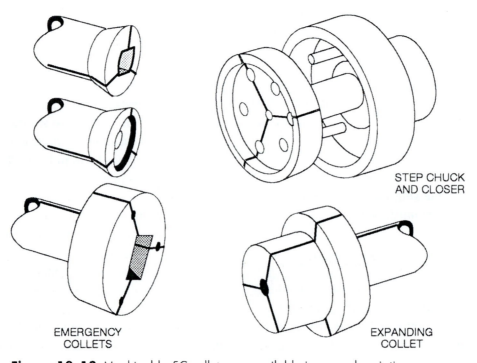

STEP CHUCK
AND CLOSER

EMERGENCY
COLLETS

EXPANDING
COLLET

Figure 18–13 Machinable 5C collets are available in several variations.

A

B

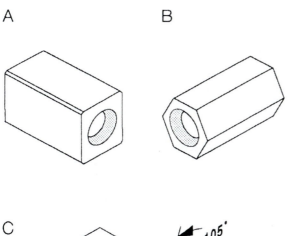

C

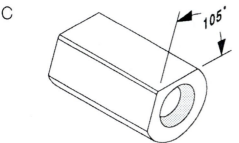

Figure 18–14 Collet blocks may also be used to mount 5C collets.

A

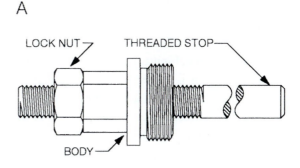

B

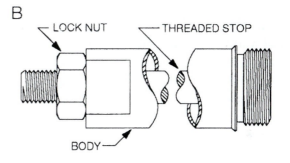

C

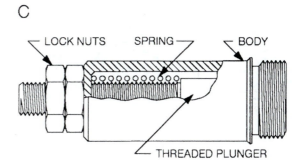

Figure 18–15 Collet stops.

available in two styles, one with four sides (view A) and one with six sides, (view B). For special operations, these collet blocks can also be made to suit other special considerations (view C).

Collet Stops. When collets are tightened, they have a tendency to pull the workpiece into the spindle. The exact amount of movement depends on the size of the workpiece, so even small size variations can have an effect on the mounted workpiece's actual position. To correct this problem, a collet stop can be installed in the collet.

Collet stops attach to the mounting end of the collet and allow the workpiece to be positioned to a preset depth in the collet. With a collet stop, the position of the workpiece is relative to the collet and not to the machine tool. The depth of the collet stop only controls the position of the workpiece with respect to the collet, and not the cutting tool.

Collet stops come in a variety of styles and forms. The specific type used is determined by the workpiece requirements. The most basic collet stop is the solid stop assembly, as shown in Figure 18–15A,

which is threaded into the end of the collet. The solid stop assembly consists of a body, lock nut, and threaded stop, and is well suited for shorter workpieces. Since this collet stop attaches directly to the collet, the depth of the collet determines the maximum chucking depth. Most standard collets permit chucking depths of up to 3.13 inches.

For longer workpieces, a long collet stop, as shown in Figure 18–15B, may replace the solid stop assembly. This unit extends the depth of the collet and the range of workpiece lengths the collet can accept. The long stop assembly is furnished with a threaded stop, a lock nut, and an extended-length threaded stop.

When mounted, a long collet stop permits a maximum chucking depth of about 7.50 inches.

A third type of collet stop is the spring stop, or ejector stop, as shown in Figure 18–15C, which is used where a spring action is required. This type can act either as a solid collet stop and a spring ejector or as a spring ejector alone. In either case, the position of the stop in the collet is preset with the threaded plunger. Unlike the other collet stops, the spring stop does not have a threaded body to adjust the threaded plunger; instead, a set of lock nuts is used for this purpose.

Foolproofing with Collet Stops.

Collets are universal workholders that require imagination for use in specialty operations. Foolproofing a collet setup for a symmetrical workpiece is one example of an operation that requires a bit of thought.

The symmetrical workpiece shown in Figure 18–16A requires multiple setups to machine different features. Here one end of the workpiece is turned and has a small center hole. The opposite end has the same outside diameter but a larger hole. Since the outside diameter is the same at both turned ends, the setup problem for the other machined features is making sure that the correct details are machined on the proper ends. Installing a foolproofing device inside the collet is one solution to this problem.

The foolproofing pin for mounting the workpiece on the end with the large-diameter hole is installed as shown in Figure 18–16B. Here the foolproofing pin is mounted in a collet stop. The diameter of this pin is made to suit the larger diameter hole, and the pin is mounted in a fixed position. With this installation, the workpiece can only be mounted with the larger hole inside the collet.

The opposite end requires a spring-loaded mount for the foolproofing pin. As shown in Figure 18–17, a stepped foolproofing pin is installed for this end of the

A

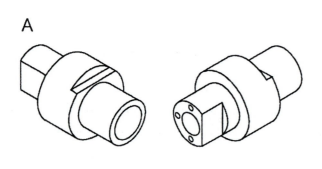

A

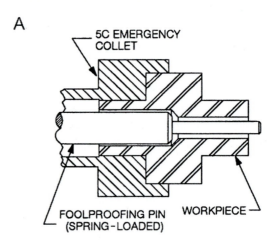

B

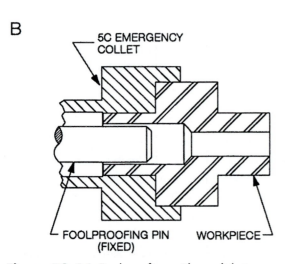

Figure 18–16 Foolproofing with a solid stop.

B

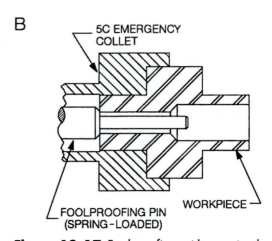

Figure 18–17 Foolproofing with a spring-loaded stop.

workpiece. The larger-diameter step is made larger than the workpiece's small hole yet slightly smaller than its larger hole. The small-diameter step is made to go completely through the center hole. The foolproofing pin is spring-loaded to prevent the workpiece from being loaded incorrectly.

As shown at view A, if the end with the large hole is installed in the collet, the end of the foolproofing pin will extend beyond the end of the workpiece. This tells the operator that the part is loaded incorrectly. When the part is mounted correctly, with the end with the small hole inserted in the collet, as shown at view B, the spring-loaded foolproofing pin is pushed back in the collet. Here nothing extends beyond the end of the workpiece.

Controlled Length Collets.
When a collet is engaged, the gripping action collapses it around the workpiece. The collet is drawn into the tapered socket with the draw bar or draw tube. While efficient, this gripping action also affects the locational accuracy and repeatability of the workpiece.

When clamped, both the collet and the workpiece are drawn into the spindle. When turning an outside diameter, this action has little or no effect on the overall accuracy, but when facing to a specific length, the positional accuracy of the workpiece within the collet is very important. As illustrated in Figure 18–18, when mounted in the collet, the workpiece is positioned as shown at view A. When the collet is closed, however, the workpiece shifts to the position shown at view B.

This would not be a problem if the movement of the workpiece were consistent, but no matter how the

workpiece is positioned in the collet, when it is clamped, its position changes. The reason for this lies in the basic design and operation of the collet.

Collets are intended to hold parts of a specific diameter. When this diameter varies even slightly, the collet grips the workpiece differently. As shown in Figure 18–19A, smaller-diameter parts will normally pull the collet further into the spindle than will larger-diameter parts, because of the fixed size relationship between the collet taper and hole and the variable hole diameter (Figure 18–19B).

The answer to this problem is the controlled length collet (Figure 18–20). This design is actually a collet within a collet rather than a one-piece unit. The outer collet acts in the normal manner, but instead of pulling the workpiece into the spindle, its gripping action forces the inner, or controlled length, collet against the face of the spindle nose. This action allows this collet arrangement to achieve extremely accurate positional

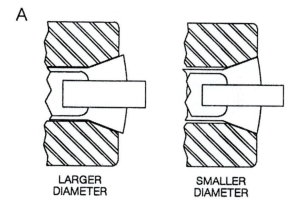

A

LARGER DIAMETER   SMALLER DIAMETER

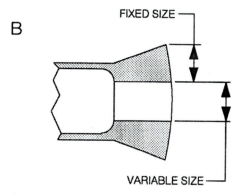

B

FIXED SIZE

VARIABLE SIZE

Figure 18–19 As the diameter of the workpiece changes, the position of the workpiece in the collet also changes.

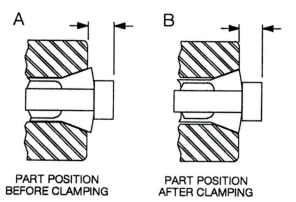

A   B

PART POSITION BEFORE CLAMPING   PART POSITION AFTER CLAMPING

Figure 18–18 When a collet is closed, the workpiece is drawn into the collet chuck.

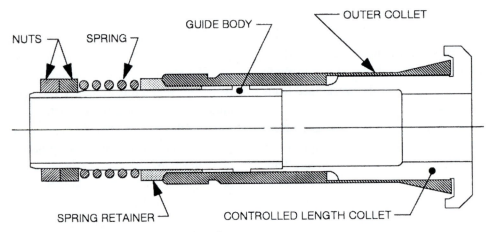

Figure 18–20 Controlled length collet.

repeatability from part to part, regardless of any slight variations in workpiece diameters.

Although the controlled length collet arrangement is very useful, it will not work for every workpiece. Often the size, shape, or other aspects of a workpiece design lend themselves to other chucking devices. For these situations, other controlled length workholding options are available to the designer.

One such arrangement is the controlled length work stop, as shown in Figure 18–21A, in which the work stop is mounted on the end of the spindle and uses a removable stop plate. Controlled length work stop units are made with either a tapered mount or a threaded mount. The stop plates are furnished as blanks and are machined to suit the requirements of the workpiece. They are held in place in the work stop with set screws with conical ends. This design ensures proper alignment and securely holds the stop plate.

As shown, with this setup a standard 5C (or other size) collet grips the workpiece, pulling it against the stop plate. The stop plate acts as a solid locator and provides consistent and repeatable location from one workpiece to the next.

A similar arrangement is shown in Figure 18–21B. Here a standard fixture plate is used in place of the controlled length work stop unit. These fixture plates are available in a variety of standard sizes and may be machined to suit whatever the application requires.

In this design, the fixture plate is bored to suit the mounting diameter of a special bushing. Machining the bushing to suit the specific workpiece shape

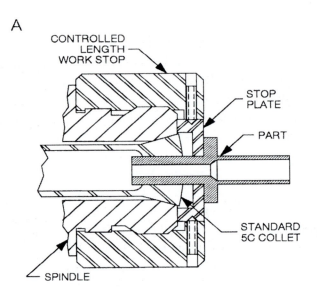

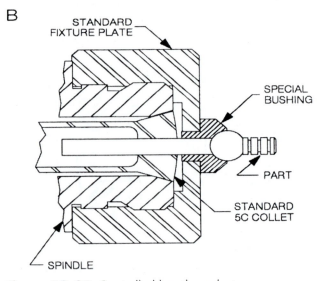

Figure 18–21 Controlled length work stop.

provides the necessary rigidity and reduces the vibration caused by the grooving operations. Once again, a standard 5C collet holds the workpiece. As before, the workpiece is pulled back into the bushing as the collet is tightened.

Specialty Collet Chucks. The 5C-style collets are a very useful form of workholder. However, not all machine tools are equipped to use them. In these cases, a specialty collet chuck may be used to suit both the machine tool and the 5C collets.

The Master Jaw System®, as shown in Figure 18–22, is another form of collet mounting device. It is well suited for machine tools where different types of workpieces are machined. The design of this unit allows a standard chuck to hold collets by simply changing the chuck jaws. So, rather than replacing the entire chuck, the operator only needs to change the jaws to switch between chuck and collet operations.

The system consists of three chuck jaws and a 5C collet adapter (Figure 18–23). This arrangement allows the collets to be installed directly on the chuck, like a set of custom soft jaws, without disturbing the

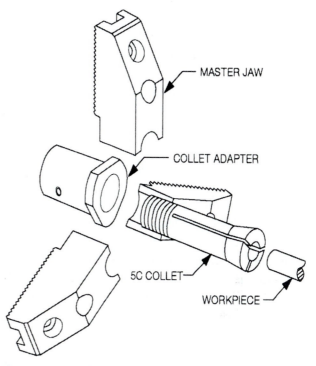

Figure 18–23 The Master Jaw System® consists of three chuck jaws and a 5C collet adapter.

Figure 18–22 Master Jaw System® (*Courtesy of EDI Tooling Systems*).

chuck mount or draw tube. The collet is automatically closed around the workpiece as the chuck is tightened.

The basic design of this system also incorporates a length control feature to help to precisely position the workpiece in the collet. When the chuck jaws close around the collet, the clamping action causes the collet to pull the adapter tightly against the back side of the jaws. This precisely controls the position of the collet with respect to the chuck. The overall accuracy of this system depends largely on the chuck and the collets, but runouts in the .0008- to .0015-inch range are typical.

Another form of specialty collet chuck is the pneumatic collet chuck, as shown in Figure 18–24, which is made to be flush-mounted to machine tables, faceplates, or any similar accessory. This design allows the chuck to be applied in a wide variety of applications where a collet is necessary. Figure 18–25 shows a typical applica-tion. Here the chuck is mounted on a faceplate and is used to hold an expanding collet in a cylindrical grinding operation.

When collets are used for turning operations, the lathes are normally equipped with a collet closer. These closers are often manually operated, but newer machine tools are equipped with power-operated closers. For lathes that do not have a power closer, the unit shown in Figure 18–26A can be installed. This is a pneumatic collet closer that mounts directly to the existing machine through the spindle. A dead-length stop assembly, as shown in Figure 18–26B, is also available for this unit to permit the complete length of the spindle to be used in setting the dead-length stop. The dead-length stop assembly mounts directly to the collet closer, as shown in Figure 18–27.

Another style of dead-length stop that may be used with either a collet or a chuck assembly is the

Figure 18–24 Pneumatic collet chuck (*Courtesy of Dunham Tool Company*).

Figure 18–25 Using the pneumatic collet chuck for a cylindrical grinding operation (*Courtesy of Dunham Tool Company*).

spindle work stop, shown in Figure 18–28. This stop is inserted into the machine spindle and set to the correct position against the workpiece. Once positioned, the unit is fixed in place by tightening the adjusting rod, which expands the clamping mechanism and securely holds the unit in the spindle.

Automatic bar feeders, as shown in Figure 18–29, are another type of accessory found on production turning machines. These devices are designed to automatically advance the bar-stock into the collet, or chuck, at the beginning of each turning cycle.

Expanding Collets

Most collet operations are performed with the standard style of collet that grips the workpiece by its external surface. However, there are times when the workpiece must be held by an internal diameter. In these cases, an expanding collet is the ideal choice.

Expanding collets, rather than collapsing in around a workpiece, expand outward against the inside of a hole. As shown in Figure 18–30, these collets are furnished with a gripping area made from aluminum or another soft material. This gripping area is typically machined to suit a specific hole diameter. Depending on the manufacturer, this type of expanding collet, with the proper accessories, can handle parts as large as 6.00 inches in diameter.

The standard range of the setup, as shown here, with the standard accessories, is .250 inch to 6.00 inches. Figure 18–31 shows a typical expanding collet application. Here the collet has been machined to the proper diameter and the part has been mounted for machining. For parts with internal diameters smaller than .250 inch, another accessory mandrel has been developed. This mandrel, as shown in Figure 18–32, can hold parts with an internal chucking diameter between .065 and .250 inch.

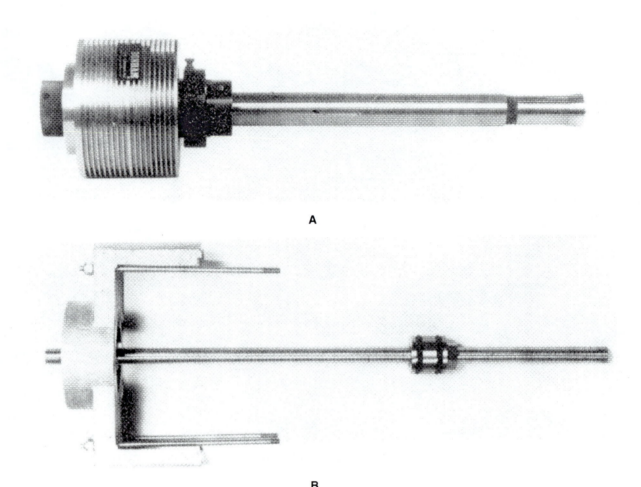

A

B

Figure 18–26 Pneumatic collet closer and dead-length stop assembly (*Courtesy of Dunham Tool Company*).

VISES AND VISE ACCESSORIES

The plain milling machine vise has always been a very popular workholder for milling, drilling, and a number of other applications. However, like other commercial components, the milling machine vise has also been improved and modified to suit a wide variety of applications.

Specialty Vises

The vise shown in Figure 18–33 has a *drilling fixture attachment* mounted to perform different, yet repetitive, drilling tasks. This attachment is a modification of a standard vise. Its solid jaw contains an end stop and three individual bushing arms. Both the end stop and the bushing arms can be adjusted to suit drilling operations in parts of many different sizes and shapes.

The Multivise™ shown in Figure 18–34 is a novel modification of the standard milling machine vise. It contains three positionable jaws that can be set to accommodate many different part shapes (Figure 18–35). Once set, the jaws are locked with the cam lever and the vise assumes the preset position for the duration of the production run. Figure 18–36 shows an example of a part set up in this three-jaw vise.

Another novel vise is the Vertivise® (Figure 18–37). This vise is made to be used vertically and has a set of stop bars positioned in the holes in the inside surfaces of the jaws. The part is then positioned as shown and machined as required.

Other forms of specialty vises are also commercially available. The Modular Vise® is quite useful for various workholding operations. The base may be used as a single piece, as shown in Figure 18–38A, or it may be substituted with two shorter bases, as shown in

Figure 18–27 The dead-length stop assembly mounts directly to the collet closer (*Courtesy of Dunham Tool Company*).

Figure 18–38B, to suit larger workpieces. The vise jaws can also be changed to suit the workpiece. While the standard jaws will suit most work, the fixed jaw can be rotated or both jaws can be replaced with specialty jaws to match the workpiece requirements. The vise and jaws may also be set up as a self-centering vise. The swivel base, rather than acting on a fixed point, can slide within the mount to permit the vise to be positioned at any point along its length.

One other type of vise ideally suited for different sized workpieces is the Bi-Lok Machine Vise™ (Figure 18–39A). In use, this vise can hold identical parts or two different sized parts (Figure 18–39B). It is totally self-compensating, and the same pressure is applied on both gripping positions regardless of the part sizes. The same principle behind the Bi-Lok Vise is applied to the Multi-Lok Vise™ unit (Figure 18–40). Here a group of vises are mounted together in units of two, four, eight, or more to hold a series of identical parts or a variety of parts for either horizontal or vertical

machining operations. The individual jaws on the Multi-Lok are also removable and may be replaced with soft jaws machined to suit specific part shapes.

The Triple Precision Mount Vise™, as shown in Figure 18–41A, is another specialty vise used for workholding. It differs from other vises in its basic clamping action. Rather than pushing the movable jaw, this design uses a pull action, which helps keep the workpiece securely seated against the base of the vise. Figure 18–41B shows this type of vise setup, with parts loaded, on a horizontal machining center. As with many newer vises, the design of the Triple Precision Mount Vise permits several vises to be mounted close together for multiple-part setups. Also, the design of its base element permits this vise to be mounted on its bottom or on either of its two sides.

When very large workpieces must be mounted on a machine table, the Maxi-Mill™ vise, as shown in Figure 18–42, is very well suited. This vise can convert the entire table of a vertical milling machine into a vise. As

Figure 18–28 Spindle work stop (*Courtesy of Five Manufacturing Company*).

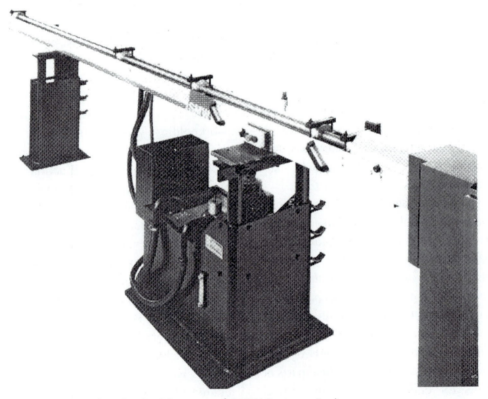

Figure 18–29 Automatic bar feeder (*Courtesy of SMW Systems, Inc.*).

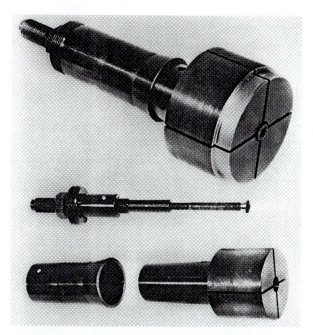

Figure 18–30 Expanding collet set (*Courtesy of ROVI Products*).

Figure 18–31 Expanding collet setup for boring a large-diameter workpiece (*Courtesy of ROVI Products*).

shown, the Maxi-Mill™ is a three-piece unit that consists of a fixed jaw, a movable jaw, and a rack gear that holds the jaws in position. In use, the fixed jaw is mounted at one end of the table and the movable jaw is mounted at the other end. The rack gear, which is positioned in the center "T"-slot, meshes with gear segments under each of the other elements and prevents the jaws from spreading as clamping pressure is applied.

Despite their wide and varied applications, many milling machine vises cannot hold odd shapes or multiple workpieces very well. One answer to holding these workpieces is the Twin Lock Workholding System™, as shown in Figure 18–43, which consists of two moving blocks that replace the movable jaw on any standard angle-locking vise. These moving blocks can hold either single parts or multiple parts and have the unique ability of applying the holding forces in both horizontal and vertical directions.

The jaws of this vise attachment pivot on a yoke and piston arrangement, as shown in Figure 18–44, and are self-adjusting to suit workpiece variations of

Figure 18–32 Expanding collet setup for turning a workpiece with a very small inside diameter (*Courtesy of ROVI Products*).

Figure 18–33 Vise-mounted drilling attachment (*Courtesy of Universal Vise and Tool Company*).

Figure 18–34 Three-jaw vises (*Courtesy of James Morton Company*).

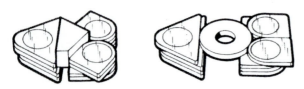

Figure 18–35 The jaws may be positioned to suit many different part shapes (*Courtesy of James Morton Company*).

Figure 18–36 Typical setup using a three-jaw vise (*Courtesy of James Morton Company*).

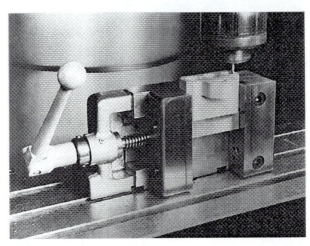

Figure 18–37 Vertivise® vertical vise (*Courtesy of Mid-State Machine Products, Inc.*).

up to .500 inch. If two different sized parts are held, both are clamped with the same force. The vertical travel of the pistons is .100 inch.

Several of the setup possibilities of the Twin Lock are shown in Figure 18–45. The setup at view A shows how this system is used to hold a rectangular part. Two parts can also be held in this unit, as shown at view B. A removable stop may also be positioned on the solid jaw to act as a reference point for mounting both parts. When the workpiece has two different clamping surfaces, as shown at view C, the unit easily accommodates the variations. View D shows how the jaws can be used to hold round parts.

The off-center design of the mounting holes in the removable jaw elements permits both large- and small-diameter parts to be held by simply flipping the

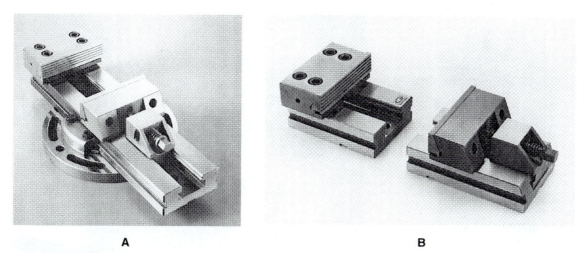

A   B

Figure 18–38 The Modular Vise™ may be used as a single unit (A) or on independent bases (B) (*Courtesy of Palmgren Steel Products*).

jaw elements over and remounting them on the moving blocks. Each of these jaw elements is keyed to the moving block and has a repeatability accuracy of .001 inch. Thus, the side of one jaw element may be used to establish a perpendicular mounting plane for smaller parts, as shown at view E.

When the clamping elements are mounted to the pistons, a completely new set of clamping options is available. Mounting a set of swivel jaws to the piston allows this unit to hold both tapered and larger round parts, as shown at views F and G. Odd shapes and contours may also be held simply by a set of swivel jaws made to conform to the part shape, as shown at view H. The setup shown at view I is very useful for machining three sides of a part in one setup. Here a set of strap clamps is mounted in the pistons to hold the part.

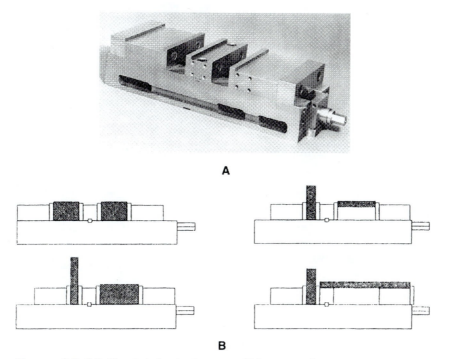

A

B

Figure 18–39 The Bi-Lok Machine Vise™ has two clamping areas (A) and can hold two identical or different sized parts (B) at the same time (*Courtesy of Chick Machine Tool, Inc.*).

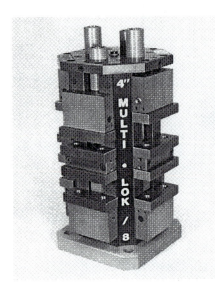

Figure 18–40 The Multi-Lok Vise™ uses the same principle found in the Bi-Lok™ Machine Vise to hold multiple workpieces vertically or horizontally (*Courtesy of Chick Machine Tool, Inc.*).

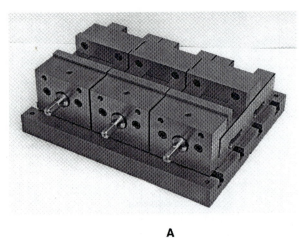

A

B

Figure 18–41 The Triple Precision Mount Vise™ uses a pulling action rather than a pushing action to hold parts (*Courtesy of Huron Machine Products*).

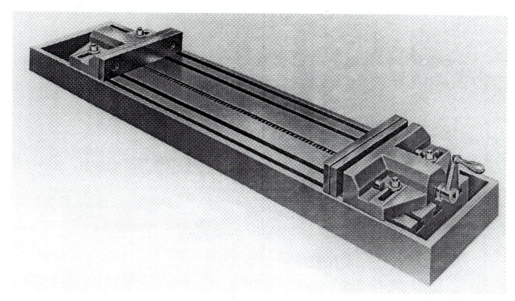

Figure 18–42 The Maxi-Mill™ vise can convert the complete table into a vise for larger parts (*Courtesy of TAYCO Tools, Inc.*).

As shown at view J, even setups involving multiple small parts are easily accomplished with another set of swivel jaws. Here the clamping action is achieved by installing an anvil plate against the sides of the parts. As the vise is tightened, the swivel jaws contact the anvil plate and push it against the parts. As the pressure is increased, the swivel jaws pivot and apply the necessary holding force to both ends of the stacked parts.

Figure 18–43 The Twin Lock Workholding System™ (*Courtesy of Discount Tool, Inc.*).

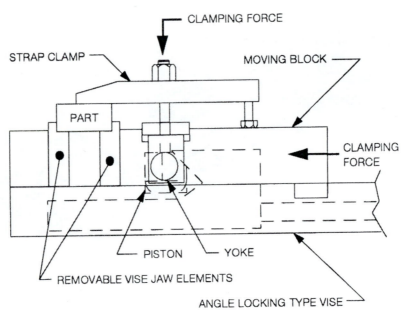

Figure 18–44 Operation of the Twin Lock Workholding System™.

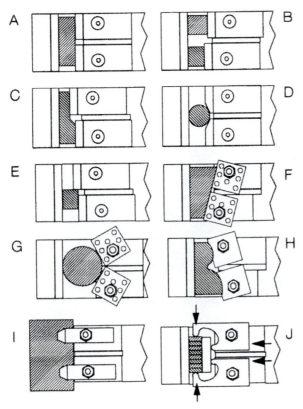

Figure 18–45 Twin Lock Workholding System™ applications.

Vise Accessories

In addition to the vises described, several vise accessories can aid in holding odd-shaped or multiple workpieces. The specific vise accessory that should be used with any vise is normally determined by the type and shape of the parts to be machined. Likewise, the accessories selected should be easily mounted and removed to return the vise to its general-purpose arrangement.

The first of these accessories is the Knollwood Vise Adapter™, which, as shown in Figure 18–46, provides a secondary clamping action, perpendicular to the normal clamping direction of the machine vise. This allows a standard vise to be used for clamping multiple parts. The Knollwood Vise Adapter helps to compensate for the small differences in part size and eliminates the problems caused when one or two parts are held securely while the others are loose.

Odd- and irregular-shaped workpieces are also a problem when workholding is involved. One especially difficult problem is holding parts with compound angles, but, as shown in Figure 18–47, with the Swivel Jaw™ mounted in the vise, even these parts are securely held. The basic unit in this vise accessory is

Figure 18–46 The Knollwood Vise Adapter™ aids in holding multiple parts by applying a sideways clamping force (*Courtesy of Royal Tools, Inc.*).

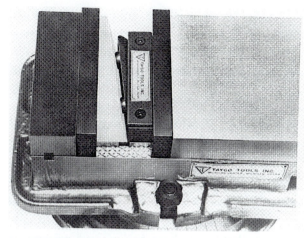

Figure 18–47 Odd and irregular parts are easily held with the Swivel Jaw™ vise accessory (*Courtesy of TAYCO Tools, Inc.*).

the rectangular block that houses the swivel unit. This block gives the device a stable mount and locates and aligns the accessory in the vise. The main swivel unit has one flat side and a semicircular form on the opposite side. This semicircular shape permits the swivel to move, in an arc, within the mounting block. Contained within this swivel element are two flattened balls that also swivel in their sockets. This permits each ball to move independently of the other and allows these accessories to conform to irregular angular or tapered forms with either simple or compound angles.

One other form of vise accessory used for holding odd and irregular shapes is the Hydra-Jaw™ (Figure 18–48). These vise jaws are made up of a series of individual clamping rods that are hydraulically operated. Each rod is interconnected with the other rods, and as one is depressed the others are extended. The hydraulic reservoir is self-contained within the

Figure 18–48 The Hydra-Jaw™ vise accessory adjusts itself hydraulically to hold odd shapes (*Courtesy of Hydra-Jaw*).

body of the attachment and requires no additional attachments. Since these jaws are hydraulically operated, the clamping pressure is equal and the jaws self-adjust to accommodate the slight differences in part sizes. The clamping rods are available with either flat pads or V-pads for holding round workpieces.

SPECIALTY CLAMPS AND WORKHOLDING DEVICES

In addition to chucks and vises, a wide assortment of specialty clamps and other devices are commercially available to hold just about any type of workpiece. Following are some of the clamps and devices in this group of workholding components.

Independent Clamping Elements

Independent vise jaws, as shown in Figure 18–49, provide auxiliary clamps anywhere they are required on the workholder. These clamps are available in several styles and may be positioned anywhere on the workholder or machine table. The jaws work on a screw thread that, when tightened, moves the clamping jaw down an incline. This movement exerts both a lateral force and a downward force against the workpiece, both clamping the part and forcing it down against the base of the workholder or machine table.

Edge clamping is another form of workholding common to many machining operations. Here the workpiece is gripped by its edges so that the top surface can be machined without obstruction. Vises are typically the most common edge-gripping clamps for these operations. If the workpiece is larger than the vise capacity or has an odd shape, two additional styles of edge clamp, the flat clamp and the slot clamp, work well.

The flat clamp, shown in Figure 18–50, is intended for low-profile clamping of flat workpieces. These clamps are designed around two heat-treated steel clamping elements. The larger piece holds the clamp to the machine table through the two mounting holes. The smaller element, the clamp, is connected to the larger element with a spring and a pivot rod. In use, the clamp screws are installed in the "T"-slot and the clamp body is slid into contact with the workpiece. The spring connector holds the clamping piece up off the table. Once positioned, the clamp screws in the larger element are tightened to lock the flat clamp to the table. The front screw is then tightened, forcing the clamping element against the workpiece. This clamping action applies force in both horizontal and downward directions to both clamp the part and hold it down against the table.

Slot clamps, as shown in Figure 18–51, operate in much the same way as flat clamps do. They also apply horizontal and downward clamping actions against the workpiece. Slot clamps are designed to hold thin workpieces and are mounted directly in the "T"-slots rather than on the machine table. This low-profile installation allows the clamps to hold parts as thin as .13 inch without distortion. In use, the slot clamp body is installed in the "T"-slot in much the same way as a T-nut is. The clamp is then moved into contact with the

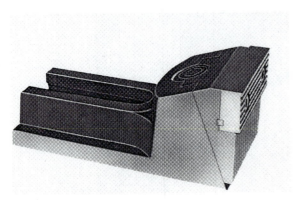

Figure 18–49 Independent vise jaw (*Courtesy of James Morton Company*).

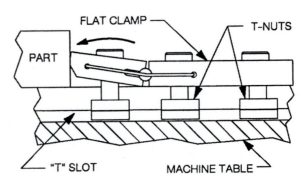

Figure 18–50 The flat clamp is well suited for edge-clamping operations (*Courtesy of James Morton Company*).

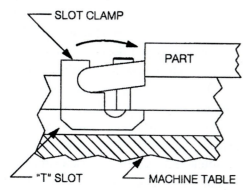

Figure 18–51 The slot clamp is used for edge-clamping thin parts (*Courtesy of James Morton Company*).

workpiece. Once the clamp is positioned, a set screw in the clamp body is tightened to hold the clamp body in the "T"-slot. The front screw is then tightened, forcing the clamping element against the workpiece.

For those edge-clamping applications where pallets or pallet systems are used, the Vector Clamping System™ workholding device provides both clamping security and a space-saving design. As shown in Figure 18–52A, this clamp combines the speed of a wedge clamp with the simplicity of the screw in an integrated clamping unit. The major elements of this clamp are

the clamp body, the clamping element, and the mounting bolts, Figure 18–52B. The two clamping elements available with this clamp, Figure 18–52C, are the solid end for greater clamping contact and the reduced end for ample cutter clearance.

In clamping situations where edge clamping is needed, the older forms of commercial edge clamp are often too large. However, with the newer MITEE-BITE™ clamping system, the problem of edge clamping in small spaces has been solved. This unique clamp combines the stability of a screw clamp with the rapid action of a cam clamp. The principal elements of the system are the eccentric socket-head cap screw and the hexagonal brass clamping element (Figure 18–53A).

The cam-action movement of the clamp is achieved by turning the screw inside the clamping element. The hexagonal clamping element has six individual clamping surfaces and is made of brass, so there is little chance of damaging the clamped workpiece. The clamp may be mounted directly to a workholder through a tapped hole or in a "T"-slot with a T-nut.

Figure 18–53B shows two applications where the MITEE-BITE™ is used to hold both a rectangular workpiece and a round workpiece. For setups where

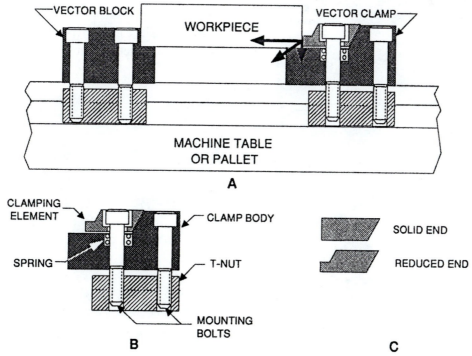

Figure 18–52 The Vector Clamping System™ allows secure clamping in limited spaces.

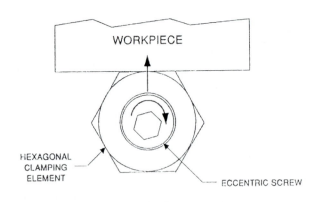

WORKPIECE

HEXAGONAL
CLAMPING
ELEMENT

ECCENTRIC SCREW

A

B

Figure 18–53 The MITEE-BITE™ clamps use an eccentric screw inside a hexagonal clamping element (*Courtesy of MITEE-BITE Products Co.*).

the workpiece must be elevated off the mounting surface, the riser variation can be used (Figure 18–54).

Another unique system that departs from conventional clamping methods is the Mono-Bloc™ clamp (Figure 18–55A). This is a self-contained clamping unit that uses a worm and wormwheel arrangement, Figure 18–55B, to perform the clamping action. Such an arrangement permits the clamps to have a wide clamping range and infinite adjustability. These clamps are available in several grades and sizes with a variety of riser elements and arm extensions. They can be used for holding parts from 0 to 12 inches tall.

The Terrific 30™ clamp, as shown in Figure 18–56A, is one other form of clamp that offers both utility and versatility for a range of applications. This is a retractable form of workholder that operates from a single point. In use, the part is positioned in front of the clamps and the handle is turned one and a half turns to achieve both the extension and clamping action, Figure 18–56B. Turning the handle the same one and a half turns in the opposite direction releases and retracts the clamp. Terrific 30™ clamps may be mounted either directly to the machine table or on a series of riser elements.

The universal milling fixture, as shown in Figure 18–57, is very useful for holding cylindrical parts that must be machined on or near their top surface. Positioning the workpiece in this type of workholder permits the top to be completely free of any clamps and allows

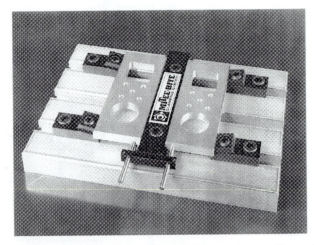

Figure 18–54 The MITEE-BITE™ riser clamp allows the workpiece to be clamped in limited areas and elevated off the table (*Courtesy of MITEE-BITE Products Co.*).

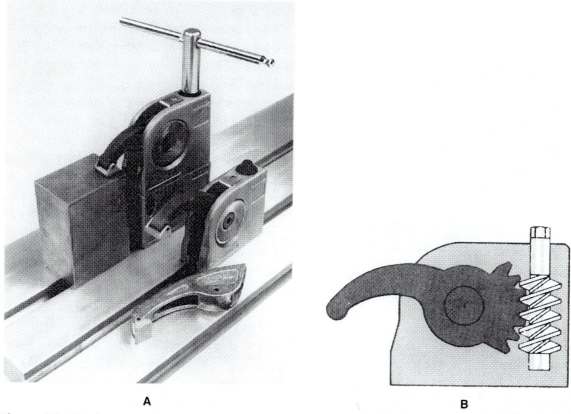

A

B

Figure 18–55 The Mono-Bloc™ clamps work on a worm and wormwheel principle (*Courtesy of Royal Products*).

three-point contact between the workholder and the part. This three-point contact ensures both accurate location and total clamping security.

Another handy device for mounting either parts or complete workholders is the Hold-All™ (Figure

18–58A). This unit raises the parts off the machine table and is well suited for applications where the area under the part must be clear for operations that go completely through the part. Figure 18–58B is an example of such a setup. Here the part is drilled and the holes are

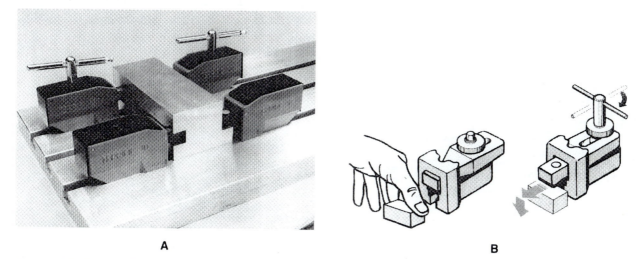

A

B

Figure 18–56 The Terrific 30™ clamps are a form of retracting clamp that operates with only one and a half turns of the handle (*Courtesy of Royal Products*).

Figure 18-57 Universal milling machine fixture (*Courtesy of Universal Vise and Tool Company*).

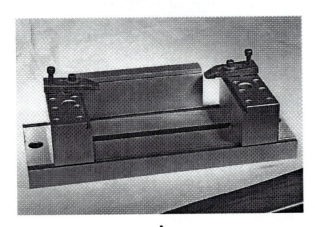

A

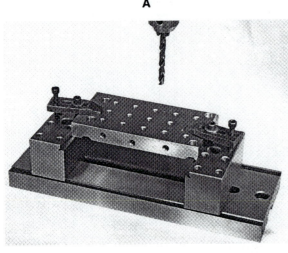

B

Figure 18-58 The Hold-All™ is used to hold parts off the table to permit operations that go completely through the part (*Courtesy of ATCO Precision Tools*).

either tapped or counterbored. The Hold-All™ consists of a base plate and three adjustable rails used to position and locate the part. In addition, clamps may be positioned in several locations in the end units to securely hold the part during machining.

A workholding system that greatly simplifies the fixturing process by reducing the number of steps needed to build a workholder is the Grip Strip Tooling System™. The basic Grip Strip unit is a laminated bar consisting of a steel mounting plate and an aluminum clamping plate bonded to a resilient intermediate layer of rubber (Figure 18-59). A set of rails on both sides of the aluminum clamping plate provide the mount for the clamping devices as well as a solid mounting area for attaching the unit to the machine tool or base plate.

This ingenious workholding alternative combines the design flexibility of custom-made clamps with the utility and lower cost of standard clamping devices. Rather than requiring separate fixture-building operations, the Grip Strip is machined on the same machine tool intended to produce the workpieces. The fixture blank is mounted to the machine tool, and the cavity, or nest, as shown in Figure 18-60, is machined using the same computer program written to machine the workpieces. The cutter path is simply set to cut the inside periphery rather than the outside.

Grip Strip fixtures are available in several standard sizes ranging from 2.00 to 8.00 inches wide in a standard length of 18.00 inches. These bars are also available with three different clamping plate thicknesses, .25 inch, .50 inch, and .75 inch, to suit practically any workholding application.

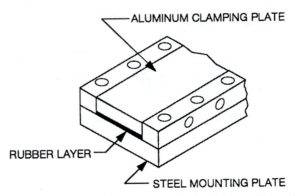

Figure 18-59 The Grip Strip Tooling System™ is a laminated bar with a steel plate and an aluminum plate bonded to a layer of rubber (*Courtesy of Morgan Enterprises*).

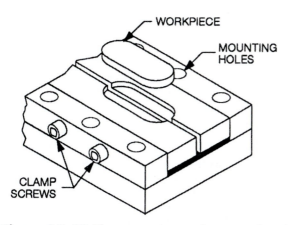

Figure 18–60 The part is mounted in a machined nest in the Grip Strip (*Courtesy of Morgan Enterprises*).

The toggle clamp is one of the most common clamp forms used throughout industry. But for all its applications, the main problem with the toggle clamp is its inability to compensate for different sized parts. With the Automatic Toggle Clamp, shown in Figure 18–61, even this problem has been solved. Instead of using fixed pivots and levers, like standard toggle

Figure 18–61 The Automatic Toggle Clamp allows different size parts to be securely held with the same clamp (*Courtesy of Carr Lane Manufacturing Company*).

clamps, the Automatic Toggle Clamp uses a self-adjusting feature to adapt the clamp to different workpiece heights. The clamp arm can handle differences in clamping heights of up to 15 degrees. This results in a total automatic adjusting range of more than 1.25 inches. Additional adjustment is permitted by manually moving the threaded spindle. Together, these adjustments result in a considerable clamping capacity. In use, the clamp is first set to the average workpiece height (Figure 18–62A). Once set, the clamp automatically adjusts to suit a range of workpiece variations (Figure 18–62B and Figure 18–62C).

Another form of clamp that is useful for specialized operations is the Claw Clamp, as shown in Figure 18–63, the basic design of which is simple yet combines several unique features. The Claw Clamp can be used to clamp in two directions. As shown, it may be used conventionally, or the clamp arm may be reversed and the unit used as a spreader clamp. When the clamp arm is reversed, a clamp shoe is attached to the rail to provide stable contact. The contact pad on the clamp arm is pivoted and permits movement to 30 degrees to accommodate minor part irregularities as well as any angular surfaces.

The clamping action of the Claw Clamp is accomplished by the screw mounted in the clamp arm. The position of the screw, outside the clamping area, allows the clamp to be used in confined spaces. This compact design, coupled with the 4.00-inch throat depth, makes this clamp useful for a range of special tasks. The Claw Clamp is available in three sizes, 12.00 inches, 24.00 inches, and 40.00 inches, and has a maximum recommended holding force of 1760 pounds.

Workholder Mounting Devices

The final clamping device discussed here is intended to clamp not the workpiece but rather the workholder. Many methods are used to hold the workholder to the machine tool during machining. However, the main problem has always been to find a way to reduce the time needed to set up and align the workholder. With the Ball-Lock™ mounting system, both position and accuracy, as well as the attachment, are handled with a single set of elements. This mounting system consists of three major parts: a flanged shank, a liner bushing, and a Ball-Lock bushing (Figure 18–64). In use, the bushings are installed in both the fixture base and the machine base or subplate. The flanged shank is then

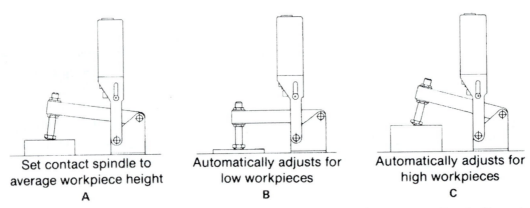

Set contact spindle to average workpiece height
A

Automatically adjusts for low workpieces
B

Automatically adjusts for high workpieces
C

Figure 18–62 When used for a variety of different sized parts, the Automatic Toggle Clamp is first set to the average part height and will then adjust itself to suit the thinner or thicker parts (*Courtesy of Carr Lane Manufacturing Company*).

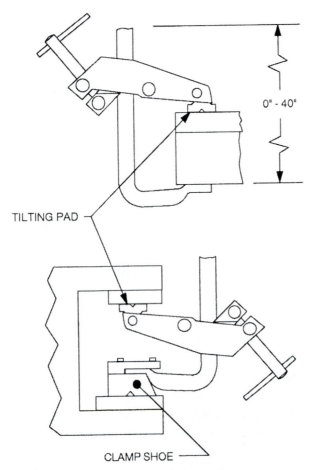

TILTING PAD

0" - 40"

CLAMP SHOE

Figure 18–63 The Claw Clamp holds parts securely with either a conventional clamping or spreader clamping action (*Courtesy of James Morton Company*).

Figure 18–64 The Ball-Lock™ system helps to quickly and accurately locate workholders on machine tables or subplates (*Courtesy of Jergens, Inc.*).

installed through the liner bushing into the bushing and is tightened down with the socket-head cap screw.

The operation of the Ball-Lock is shown in Figure 18–65A. As the socket-head cap screw is tightened, it forces the larger ball against three small balls. This moves the three locking balls outward in their sockets against the chamfered, or conical, sides of the Ball-Lock bushing. The three balls then contact the chamfer, which not only locks the balls in position and prevents the flanged shank from moving, but also applies a downward force against the shank, pulling down the flanged shank, locking it against the fixture base. Figure 18–64B shows how these units are positioned for attaching both small and large workholders.

A quick-change pallet system that incorporates the Ball-Lock is also available. This is the Ball Lock™ Quick-Change Kit (Figure 18–66). The kit consists of a steel subplate with four receiver bushings installed, two aluminum fixture plates with two liner bushings installed, and four 20-millimeter Ball-Lock shanks. The subplate is available in a 16 × 16 × -3/4-inch size, and there are three fixture plate sizes ranging from 12 × 14 × 3/4-inch to 16 × 16 × 3/4-inch. An additional 16 × 25 × 1.125-inch dual-station subplate, shown in the figure is available for mounting a single 16 × 16-inch plate, a single 14 × 14-inch plate, or two 12 × 14-inch fixture plates side by side.

The fixture plate may be used as a base for the workpiece fixture or as a secondary base for an existing fixture. The steel subplate may be mounted directly to the tombstone or other mounting device. The complete unit then functions in much the same way as a pallet arrangement. The individual subplates are mounted and removed with the Ball Locks.

SUMMARY

The following important concepts were presented in this unit:

- The two major groups of low-cost workholding devices used in fixturing workpieces today are:
  - Chucks and vises
  - Specialty clamps and workholding devices

- The benefits of using these low-cost components and devices include:
  - Lower tooling costs
  - Simplified designs by combining commercially available elements with specialty items
  - Increased versatility

REVIEW

1. What devices are primarily used in low-cost jigs and fixtures?
2. What two general soft-jaw styles are used for holding parts in chucks?
3. Which method of turning soft jaws allows production machines to keep working rather than being used for making jaws?
4. What three general styles of machinable collets are used for workholding?
5. What setup advantage does the Master Jaw System® offer?
6. With regard to the way they hold a part, how does an expanding collet differ from a normal collet?
7. Which vise uses three positionable jaws to accommodate a variety of part shapes?
8. What vise is made to be used on its side to hold parts?
9. Which vise is capable of holding two different sized parts?
10. What vise is capable of being mounted on any of three sides and has a pull action rather than a push action?
11. What is the benefit in using the Maxi-Mill™ vise for holding large parts?
12. Which vise accessory applies the holding forces in both horizontal and vertical directions?
13. List three vise accessories that are used for specialized parts.
14. Which clamping device uses an eccentric socket-head cap screw and hexagonal brass clamping element?
15. What two types of clamps are very well suited for work in small spaces?
16. List two types of clamps used to elevate a workpiece off the machine table.

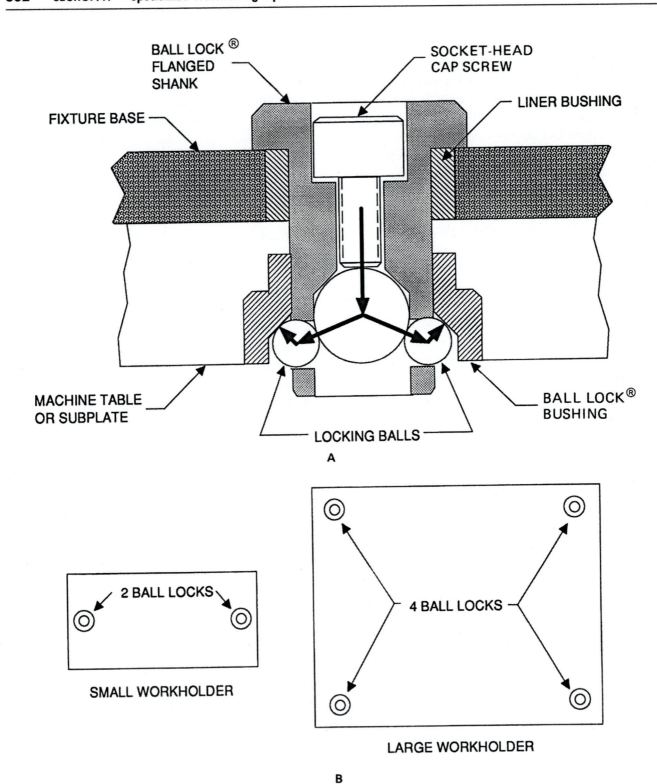

Figure 18–65 Cutaway view of the Ball-Lock™ system (*Courtesy of Jergens, Inc.*).

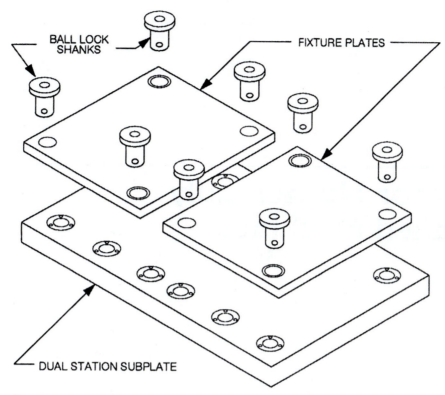

Figure 18–66 Ball-Lock™ Quick-Change Kit.

17. What type of clamp works with a worm and wormwheel arrangement?

18. What type of clamping action does a Terrific 30™ clamp use?

19. What advantage does the Automatic Toggle Clamp offer?

20. What is the purpose of the Ball-Lock™ device?

UNIT 19

Tooling for Numerically Controlled Machines

OBJECTIVES

After completing this unit, the student should be able to:

- Identify the meaning of the term *numerical control.*
- Identify and name the three principal axes of numerical control.
- Explain the Cartesian coordinate system.
- Describe the meaning of incremental and absolute programming.
- Specify the types of input media used with numerical control.
- Specify the most common numerical control machines.
- Identify the types of workholder used with numerically controlled machines.

INTRODUCTION

Recently, manufacturing has undergone dramatic changes in how parts are made. The demand for more production at a reduced cost has forced industry to be more efficient. One means manufacturers have used to accomplish this is the use of numerically controlled (N/C) machine tools. The ever-increasing use of numerically controlled machines has provided the tool designer with both a challenge and an opportunity to develop new tooling ideas and designs.

BASIC N/C OPERATION

Numerically controlled machines operate similarly to conventional machines, but they are much faster and without the human element at the controls. The term *numerical control* describes the means by which the movements of the machines are governed. The desired machine movements are first determined from the part print. Each movement is then assigned to a part axis (Figure 19–1). As shown, these axes are identified by the letters X, Y, and Z. Typically, on a vertical milling machine or drill press, the X-axis controls longitudinal, or left and right, movements of the table; the Y-axis controls traverse, or in and out, movements; and the Z-axis controls vertical movements of the spindle. On other machine tools, such as machining centers or numerically controlled lathes, the placement of the axes varies, so you should check each machine before designing a jig or fixture for the machine tool.

During programming, the direction and extent of all movements along each axis are recorded on one of the input media used to convey the specific instructions to the machine tool. The input media are dependent on the type of system used for the machine tool.

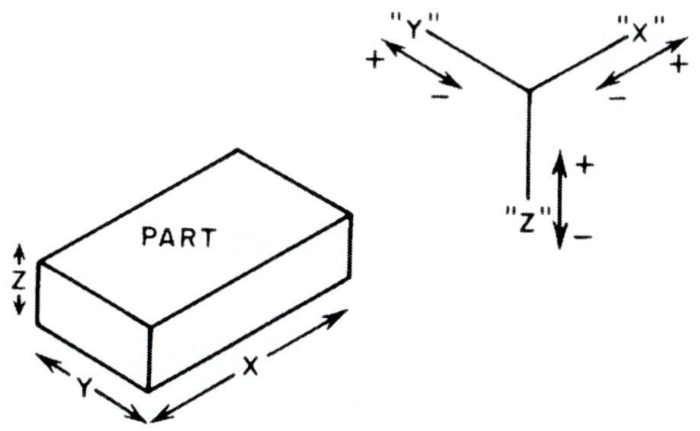

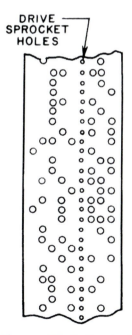

Figure 19–1 Typical arrangement of axes for numerical control.

The two forms of numerical control in use today are N/C (numerical control) and CNC (computer numerical control). The input media most frequently used with N/C systems are punched paper or mylar tape (Figure 19–2). The newer CNC systems typically use floppy disks, magnetic tape, or direct computer interfaces. Regardless of the input medium used, every numerically controlled machine tool works in basically the same way. That is, each accepts the information needed to machine the required part as coded numerical data the machine control understands. This coded information is then translated by the machine tool

Figure 19–2 One-inch perforated tape.

into specific machine motions and movements required to machine the specified part feature.

Figure 19–3A is a relatively simple diagram of how a typical N/C system and CNC system operate. In the N/C system, the basic process flow starts with the part drawing that is used to program the machine tool. Once the information from the part drawing is programmed and properly coded, the tape or other input medium is prepared. This input medium provides all the necessary instructions for the part and is fed into the machine tool that is used to make the part.

The same basic process is used with CNC, with the exception that the computer, rather than a punched tape, is used to control the machine tool. As shown in Figure 19–3B, the part design, tool design, and programming operations can all be accomplished with the same computer. The part is first designed using the CAD software. The fixturing elements are then added to design the workholder using the fixturing software. Finally, the CAM software is used to program the movements of the machine tool. Paper copies of the designs and the programs can then be prepared using either a plotter or a printer, or both.

The direct computer interface, or DNC as it is known in industry, offers greater security for the part program. Approved changes in tooling and tool path are maintained for incentive standards, and speeds and feeds are documented. Part programs can be stored off-site from the machine and downloaded when production requirements are scheduled. Accompanying mechanical detail sheets that describe each tool assembly and what each tool will machine in the part program are also available as printed material and through computer-generated copies.

THE CARTESIAN COORDINATE SYSTEM

The principle of locating points or features with numerical control is based on rectangular coordinates. With rectangular coordinates, any point or position can be specified by its distance, expressed in horizontal and vertical units, from a known point. The basic system used to plot these rectangular coordinates is the Cartesian coordinate system (Figure 19–4). Here each point is located by expressing its position from the central point, marked "0," in either plus or minus

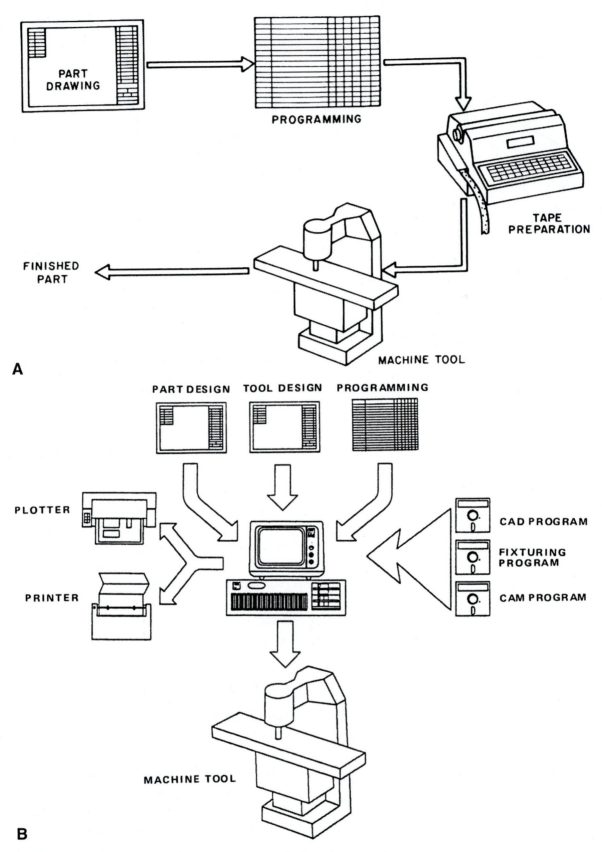

Figure 19-3 Process flow of numerically controlled machining systems.

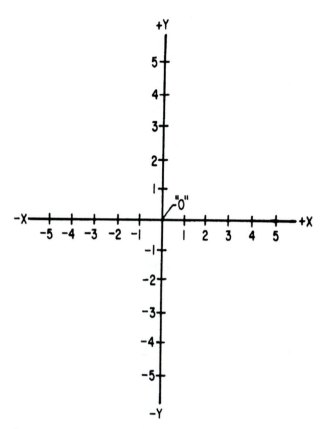

Figure 19–4 Cartesian coordinate system.

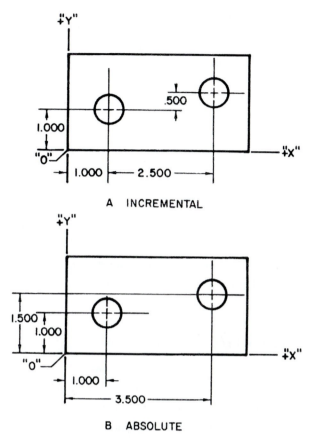

Figure 19–5 Incremental and absolute dimensioning.

movements along the longitudinal axis, X, and the traverse or cross axis, Y.

INCREMENTAL AND ABSOLUTE PROGRAMMING

The two basic methods used to locate points along any part axes are *incremental* and *absolute*. These terms, in addition to describing the methods used to locate a specific point, also describe the type of dimensioning method used to locate the feature on the part print. As shown in Figure 19–5A, incremental programming or dimensioning establishes the location of a feature based on the location of the feature immediately preceding it. Thus, if the second hole must be drilled, its location on the part is established by its position relative to the first hole. To program this location, the first hole position must first be determined. In this case, using the Cartesian coordinate system and incremental programming, the first hole is

located by positioning it at + X = 1.000 inch and +Y = 1.000 inch. The second hole is then referenced relative to the first hole and becomes + X = 2.500 inches and + Y = .500 inch. Absolute programming or dimensioning, as shown in Figure 19–5B, conversely, establishes each individual point relative to a central location. Assuming, then, that the same parts were programmed or dimensioned in this system, the points would be expressed as + X = 1.000 inch and + Y = 1.000 inch for the first hole, and + X = 3.500 inches and + Y = 1.500 inches for the second hole, as shown.

It is acceptable to switch back and forth between incremental programming and absolute programming. This would not result in the tool path being any different. The machine tool will run the same and the tool path will look the same. The justification for switching lies in the ease of programming that it provides. Patterns of holes might be common among a family of parts that share a common fixture. Once the pattern has been defined, it can be called up and

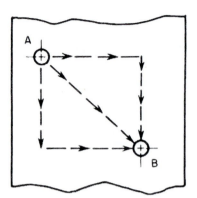

Figure 19–6 Point-to-point N/C path.

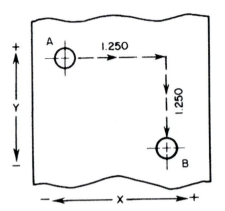

Figure 19–7 Programmed tool path.

inserted into the program. This also makes the pattern easier to identify in the printout of the program.

TYPES OF N/C SYSTEMS

The two primary types of N/C system used today are called *point-to-point* and *continuous path*. The principal difference between the two is the method or control used when moving the cutting tool between specific locations on the part.

Point-To-Point Systems

A point-to-point system is the simplest form of numerical control. This system, or machine tool, is used when the specified operations must be precisely located and the path the tool takes is of little or no importance. Typically, drilling or punching machines use this type of system. For example, the holes in the plate shown in Figure 19–6 must be accurately located, but the path the tool takes from hole A to hole B can be any of those shown.

In the operation shown, the tool starts at the reference point "0." By following the instructions on the prepunched tape, the cutting head moves to hole A (Figure 19–7). The tool descends along the Z-axis and drills the hole. The tool is retracted (up the Z-axis) and then moves 1.250 inches along the +X-axis and 1.250 inches along the –Y-axis to the second hole. When the hole is properly positioned, the program repeats the drilling cycle.

Continuous Path Systems

Continuous path systems and machines are more detailed in design than the typical point-to-point system. These systems control the tool position continuously throughout the machining cycle. Continuous path numerical control is used mainly on machine tools such as milling machines and lathes, where the path of the tool is very important to the end product. The cutter path in these systems may be plotted by hand or, as is more often the case, by a computer from the approximate rectangular coordinate locational points provided by the programmer. Using a computer not only reduces errors but also increases the speed of the required calculations. Figure 19–8 illustrates the rectangular coordinate positions that must be plotted around a typical small radius.

Retrofitted N/C Systems

In addition to the complete numerically controlled machines and systems found in shops today, a variety of retrofit systems are commercially available. These systems are intended to convert existing manual machine tools into semiautomatic CNC machines at minimal cost.

One such system is shown in Figure 19–9. Here a retrofit kit is installed on a vertical milling machine and controls all three table movements as well as the accessory rotary table. The major components of this system are the direct-drive motors and the programmable teach pendant. The motors drive the table, saddle, knee, and optional rotary

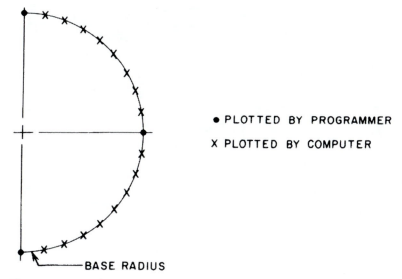

• PLOTTED BY PROGRAMMER

X PLOTTED BY COMPUTER

BASE RADIUS

Figure 19-8 Plotting coordinates for a radius.

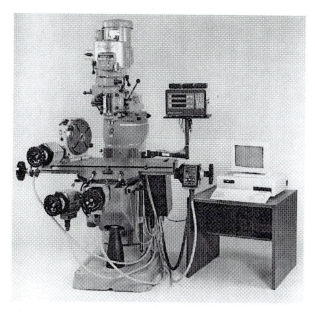

Figure 19-9 A retrofit system is ideal for converting a manual machine tool to a functional numerically controlled tool at minimal cost (*Courtesy of SERVO Products*).

table, and are controlled by the teach pendant (shown by the right table handwheel).

In use, the system is programmed either with a computer or by stepping the machine through the machining cycle with the teach pendant. Once programmed, the machine operates in the same fashion as a standard CNC machine and makes every part identical.

TOOLING REQUIREMENTS FOR NUMERICAL CONTROL

Numerically controlled machines have not removed the need for workholders, only the need for complex and expensive jigs and fixtures. The trend is away from single-purpose tooling and toward multipurpose, universal workholders.

Numerically controlled machines have forced the tool designer back to the basics of jig and fixture design. Whereas with other machines only a single surface can be machined at one time, numerically controlled machines can machine several surfaces in each setup. Whereas conventional machines require detailed and repeated tool referencing, numerical control requires only initial referencing to the workholder. Whereas the goal of conventional tool design was to transfer the needed accuracy from the operator to the fixture, numerically controlled machines transfer the required accuracy from the fixture to the machine tool.

All of this, coupled with the speed, precision, and repeatability of numerical control, has challenged the tool designer as never before.

TYPES OF WORKHOLDERS

Since the primary function of workholders for numerically controlled machines is holding and

Figure 19-10 Typical N/C fixture.

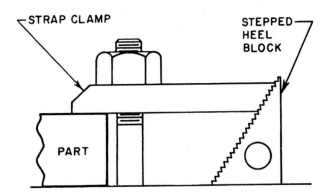

Figure 19-11 Strap clamp and block.

locating the part, there is no need for elaborate, expensive tools. Figure 19–10 illustrates a simple fixture for a typical machining center.

The most popular form of workholder used with N/C machines is the strap clamp and block (Figure 19–11). This form is the most versatile and the least expensive. Strap clamps used for numerical control applications are usually hand-operated rather than power-operated. For automated production lines, however, power clamps reduce the clamping time and speed up production.

In dedicated fixturing that runs only one part, a fixed heel block is used. This expedites the handwork to set the clamp properly and eliminates loss of machining time due to repositioning of a heel block. The physical configuration for a production clamp

might look like the clamp shown at the front of the fixture in Figure 19–10.

Torque wrenches are used in situations where excessive clamping pressure might cause part distortion. Designers must consider this possibility of distortion from clamping pressure when designing fixturing for thin-walled parts. The torque wrench delivers the same clamping pressures regardless of the strength of the machining technician. This eliminates or minimizes the potential for out of tolerance sizes due to distortion. Designing the fixture is a onetime event, whereas fighting a bore size may occur every time the part geometry is run. Fixtures that might use holes that are drilled and tapped into the part itself for attachment to the fixture provide a distortion-free setup. Reviewing these possibilities for potential incorporation into the fixture may result in a competitively better design.

Other types of workholding or positioning devices sometimes used with numerically controlled machine tools, as well as with some conventional machine tools, are the subplate and tombstone shown in Figure 19–12. These units are used to mount tooling components, fixtures, or workpieces onto the machine tool. In use, they are frequently mounted permanently on the machine tool and are used to accurately align and mount the workpieces themselves or the tooling required to machine the workpieces.

A

B

Figure 19–12 Subplate (A) and tombstone (B) used for fixturing N/C machines (*Courtesy of Mid-State Machine Products, Inc.*).

Another variation of this tooling method for numerically controlled machines is the universal holding fixture (Figure 19–13). This tool is used to provide a permanent mount for jigs and fixtures used on N/C machine tools or machining centers. The universal holding fixture consists of two major parts: the cast body of the fixture and the tooling plate (Figure 19–14). The cast body is accurately machined and provides the required support for the machining operation as well as orienting the fixture either horizontally or vertically. The tooling plate is used as a base plate for the jig or fixture and is accurately aligned and mounted on the universal holding fixture by means of the two dowel-pin locators and the socket-head cap screws. In use, the only part of these fixtures

Figure 19–13 Universal holding fixtures (*Courtesy of Integrated Systems Division, Varo, Inc.*).

that must be changed to suit a new job is the tooling plate on which the new jig or fixture is mounted. Another feature of the universal holding fixture is removable wear blocks, which are used to locate the fixture on the machine tool and may be easily replaced when they wear or if they are damaged.

Other forms of workholder are finding increased use with numerically controlled machines. Modular tools, cast workholders of plastic resins or low-melt alloys, and vises and chucks with modified jaws are also finding greater use. The key elements in designing any workholder for numerical control are simplicity and workability. To this end, the designer must rely on the basics of tool design.

Since numerical control offers so many possibilities in workholder design, many ideas are being incorporated into the basic design of the machines, including features such as single and double indexing tables, multifixture tool posts, automatic tool changers, and special vertical vises. Figure 19–15 illustrates a complex multistation numerically controlled machine in operation.

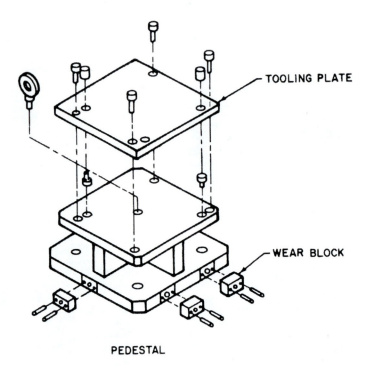

PEDESTAL

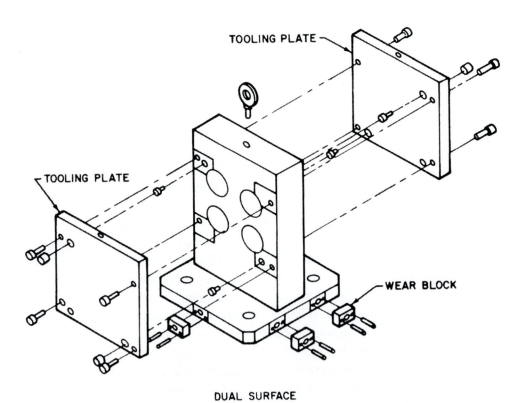

DUAL SURFACE

Figure 19-14 Parts of the universal holding fixture.

- The two forms of numerically controlled system are point-to-point and continuous path.
- One of the most popular clamps used with these machine tools is the strap clamp.
- Other workholding devices used with numerically controlled machines include tombstones, subplates, universal holding fixtures, and pallet systems.

REVIEW

1. What term describes how the movements of an N/C or CNC machine are controlled?
2. How are the machine axes identified with relation to the part?
3. Which input medium is the most commonly used today?
4. Identify the input media used most frequently with N/C. Identify two forms of input media for CNC.
5. What system is used to specify the rectangular coordinates of any location on a part?
6. Which programming method always starts from a single known reference point for a dimension?
7. Which programming method relies on the position of the last programmed location for the location of the next detail?
8. What do the + and – signs before the axis dimension specify?
9. Briefly explain what each of the following terms means with regard to N/C operation: point-to-point; continuous path; X-, Y-, and Z-axes.
10. List three standard tooling items used with numerically controlled machines.

UNIT 20

Setup Reduction for Workholding

OBJECTIVES

After completing this unit, the student should be able to:

- Identify the benefits of reducing setup costs.
- Describe techniques for simplifying workpiece design.
- Describe techniques for simplifying workpiece processing.
- Specify methods and techniques of designing cost-effective workholders.

Setup reduction for workholding may seem to be simply finding ways to get workpieces on and off a machine tool more quickly, but it actually involves much more. In its most basic sense, setup reduction for workholding is a means of reducing the time and cost of manufacturing products by reducing the time required to perform the necessary operations.

Setup reduction starts with a detailed analysis of the basic workpiece design. Once the workpiece is carefully studied, the proposed processing is then evaluated. Every production step from the design through the final inspection must be examined to find better ways of producing products. When the best combination of workpiece design and processing is achieved, the tool design is initiated.

BENEFITS OF SETUP REDUCTION

Workholding, unfortunately, has not received as much attention as other areas of manufacturing have. The cost of setting up jigs or fixtures is viewed in some companies as having a fixed value. Other companies estimate setup costs based on outdated methods and data. Seldom do companies actually study the time it takes to perform these setups. Although countless hours are spent trying to improve the cost-effectiveness of operations, many times the one area that can save the most is overlooked.

Reducing setup expenses offers a manufacturer many benefits. Some of these benefits, such as reducing overall production costs, are quite clear, but others are not as obvious. By following just a few basic cost-reduction practices the tooling expenses can also be greatly reduced. Likewise, both the production speed and the production volume can be increased by reducing the lead times required for each machining operation. This permits considerably faster production changeovers.

Probably the single largest benefit of setup reduction in all areas of manufacturing is increased productivity, which has a dramatic effect on all of us. In most cases, increased productivity makes a company more competitive and allows it to make its products less expensively. This affects the consumer by allowing the company to sell its product at a lower price. The company becomes more profitable and benefits from increased sales.

THE SETUP REDUCTION PROCESS

The processes or procedures used in reducing setup expenses represent not a completely new design method but rather simple refinements of how jobs are fixtured today. Just about anything that reduces the time required for a job will have a positive effect on the complete process. Setup reduction should be considered in every area of the workpiece production cycle. The three general areas that should be studied for ways to reduce setup costs are workpiece design, workpiece processing, and workholder design.

Workpiece Design and Setup Reduction

The starting point for reducing workholding setup expense must be the workpiece being fixtured. Simplifying workpiece designs will make manufacturing easier and reduce overall tooling and production expenses. The first place to look for ways to improve jig and fixture designs is the initial product design.

Product designs too often are developed with little or no concern for how the parts are made. Simply designing products so that they are easier to manufacture will go a long way in improving the workholder designs. Occasionally, ideas that look very simple on paper are actually quite complicated when applied to the product.

Ideally, the product designer should have a working understanding of how the proposed parts will be made and the capabilities of the manufacturing departments. The product designer is normally the best person to determine how the workpiece must be located and clamped to meet the design requirements. Unfortunately, the designer may not have manufacturing experience, and this is where concurrent engineering becomes critical to the success of a business. Manufacturing has expanded its capabilities. Machine

tools and computers have married products to produce a new breed of technology that is difficult to comprehend if the individual is not exposed to it on a continual basis. Specialization is common and cross-training is expensive, although not impossible. In the past, designers have typically been exposed to years of manufacturing and are familiar with machine tools, fixturing, and the tool used to produce components. This is not always true of the young designer, who may be heavily credentialed with a four-year engineering degree but have little or no experience with manufacturing. The concurrent engineering approach to product design and production brings together all of the experiences and skills that are necessary to find solutions to problems. This team of individuals provides the information and feedback necessary for all areas to benefit from one another.

One continual problem for tool designers is locating the workpiece. This is especially true when datums are incomplete or missing from the design drawing or when the datums are placed in areas that cannot be easily located. Simply double-checking a design can often greatly reduce overall production costs. Making sure that all the workpiece specifications are reasonable and within manufacturing capabilities can simplify the necessary workholders.

The product designer should also question each feature or detail of a design. Here the three questions that should be asked are (1) Is the feature or detail necessary, or can it be eliminated? (2) Can the feature or detail be simplified? and (3) Can the feature or detail be manufactured economically? Questioning each aspect of a proposed design not only validates and verifies the design but also ensures that each feature can be produced economically.

Overly tight workpiece tolerancing is another problem in designing workholders. Too often the tolerances specified are either unnecessary or almost impossible to hold. When specifying tight tolerances, first make sure that there is a valid reason for using a tight tolerance, and then make sure that manufacturing has the capability to hold the tolerance. Question every tight tolerance. Look for every opportunity to open up a tight tolerance. Remember that the tolerance specifications listed for a workpiece are reduced even further

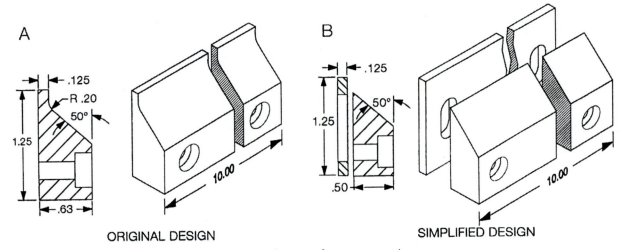

A

ORIGINAL DESIGN

B

SIMPLIFIED DESIGN

Figure 20-1 Redesigning a cutoff blade to make manufacturing simpler.

for the workholder. A tight tolerance on the workpiece is actually even less when applied to the jig or fixture.

Parts that have been geometrically toleranced are examined closely for mating part assembly. There is a complete mathematical background for determining the tolerances and, when followed, they produce more good parts that assemble and interchange than the rectangular coordinate dimensioning system. Stack-up of tolerances are calculated and the maximum tolerance is determined. Be advised that the designer may never give you all of the calculated tolerance. It is sometimes wise to reduce the tolerance. For example, this provides the engineer with an opportunity to allow for deviation to the tolerance specified in situations where a foundry error may not have allowed sufficient material for cleanup. When these situations will not negatively impact the function of the product, a reliability deviation letter will be issued to accept the out-of-tolerance material.

When studying a product design, the designer's primary goal should be simplifying the product. Here the function of the part often determines the best design. Simply determining what a workpiece does sometimes leads a designer to a more efficient approach.

An example of redesigning a workpiece to make manufacturing easier is shown in Figure 20-1. As shown at view A, the cutoff blade was first designed as a one-piece unit machined from 1.125 × .63-inch rectangular bar-stock. The blade was to be hardened, so the material specified was an A2 tool steel.

The initial tooling estimates called for a mill fixture, a grinding fixture, a drill jig, and special form cutters. The mill fixture and a special form cutter were required to machine the radius and angled area of the part. The grinding fixture was used to dress the top and the long side of the cutting edge. The drill jig was needed to drill and counterbore six mounting holes. To reduce possible cracking in the counterbored area, a special counterbore with radiused corners was also specified.

By studying the function and operation of the part, the designer found that the area of the workpiece that did all the work was the .125-inch-wide cutting edge. The cutting action required the blade to move less than .063-inch vertically. The rest of the workpiece was simply a carrier and support for this cutting edge.

Starting with this information, the designer looked for ways to simplify the design to suit the operations actually performed by the workpiece. The result was the design shown at view B. Here the part was a two-piece unit. Since only the cutting edge needed to be hardened, a piece of .125 × 1.125-inch, A2 precision-ground stock was selected for the cutting blade. Using this material reduced both the machining requirements and the grinding requirements. The supporting bar was changed to 1018 carbon steel.

Both the slots in the blade and the angle on the supporting bar were machined in vise-jaw fixtures using standard end mills. Since the supporting bar was not hardened, standard drills and counterbores were used for the mounting holes in the supporting bar. The cutting blades were ground on a surface grinder using parallels and clamps.

The slotted design of the cutting blade allowed the blade to be flipped over when the first side became dull. This doubled the life of the cutting blade. When the complete cutoff blade required changing, only the cutting blade was changed; the original support bar was reused.

Using replaceable components is a common practice. Vendor companies produce many interchangeable components that assemble into and out of a fixture easily. Component libraries are readily available on CD Rom and through Internet sources. Inspection of fixtures after they have finished their required production should be conducted on a regular basis to determine whether replacement of components is needed. Wear characteristics can also be established and, when done in conjunction with statistical process control (SPC), may result in better materials or sources of materials being identified. SPC is a statistical method that helps to identify problem areas in manufacturing and has been overlooked when examining fixture credibility, setup, and teardown. Many problems in this area are still considered just part of the setup.

Another example of simplifying a design is the drive shaft assembly shown in Figure 20–2. Here the original design called for a shaft with a machined spline to be mated with a splined hole. To produce these splines, the shaft was first turned to an exact diameter and milled with a special cutter. The hole was produced with a special broach. As a result of analyzing the operation of these parts, a simplified design using a standard

size of hex stock, matched with a broached hexagonal hole, was used instead of the much more expensive splined arrangement.

The workpieces shown in Figure 20–3 also point out how simple changes made in a design can save many hours of work in manufacturing. The original design, shown at view A, required milling three different levels. However, after the function of the workpiece was studied, a single surface was found to work just as well. This simplified machining the surface and eliminated two steps. The workpiece shown at view B required a hole drilled on an angled surface. Simply

A

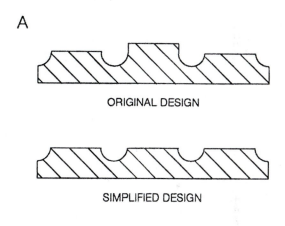

ORIGINAL DESIGN

SIMPLIFIED DESIGN

B

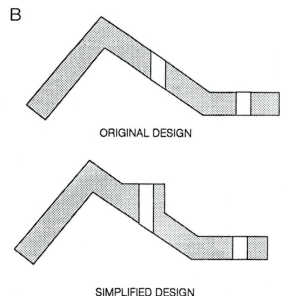

ORIGINAL DESIGN

SIMPLIFIED DESIGN

Figure 20–3 Simple design changes can save many hours of work in manufacturing.

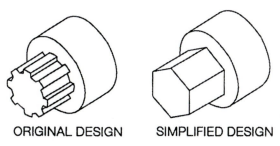

ORIGINAL DESIGN   SIMPLIFIED DESIGN

Figure 20–2 Simplifying the design of a drive shaft assembly.

adding additional material to make the top of the hole flat reduced the required machining time by two thirds.

Cast and forged workpieces are other workpiece styles that require jigs and fixtures. Although widely used, cast and forged parts are some of the most difficult to fixture, largely because of the lack of uniform and consistent locating and clamping surfaces. But with just a little foresight during the initial design, even these parts can be made more manufacturable.

The first step in simplifying the design of cast or forged parts is specifying the locating points. Merely specifying how the part should be located eliminates much of the guesswork for the tool designer. Making these locating decisions during the initial part design also allows the designer to find and fix any design problems that might affect how a part is fixtured.

The casting designer will identify "datum targets" that the tool designer will use to establish a reference frame for manufacturing. Using these datum targets to stabilize the part for manufacturing should be the concern of the tool designer. Metrology may inspect the casting before manufacturing to determine whether the casting will locate into the designed fixture based on this information. The foundry will hold these targets to a closer tolerance to provide the necessary control of features related to these datums. Agreement on these points used to locate the casting is critical to the success of the fixture. As always, there are limitations and concessions that must be made. The casting, while capable of taking many shapes to conform to a prescribed function, may not present a clean pattern for locating points or areas.

One method to simplify the location of complex cast parts is using tabs to locate the workpiece. Here the tabs are cast or forged directly on the workpiece, as shown in Figure 20–4. They are then machined to suit the locational methods and used for locating and clamping the workpiece throughout the machining. Once the workpiece is complete, the tabs are removed. This method is well suited for parts that are not expected to require remachining during their service life.

For parts that may need to be remachined or where the part design does not permit tabs to be added, special tooling holes may be used to locate the workpiece. Here the locating holes are drilled and reamed in nonfunctional areas of the workpiece.

These holes are then used to locate the workpiece for either initial machining or remachining operations.

The tooling holes should be clearly designated on the drawing and on the workpiece to identify them as locating points. One method of identifying these locating holes in the workpiece is with a shallow counterbore at the top and bottom of each one (Figure 20–5).

Workpiece Processing and Setup Reduction

Simplifying the manufacturing processes is another important step in reducing the overall cost of any product. By studying the existing or proposed processes, the designer can often uncover significant savings. In most companies, the part processing is determined by a process engineer rather than by a tool designer. Generally, when the tool designer receives the part print and process sheet, the specific processing has already been determined. Despite this, the tool designer should also confirm the processing specified for the workpiece, and if a better method is found, the process engineer should be informed.

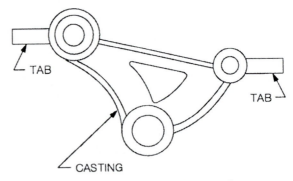

Figure 20–4 Cast or forged tabs can often serve as locational and clamping points on some complex workpieces.

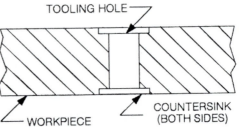

Figure 20–5 Tooling holes may also be used to locate the workpiece.

There is no one correct method to do any job. Almost any machining operation can be performed in any number of ways. The ideal method is the one that gives the desired result at the lowest possible cost. The tool designer and process engineer are the key individuals in determining the tooling and processes to manufacture a part. They are closer to the manufacturing process than the design engineer is, and they have more concentrated expertise in these areas. Downstream from both the process engineer and the tool designer are the programmers of CNC equipment. Production department supervisors and machining technicians also provide insights based on their level of expertise. Together this team of collaborators can have a positive impact on optimizing the design and manufacturing process. Therein lies the value of concurrent engineering.

The team should always look for processing alternatives that will save time or processing steps. If a simple operation can be eliminated or combined with another step, a considerable amount of time can be saved over the complete production run. Two or three minutes saved on each part may not seem like very much, but when you consider a run of several thousand workpieces, the savings are dramatic.

The following are some general points to remember when designing a part for easier manufacturing. These general guidelines will not only simplify manufacturing but also help to reduce the overall cost of making any product.

Maximize Design Simplicity. Every design should be made as simply as possible. Only the elements necessary to the function of the workpiece should be included. A product should be designed for maximum simplicity in both how it is made and how it functions. All unnecessary workpiece or process complexity should be reduced or eliminated at every opportunity.

As shown in Figure 20–6, altering how a feature is made can reduce the cost of making it. Each of these modified features, while much less expensive to make, will still perform the intended function. In the shaft shown at view A, the radius between the two diameters is changed to an undercut. The radius in the initial design makes the part more difficult to grind accurately because both the diameter and the radius must be

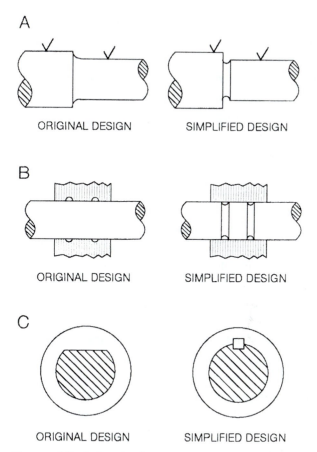

Figure 20–6 Simple changes to processing can reduce the cost of making the feature.

ground together. Using an undercut eliminates this problem and makes grinding the shaft faster and more accurate.

The features shown in the initial designs in the assemblies at views B and C of Figure 20-6 are also very difficult to machine. The "O"-ring grooves are much easier and less expensive to machine on the outside of the shaft than on the inside of the hole for this assembly. Changing the *D*-shaped hole and shaft design to a keyed assembly simplifies this unit and eliminates the need for a special broach.

The complete design should be tailored to the specific functional requirements of the product. Simplifying the product design reduces workpiece complexity and makes the product less expensive to build by reducing both manufacturing and fixturing requirements.

Select Suitable Materials. Workpiece materials should be selected on the basis of functionality, availability, and cost. When possible, specialized or very expensive materials should be avoided. Material selection should be based on the suitability of the material for the expected tasks.

Other considerations, such as machinability ratings, heat-treating operations, and general properties of the materials must also be evaluated. As a rule, where cost is involved, choose the materials that have the lower processing cost, not those that are simply the least expensive. For example, some alloy steels, although initially more expensive than carbon steels, are actually less expensive in the long run. Carbon steels often require additional processing to achieve the same characteristics of alloy steels, thus, where hardening an alloy steel is relatively simple, case-hardening a carbon steel part can be more expensive.

Use the Widest Possible Tolerances. Tight tolerances add only cost to the product—not quality. Tolerances should always be based on part function and requirements. They must also be within the capability of the specified processes. Any extremely close tolerance should be examined to make sure it is really needed.

The tolerances shown in Figure 20–7 are an example of tight tolerances that have questionable

value. Here a chamfer and a radius are dimensioned with very tight tolerances, but in most cases this type of feature does not require them and they should be removed. If a feature requires tight control, then the tolerance should be applied, but remember that tight tolerances are very expensive and should be specified only where they are really needed.

The general tolerances listed in the title block should also be carefully analyzed to make sure the stated values are realistic. Ensure that all surface texture specifications are within the capability of the process and are actually necessary to the function of the part.

Standardize Where Possible. Standardize features, methods, and hardware wherever practical. Always use standard components instead of specialized components to simplify designs and reduce costs. Reduce the inventory of common hardware items such as screws, bolts, and washers by deciding on a limited number of standard sizes and styles of each fastener. Likewise, material selections should be limited to a range of standard types, sizes, and shapes.

Standard size cutters and tooling items should always be used in place of specialty items. Standard cutters are much less expensive than special cutters. Limit the tooling inventory by deciding on standard tap drill sizes, body diameter drills, and counterbores for each thread size. Having one standard set of cutting tool sizes for each thread size allows larger quantities of the same tools to be purchased, thus reducing the per-tool cost.

Larger corporations with more than one manufacturing facility receive a maximum benefit purchasing tooling when they combine their quantities and let bids to their potential providers. Contracts are issued to the most economical provider with multiple release dates based on a needs analysis. Cost is not the only criterion for the purchasing engineer. Quality is an issue that must be considered in the purchasing plan. If production is up and consumption is ahead of schedule, the order can be expedited to bring tools in earlier than previously planned. Agreements for unused surpluses are also negotiated before contact acceptance.

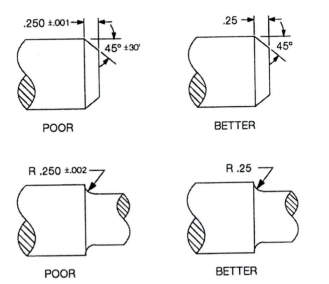

Figure 20–7 Chamfers and radii are part features that seldom require tight tolerances.

Common and repetitive machining operations that do not have a direct bearing on part function should also be standardized. Processes such as chamfering should have standard drawing specifications to reduce the problems encountered with unnecessarily tight chamfer specifications. Figure 20–8 shows how similar features can be standardized where possible. With the chamfered workpiece shown at view A, the 35-degree angle is important and the 45-degree angle is only intended to break the corner. By changing both angles to 35 degrees, the same function is accomplished and a machining operation and a tool change are eliminated. The same change was made to suit the .13-inch decorative radius on the second example, shown at view B.

Internal radii are often specified for recesses, slots, or similar features. Here, if practical, the radii specified should match the cutter size intended to machine the feature (Figure 20–9). This eliminates unnecessary secondary operations required to produce radii of other sizes. Other machining specifications that should be standardized when practical are:

- Undercuts, grooves, "O"-ring seats, and thread runouts
- Countersinking, counterboring, and spotfacing operations
- Decorative or press-fit knurling operations

Reduce Production Steps and Minimize Handling.
Where practical, do as many operations as possible from each setup. When the part design permits, gang or group operations together. Design workpieces so that individual operations can be minimized. If volume allows and standard cutters are unsuitable, special cutters may be used to combine machining operation details.

The workpiece shown in Figure 20–10 illustrates how different operations can be combined to simplify processing. Shown at view A as originally designed, it requires the holes to be drilled and the flange to be milled to provide a seat for the mounting bolts. However, as shown at view B, the milling operation can be replaced with a simple spotfacing operation. This is performed in the same setup as the drilling.

Another example of simplifying the processing is shown with the workpiece in Figure 20–11. As shown at view A, the cast ring has eight drilled and counterbored holes. However, by facing the complete surface while turning the rest of the workpiece, as shown at view B, the counterboring step is eliminated.

Where possible, reduce or eliminate secondary machining or finishing operations. Milling cutters, for example, may sometimes be used for cutting operations instead of sawing and machining the cut surface. Special machining or finishing operations on unimportant areas of the workpiece should also be eliminated.

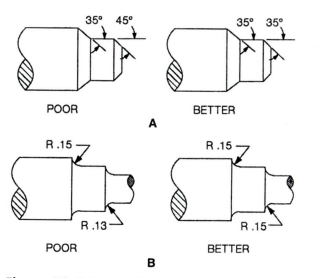

Figure 20–8 Standardizing machining operations can also reduce the cost of processing.

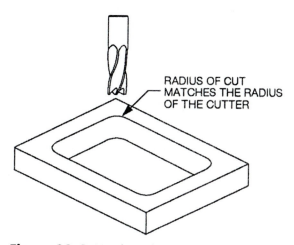

Figure 20–9 Matching the size of specified features to the size of the cutters can eliminate unnecessary secondary operations.

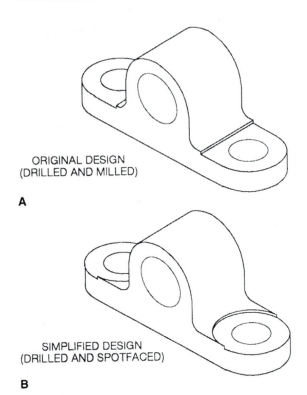

ORIGINAL DESIGN
(DRILLED AND MILLED)

A

SIMPLIFIED DESIGN
(DRILLED AND SPOTFACED)

B

Figure 20-10 Combining drilling and spotfacing to eliminate a milling operation.

The following are a few ideas that should be considered to reduce the production steps and parts handling:

- Design parts for bar-stock where possible. The bar will hold the part during machining.
- If practical, mount parts on fixturing plates.
- Consider using palletized arrangements.
- Where possible, use automatic feeding devices.

Always balance the precision and production needed against the cost and capability of the production methods. Ensure that the production methods are compatible with the specified requirements.

Workholder Design and Setup Reduction

Once the workpiece design and processing are evaluated, the designer looks at ways to design the workholder. Here the key to setup reduction is simplifying tool designs. Where possible and practical, use commercial components to reduce the cost of building the jig or fixture. For components that must be custom-made to suit a workpiece, always keep the design as simple as possible.

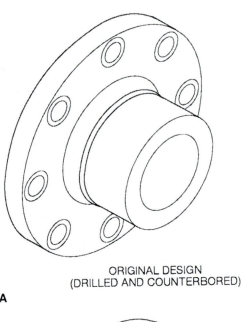

ORIGINAL DESIGN
(DRILLED AND COUNTERBORED)

A

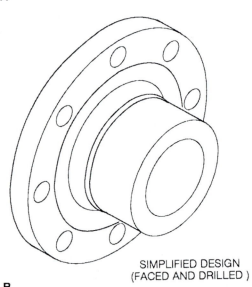

SIMPLIFIED DESIGN
(FACED AND DRILLED)

B

Figure 20-11 Combining a facing step with the required turning operations helps eliminate eight counterbored holes.

In any workholder design, always try to follow the *Golden Rule of Workholding: Grab the workpiece one time, and don't turn it loose until you're finished with all operations.* This may not be possible with every workpiece, but where you can, design a jig or fixture to perform all the required operations without removing the workpiece. This may require a palletized design where the workpiece is moved between machines on pallets.

The goal here is to prevent inaccuracy by maintaining the precise position of the workpiece. Each time a workpiece is removed from one workholder and placed in another, the chance for errors increases. To reduce this possibility, the best practice is to do all operations in one setup.

Workholder Classifications. All workholders can be grouped into three basic classifications. These are dedicated workholders, temporary workholders, and modular workholders. Each category was covered in detail in earlier chapters. The specific selection of which workholder to use is determined by several general factors:

- Type of workpiece to be machined
- Type of operations being performed
- Number of parts per run
- Number of runs expected
- Expected life of workpiece runs

Dedicated workholders are intended mainly for a specific workpiece or a family of similar parts. They are usually permanent, constructed from a series of specialized parts and components, and intended primarily for long production runs of many parts. Dedicated jigs and fixtures are commonly the most expensive type of workholder, but they are typically the most efficient for specialized workpieces.

Temporary workholders are a variation of the dedicated category. They are intended mainly for general-purpose applications and are often built around general-purpose commercial components such as vises, chucks, or collets. Temporary workholders are designed primarily for short production runs of a few parts. They are usually the least expensive form of workholder and work efficiently for short-lived, limited numbers of specialized parts.

Modular workholders are specialized workholders constructed from a series of standard parts. Like dedicated workholders, they are often intended for a specific workpiece or a family of similar workpieces. They may be used for any size production run as either permanent or temporary. Many times the modular form of tooling is the most cost-effective and efficient for infrequent production runs.

Guidelines for Optimal Workholder Design. Many times more than one type of workholder may be used for a particular workpiece. The specific type usually should be based on the expected production volume. Here the two factors to consider are the number of parts expected to be made and the number of times the job will be repeated.

The following factors should be considered in selecting a specific form of workholder to achieve the maximum benefit. It is doubtful that a single workholder design will be best for all these categories. However, the ideal workholder is the one that will meet most of these conditions for a particular workpiece.

- The workholder should have a universal design where practical to be usable for more than one workpiece.
- Workholder design should cut lead time in design and construction of the workholder.
- Workholder design should cut the cost of designing and building the workholder.
- Where possible and practical, the workholder should be reusable.
- The workholder should not become obsolete.
- The workholder design should be easily modified to suit future design changes to the workpiece.
- The workholder design should be economical and pay back the investment within a short time.
- The workholder design must provide and maintain necessary accuracy for the life of the production run.
- The workholder should be economical for any number of workpieces.
- The workholder should be capable of being easily built by both skilled and semiskilled workers.

SUMMARY

The following important concepts were presented in this unit:

- Setup reduction for workholding is a means of reducing the time and cost of manufacturing products by reducing the time required to perform the specified operations.
- Reducing setup expenses offers a manufacturer many benefits. In addition to lower overall

production costs, benefits include lower tooling expenses, reduced lead times, increased production speed, higher production volume, and faster production changeovers.

- The first way to improve jig and fixture design is with the initial product design. Simplified workpiece designs make manufacturing easier and reduce overall tooling and production expenses.
- Studying the manufacturing processes is also an important step in reducing the overall cost of any product.
- Process engineers normally decide the processing, but the tool designer should confirm the processing specified for the workpiece, and inform the process engineer if a better method is found.
- The *Golden Rule of Workholding* is *Grab the workpiece one time, and don't turn it loose until you're finished with all the operations.*
- All workholders can be classified as dedicated, temporary, or modular.
- The specific type of workholder selected for any workpiece should usually be based on the number of parts expected to be made and the number of times the job will be repeated.

REVIEW

1. List four benefits of reducing setup expenses.
2. The largest benefit of setup reduction is increased productivity. What effects does this increased productivity have?
3. What three areas should the tool designer consider when reducing setup expenses?
4. What is the primary goal in studying a product design before designing a workholder?
5. What two methods may be used to simplify the location of complex cast or forged workpieces?
6. List three general guidelines to simplify manufacturing and reduce the overall cost of making a product.
7. What is the *Golden Rule of Workholding*?
8. What are the three basic classifications of workholders?
9. Which form of workholder is intended for general-purpose work and is the least expensive?
10. Which form of workholder is constructed from standardized components to suit specialized workpieces?
11. Which form of workholder is typically the most efficient for specialized workpieces and usually the most expensive?
12. List five factors to consider in selecting a specific form of workholder to achieve the maximum benefit.

UNIT 21

Tool Materials

OBJECTIVES

After completing this unit, the student should be able to:

- Describe the properties and applications of carbon and tool steels.
- Define the characteristics and applications of nonferrous metals.
- Describe the properties and uses of nonmetallic tool materials.

In addition to designing jigs and fixtures, the tool designer is responsible for selecting the proper materials used to build these tools. In many cases, the materials chosen can influence how the tool is constructed.

Adaptability, durability, and economy must be considered before a material is selected for a particular tool. However, before any choice can be made, the designer must have a working knowledge of the properties and characteristics of the materials common to tool construction.

PROPERTIES OF TOOL MATERIALS

The properties of tool materials can best be described as those features that directly influence the behavior of the material while in use. These properties can have either a positive effect or a negative effect, depending on the intended use of the material.

The properties of tool materials that concern the tool designer are hardness, toughness, wear resistance, machinability, brittleness, tensile strength, and shear strength.

Hardness is the ability of a material to resist penetration or indentation. It is also a means to measure one material against another. Normally, the harder the material, the greater its tensile strength. The most widely used methods of measurement are the Rockwell and the Brinell hardness tests. A comparison of these two systems is shown in Figure 21–1.

Toughness is the ability of a material to absorb sudden applied loads or shocks repeatedly without permanent deformation. Hardness controls toughness to approximately Rockwell C44-48 or Brinell 410-453. After these values, brittleness replaces toughness.

Wear resistance is the ability of a material to resist abrasion either by a nonmetallic material or by constant contact with a material of equal hardness. Hardness is also a prime factor in wear resistance. Wear resistance usually increases with hardness.

Machinability is the measure of how well a material can be machined. Factors concerning machinability are cutting speed, tool life, and surface finish. A comparative chart of machinability ratings is shown in Figure 21–2.

APPROXIMATE HARDNESS CONVERSION					
Rockwell "C"	Brinell BHN	Tensile[1] Strength	Rockwell "C"	Brinell BHN	Tensile[1] Strength
65	740[2]	—	41	381	189
64	723[2]	—	40	370	182
63	705[2]	—	39	361	178
62	690[2]	—	38	352	171
61	670[2]	—	37	344	166
60	655[2]	—	36	335	163
59	635[2]	—	35	325	160
58	615[2]	—	34	318	154
57	596[2]	—	33	310	149
56	577[2]	—	32	300	143
55	560[2]	—	31	294	140
54	544[2]	283	30	285	137
53	525[2]	275	29	279	133
52	504	268	28	270	130
51	490	259	27	265	127
50	479	250	26	258	124
49	467	244	25	253	122
48	453	235	24	247	119
47	445	228	23	243	116
46	434	220	22	237	113
45	421	214	21	231	110
44	410	207	20	226	108
43	400	201	(18)[3]	219	106
42	390	195	(16)[3]	212	102

[1] Tensile strength expressed in 1000 psi

[2] Tungsten Carbide Ball (other values use standard steel ball)

[3] Values beyond normal range, for reference only

Figure 21-1 Hardness value comparison.

MACHINABILITY COMPARISON					
Material Type	**Number**	**Mach Rating**	**Material Type**	**Number**	**Mach Rating**
Carbon Steels	1010	65	Nickel-Chromium-Molydbenum Steels	4320	63
	1015	72		4340	57
	1020	72	Nickel-Molybdenum Steels	4620	64
	1030	70		4640	66
	1040	64		4820	53
	1050	55	Chromium Steels	5045	70
	1060	51		5120	75
	1070	49	Carbon Chromium Steels	5140	70
	1080	42		50100	45
	1095	42		52100	40
Free Machining Steels	① 1112	100	Chromium-Vanadium Steels	6102	57
	1115	81		6150	60
	1125	81	Nickel-Chromium Molybdenum Steels	8640	66
	1140	72		8750	60
	1151	70		9310	48
Nickel Steels	2330	55		9747	64
	2340	57	Manganese-Silicon Steels	9260	51
	2515	52	Stainless Steels	303	60
Nickel-Chromium Steels	3120	66		440	37
	3150	60	Aluminum Alloys	1050	300–1500
	3310	51		2024	500–1500
Molybdenum Steels	4037	73	Brass	General	100
	4063	52	Bronze	General	60
Chromium Molybdenum Steels	4130	72	Copper	General	70
	4140	62	Nickel	General	40
	4150	59	Magnesium	General	500–2000

[1] Based on 1112 steel as 100%

Figure 21–2 Machinability comparison.

Brittleness is the opposite of toughness. Brittle materials have the tendency to fracture when sudden loads are applied. For the most part, materials that are very hard are also very brittle.

Tensile strength is the measure of a material's resistance to being pulled apart (Figure 21–3). It is the primary test used to determine the strength of a material. Tensile strength increases proportionally with hardness to approximately Rockwell C54 or Brinell 544. Beyond this point, brittleness makes tensile strength values inaccurate.

Shear strength is the measure of a material's resistance to forces applied in opposite and parallel directions (Figure 21–4). Shear strength is approximately 60 percent of tensile strength.

The primary factors that control the properties of a metallic material are alloying elements and thermal or mechanical treatments. The properties of nonmetallic materials are controlled either naturally or by processing during manufacture. These properties cannot normally be modified after manufacture.

Knowing what the properties are and how one relates to another, while important, is not enough. The tool designer must also know which materials to use to achieve the desired results. The three general forms of tool material are ferrous, nonferrous, and nonmetallic.

FERROUS TOOL MATERIALS

Ferrous tool materials include cast iron, carbon steel, alloy steel, and tool steel. These metals have a base of

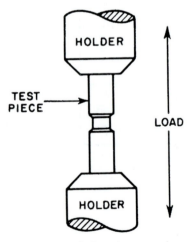

Figure 21–3 Tensile strength.

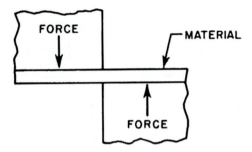

Figure 21–4 Shear strength.

iron. Ferrous metals make up the largest group of tool material in common use. *Cast iron*, *alloy steel*, and *tool steel* are general terms that cover a wide range of materials. A more specific study of each group is needed to fully understand its limitations and capabilities.

Cast Iron

Cast iron is used for tool bodies and some commercial jig and fixture components. It is generally being replaced by other materials that are less expensive and take less time to fabricate into tools. The chief disadvantage of cast iron is the large amount of lead time required. Before a cast tool body can be used, it must be made into a pattern and then a mold, and then poured. These added steps not only take longer but also cost more. Other materials, such as preformed sections, are less expensive and more efficient for the majority of jig and fixture work.

Carbon Steel

Carbon steel is the primary material of jig and fixture tooling. Its ease of fabrication, low cost, availability, and versatility have made it popular for tool construction. The three main types of this steel are low carbon, medium carbon, and high carbon.

Low carbon steels are used mainly for structural parts of a jig or fixture. They should be used only in areas where mass is required and no wear or stress will occur, such as base plates or supports. The carbon content of this steel is between .05 and .30 percent. Low carbon steel can be case-hardened to resist wear for low-production tools. It is also easily welded or joined by any standard process.

Medium carbon steels are used in much the same way as low carbon steels, but in areas of tooling that require more strength. Medium carbon steels work well as clamps, studs, nuts, and in almost any area where toughness is required. The carbon content of medium carbon steel is between .30 and .50 percent, which allows the material to be hardened by case-hardening or other conventional hardening processes. Medium carbon steel is more difficult to machine, case-harden, and weld than low carbon steel. In other characteristics, it is almost the same. Because of the increased cost of this material, it should be limited to areas where it would be most effective.

High carbon steels are generally limited to tool construction in areas that are subject to the most wear. Parts such as drill bushings, locators, wear pads, and supports can be made of this material. High carbon steels have a carbon content between .50 and 2.0 percent. They are easily hardened by the conventional processes, but do not resist wear as well as most tool steels do. For this reason, the use of high carbon steel for high-wear parts should be carefully analyzed before the parts are made. This steel welds poorly except under very carefully controlled conditions.

Generally, carbon steels are versatile, adaptable, and cost-efficient to use for tool construction. While the machinability and weldability of these materials decreases with increased carbon content, carbon steels are the best overall tool material.

Carbon steels are available in bars, strips, sheets, and many special shapes. They are also available in a variety of conditions such as cold-rolled, cold-drawn, hot-rolled, or ground. Normally the intended use of the material will decide the shape and condition of the steel selected. If the part is to be used as furnished, then the ground surface might be best. For parts to be machined, the surface condition is generally unimportant. In most cases, carbon steels are supplied in either an annealed state or a normalized state unless otherwise specified by the buyer.

Alloy Steels

Alloy steels are not generally used for tool construction because of their added cost. However, a brief discussion of the effects alloying elements have on steel will provide a better understanding of tool steels.

Alloying elements are elements that, when added to a material, change or modify the material and bring about a predictable change in its properties. The most common alloying elements and their effects are:

Carbon	— main hardening element
Sulfur	— easier machining
Phosphorus	— easier machining
Manganese	— controls sulfur effects; increases hardness
Nickel	— toughness
Chromium	— corrosion resistance (over 15 percent chromium); depth hardness (less than 15 percent chromium)
Molybdenum	— depth hardness; strength at elevated temperature; toughness
Vanadium	— grain refiner
Tungsten	— hardenability; strength at elevated temperature; hardness
Silicon	— fluidity; deoxidizer
Silicon and Manganese	— work hardenability
Aluminum	— deoxidizer
Boron	— hardenability
Lead	— machinability
Copper	— corrosion resistance
Cobalt	— hardness

As an aid in identifying and standardizing the great variety of carbon and alloy steels, a universal numbering system has been adopted that was developed by the Society of Automotive Engineers (SAE) and the American Iron and Steel Institute (AISI). It is used throughout the manufacturing industry as a standard for metals identification.

The basis of this numbering system is a four-digit code that indicates specific information about the metal. The first digit indicates the type of metal. For example, 1 indicates carbon steel; 2, nickel steel; 3, nickel-chromium steel; and so forth, Figure 21–5. The second digit indicates either the percentage of major alloy in the metal or a code to denote a specific alloy. The last two digits when read together express the carbon content in hundredths of a percent.

```
            BASIC NUMBERING SYSTEM FOR STEELS
  1XXX . . . . . . . . . . . . .CARBON STEELS
    1xxx . . . . . . . . . . . . .Plain Carbon
    11xx . . . . . . . . . . . . .Resulfurized Free Machining
    12xx . . . . . . . . . . . . .Resulfurized, Rephosphorized Free Machining
    13xx . . . . . . . . . . . . .Manganese Steels
    15xx . . . . . . . . . . . . .High Manganese Steels
  2XXX . . . . . . . . . . . . .NICKEL STEELS
    23xx . . . . . . . . . . . . .3.5% Nickel
    25xx . . . . . . . . . . . . .5.0% Nickel
  3XXX . . . . . . . . . . . .NICKEL-CHROMIUM STEELS
    31xx . . . . . . . . . . . . .1.25% Nickel; .60% Chromium
    32xx . . . . . . . . . . . . .1.75% Nickel; 1.0% Chromium
    33xx . . . . . . . . . . . . .3.50% Nickel; 1.5% Chromium
    30xxx. . . . . . . . . . . . .Corrosion and Heat Resisting Steels
  4XXX . . . . . . . . . . . .MOLYBDENUM STEELS
    40xx . . . . . . . . . . . . .Carbon-Molybdenum
    41xx . . . . . . . . . . . . .Chromium-Molybdenum
    43xx . . . . . . . . . . . . .Chromium-Nickel-Molybdenum
    46xx . . . . . . . . . . . . .Nickel-Molybdenum
    48xx . . . . . . . . . . . . .Nickel-Molybdenum
  5XXX . . . . . . . . . . . . .CHROMIUM STEELS
    51xx . . . . . . . . . . . . .Low Chromium
    51xxx. . . . . . . . . . . . .Corrosion and Heat Resisting Steels
    52xxx. . . . . . . . . . . . .Medium Chromium
  6XXX . . . . . . . . . . . .CHROMIUM-VANADIUM STEELS
    61xx . . . . . . . . . . . . .1.0% Chromium; .15% Vanadium
  86XX. . . . . . . . . . . .NICKEL-CHROMIUM-MOLYBDENUM STEELS
  87XX. . . . . . . . . . . .NICKEL-CHROMIUM-MOLYBDENUM STEELS
  93XX. . . . . . . . . . . .NICKEL-CHROMIUM-MOLYBDENUM STEELS
  98XX. . . . . . . . . . . .NICKEL-CHROMIUM-MOLYBDENUM STEELS
  92XX. . . . . . . . . . . .MANGANESE-SILICON STEELS
  94XX. . . . . . . . . . . .MANGANESE-NICKEL-CHROMIUM-MOLYBDENUM STEELS
```

Figure 21–5 Basic numbering system for steels.

Figure 21–6 shows a few examples of how this system works.

In cases where there is more than .99 percent carbon, a fifth digit is added. Figure 21–7 shows how this digit is read. A fifth digit is also applied for distinction between some of the corrosion-resistant alloys.

Tool Steels

Tool steels are steels that are made to exact standards for specific types of service. Tighter control during manufacture produces steels that are very predictable and reliable.

Tool steels are used in jig and fixture work for parts that are highly stressed or that must have a higher wear resistance. While carbon or alloy steels can be used, the superior properties of tool steel make

them the best choice. Figure 21–8 shows a comparison of the characteristics of tool steels.

Tool steels are generally identified by a letter and number code developed by the AISI. This system establishes the standard composition for each steel within a group and assigns each a number. Tool steel manufacturers conform to the specifications for each tool steel they make. In other words, an 01 tool steel made by one company should have the same composition as an 01 tool steel made by any other company. Many companies have names for each of their tool steels that, in some cases, have no relation to their composition. Tool steel selection for the most part is finding a steel that works well for its intended use.

The AISI identification system groups tool steels into seven general classes (Figure 21–9). Each type of steel within the class is further identified by letter

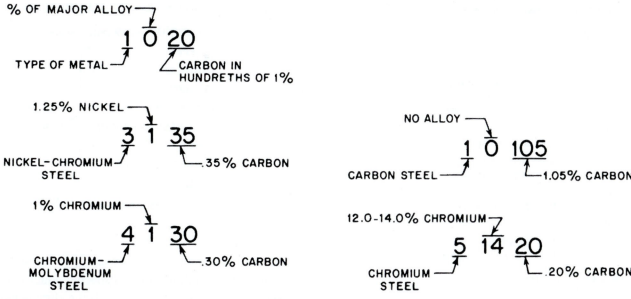

Figure 21-6 Reading the four-digit code for steel.

Figure 21-7 Reading the five-digit code for steel.

TOOL STEEL CHARACTERISTICS						
TYPE	Toughness	Wear Resistance	Resistance to Heat	Non-Deforming Properties	Machin- ability	Cost
T	P	G	G to E	G	F	5
M	P	E	G	G	F	4
H	G	F	G	G	F	4
D	P to F	E	F	G to E	P	3 to 4
A	F	G	P to F	G to E	F	3
O	F	G	P	G	G to E	2
S	G to E	F	P to F	P to F	F	3 to 4
P	G	F	P	P	G to E	3
L	F	G	P	F	F	2
F	P	E	P	P	F	2
W	G	G	P	P	E	1

Legend: E = Excellent; G = Good; F = Fair; P = Poor
5 = Highest; 4 = Med-High; 3 = Medium; 2 = Low; 1 = Lowest

Figure 21-8 Tool steel characteristics.

TOOL STEEL CLASSIFICATIONS

Water Hardening — W

Shock Resisting — S

O — Oil Hardening

Hot Work — H — Chromium Base
Tungsten Base
Molybdenum Base

Cold Work — A — Air Hardening
Medium Alloy

D — High Carbon
High Chromium

High Speed — T — Tungsten Base
M — Molybdenum Base

Mold Work — P

Special Purpose — L — Low Alloy
F — Carbon-Tungsten Base

Figure 21-9 Tool steel classifications.

BASIC NUMBERING SYSTEM FOR ALUMINUM

1XXX . . . . .ALUMINUM — 99% pure and greater

2XXX . . . . .COPPER

3XXX . . . . .MANGANESE

4XXX . . . . .SILICON

5XXX . . . . .MAGNESIUM

6XXX . . . . .MAGNESIUM-SILICON

7XXX . . . . .ZINC

8XXX . . . . .OTHER ELEMENTS

9XXX . . . . .UNUSED

MAJOR ALLOY

Figure 21-10 Basic numbering system for aluminum.

and number. These numbers are in sequence, and they have no relation to the composition of the steel. To determine the composition of tool steels, handbooks or manufacturers' data should be consulted.

NONFERROUS TOOL MATERIALS

Nonferrous tool materials such as aluminum, magnesium, and bismuth alloys are metals that have a base metal other than iron. While not as widely used for jigs and fixtures as ferrous materials are, nonferrous tool materials have the advantages of weight, stability, and workability. Because of these advantages, the use of nonferrous tool materials is increasing as the demand for low-cost versatile tooling increases.

Aluminum

Aluminum is the most widely used nonferrous tool material. The primary reasons for this are machinability, adaptability, and weight. Aluminum is available in a wide variety of forms, which further increases its usefulness.

The primary forms used for jig and fixture work are aluminum tool plates and extrusions. Aluminum tool plate is available in a variety of sizes. It is accurate to within ±0.13 millimeter over a span of 2500 millimeters. Aluminum extrusion is also quite accurate, often within ±0.05 millimeter of a specific dimension.

Another advantage of aluminum is the elimination of heat treatment or processing to increase its hardness or stability. Aluminum is normally ordered in the condition needed. Further treatments are unnecessary, saving both time and money. Aluminum can be welded or joined by mechanical fasteners.

Aluminum is identified by the Aluminum Association (AA) numbers system, which closely resembles the SAE/AISI system for steels. To identify aluminum and its alloys, the AA system uses the number codes shown in Figure 21–10. The first digit indicates the major alloy type: 1 is pure aluminum; 2 is copper; 3 is manganese; and so forth. The second digit for pure aluminum indicates the impurities that are under special control. For alloy aluminum, the second digit indicates the number of modifications. The last two digits are read together for pure aluminum. They indicate the percentage of purity in hundredths of 1 percent, over 99 percent. With alloys, the last two digits are simply numbers in a series to indicate the specific alloy. Figure 21–11 shows a few examples and how each is read.

Aluminum and its alloys have temper designations, which should be read along with the four-digit codes. These designations, as shown in Figure 21–12, indicate the exact condition of the alloy being considered. Further information about aluminum identification can be obtained from handbooks or manufacturers' publications.

Magnesium

Magnesium is another nonferrous tool material that is gaining popularity as a material for jigs and fixtures. This metal is very lightweight and versatile, and it has a high strength-to-weight ratio.

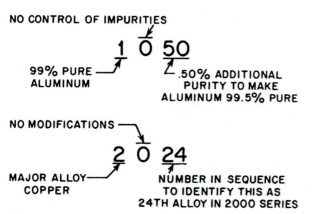

Figure 21-11 Reading the four-digit code for aluminum.

Magnesium can be machined much faster than aluminum or steel. A comparison of magnesium and other materials is shown in Figure 21–13. Magnesium can be welded or joined by mechanical devices.

The one drawback to using magnesium is the potential fire hazard. However, as long as the chips are coarse and the proper cutting fluid is used, fire hazard is greatly reduced. When machining any form of magnesium, it is a good practice to keep a supply of dry powder or sand on hand to put out fire.

Magnesium alloys are identified by a system developed by the American Society for Testing Materials (ASTM) that uses the alloy composition as the basis for identification. Figure 21–14 lists the principal alloying elements. The first one or two letters tell which alloys are used. The numbers indicate the approximate percentage of each alloy in the metal. The final letters indicate the alloy's position with regard to the other metals in the group. For example,

TEMPER/CONDITION DESIGNATIONS FOR ALUMINUM

F — As fabricated

O — Annealed

W — Solution heat treated

H — Strain hardened

    H1 — Strain hardened only

    H2 — Strain hardened and partially annealed

    H3 — Strain hardened and stabilized

    When a second digit follows an 'H' designation it indicates the degree of strain hardness. 2 = 1/4 hard; 4 = 1/2 hard; 6 = 3/4 hard; and 8 = full hard.

T — Thermally treated to produce a temper other than F, O, W, or H.

    T2 — Annealed (cast products only)

    T3 — Solution heat treated and cold worked

    T4 — Solution heat treated and naturally aged

    T5 — Artificially aged

    T6 — Solution heat treated and artificially aged

    T7 — Solution heat treated and stabilized

    T8 — Solution heat treated, cold worked, and artificially aged

    T9 — Solution heat treated, artificially aged, and cold worked

    T10 — Artificially aged and cold worked

Figure 21-12 Temper/condition designation for aluminum.

TOOL MATERIAL COMPARISON			
Material	Cutting Rate m/min	Weight[1] kg/m^3	Power Requirements
Magnesium	213	1746	100%
Aluminum	122	2707	180%
Steel	9-31	7849	630%

[1] Based on volume of 1.m^3 of magnesium

Figure 21-13 Magnesium as compared with other tool materials.

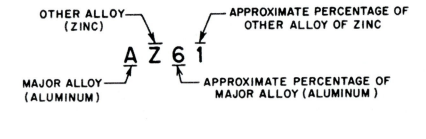

A - ALUMINUM   M - MANGANESE
B - BISMUTH   N - NICKEL
C - COPPER   P - LEAD
D - CADMIUM   Q - SILVER
E - RARE EARTHS   R - CITROMIUM
F - IRON   S - SILICON
G - MAGNESIUM   T - TIN
H - THORIUM   Y - ANTIMONY
K - ZIRCONIUM   Z - ZINC
L - BERYLLIUM

Figure 21-14 Magnesium identification system.

in the designation A 293A, *A* means it is the first alloy to receive the A293 designation, B is the second, and so forth.

Bismuth Alloys

Bismuth alloys are primarily used in jig and fixture work in the form of low-melt alloys. They are used for many special holding devices, such as nests or vise jaws.

The principal alloying elements commercially used with bismuth are lead, antimony, indium, cadmium, and tin, all of which are strong and capable of forming exact cast shapes. This makes them useful for intricate nests and special workholders.

NONMETALLIC TOOL MATERIALS

Nonmetallic tool materials have become an important part of jig and fixture work. Many times tools that are intended for a limited production run can be made faster, less expensively, and better with such materials. The materials used in jig and fixture work are wood, urethane, and epoxy or plastic resins.

Wood

Wood is used for limited production tools that do not require precise accuracy. It is normally used in several forms, including plywood, composition board (chipboard), impregnated wood, and natural wood.

Each of these forms performs well for tooling when properly used. For tools that are subject to excessive moisture, wood should be sealed and treated to prevent swelling. Natural wood should be laminated with the grain, as shown in Figure 21–15, so that warping forces counteract each other.

Many special bushings and inserts are available for use with wood tools. These bushings have external serrations that hold well when the bushing is either pressed or glued into a wood tool.

Urethane

Urethane is used in jig and fixture work for nonmar applications or secondary clamping. Its main benefit is its controllable deflection. For example, the clamp shown in Figure 21–16 must hold the part firmly without damaging the surface. A urethane pad will transfer the force while preventing the clamp from scratching the part. Secondary action clamps use the deflection capabilities of urethane to good advantage (Figure 21–17).

Epoxy and Plastic Resins

Epoxy and plastic resins are finding increasing use in jig and fixture work for special workholders. Nest and chuck jaws made of epoxy or plastic resins are tough, versatile, and inexpensive.

These compounds are normally available as ready to mix and use. Some are intended for casting, while others are in paste form. These materials can be used as is or with fillers mixed in for strength or wear resistance. Filler materials include glass beads, ground glass, steel shot, steel filings, stones, or ground walnut shells.

Resins are lightweight, strong, tough, and normally nondeforming. They also reproduce shapes with little or no shrinkage. As with woods, special serrated

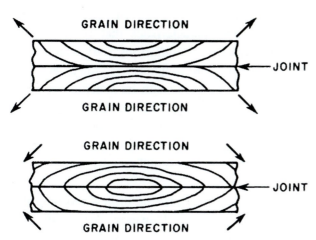

Figure 21–15 Proper grain relation to prevent warping.

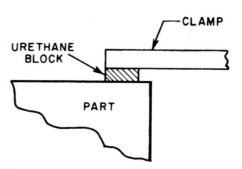

Figure 21–16 Using urethane to prevent scratching work.

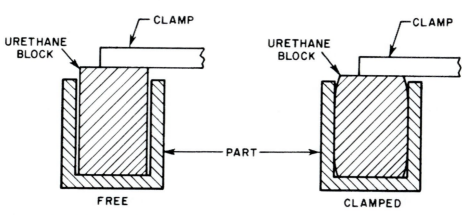

Figure 21–17 Secondary action urethane clamp.

bushings and inserts are available to further increase the versatility of these compounds. However, serrated inserts are not always needed. Some compounds, when completely set, can be used for drilling and tapping with satisfactory results.

Specialty Materials

The wide array of materials used for today's manufactured products often require that specialty materials be used when constructing special tooling devices. Because of their basic construction or processing operations, these products can call for tools with attributes that are not easily achieved with the selection of tooling materials presently available. Whether these applications require nonsparking features, nonmagnetic properties, or specific electrical or thermal conductivity ratings, other types of tool materials are often required.

In these situations, most tool designers have relied on the variety of existing nonmetallic materials. From plastics and wood to a wide array of composite and specialty materials, designers have attempted to incorporate a host of unconventional materials into their designs. Unfortunately, despite their nonmetallic composition, these materials often present a new series of problems. Quite often many of the older nonmetallic materials are very brittle; can be difficult to machine; or may lack sufficient strength, rigidity, and wear resistance to perform as required for jig and fixture applications. Likewise, in addition to lacking many of the necessary physical and mechanical properties needed to perform well for tooling applications, these materials may not stand up well in a normal machining environment.

Fortunately, today there are a selection of materials that offer both a practical and a workable alternative to either metallic and conventional nonmetallic materials for tooling applications. One example of this material is *Richlite*® *Fibre Laminate*. This material is a versatile phenolic wood fiber laminate composite that incorporates many of the best attributes and qualities of wood, metal, and plastic into a single nonmetallic material. *Richlite* is very well suited for many workholding applications, special-purpose tools, and tooling elements. In fact, just about any jig, fixture, or other specialty tool that would normally be

built with metal can be made from this very versatile and adaptable nonmetallic material.

Richlite offers excellent machinability, a high strength-to-weight ratio, very low thermal expansion, excellent electrical and heat insulation properties, and high abrasive chemical and corrosion resistance. It is also very hard, yet easy to machine with the standard tools and cutters found in the shop. Either normal metalworking or woodworking equipment may be used to cut the material to size and machine the various details. Although carbide-cutting tools are usually recommended for longer tool life, high-speed steel tools can also be used. Unlike many other nonmetallic materials that produce a fine dust during machining, using sharp tools and the proper speeds and feeds with this material creates chips that resemble a coarse sawdust. This all but eliminates the need for special respirators or expensive dust removal equipment.

In addition to these general properties, *Richlite* possesses many characteristics that compare quite favorably with most nonferrous tool materials. It offers good strength, toughness, rigidity, and wear resistance. Furthermore, with its specific gravity of 1.24, this material is approximately half the weight of aluminum. Depending on how the sheets are cut, the material offers tensile strength ratings between 12,770 psi and 25,000 psi and compressive strengths of up to 30,650 psi.

The material may also be drilled and tapped to suit threaded fasteners. However, despite its good wear resistance and overall strength, thread inserts should be installed if the screws are intended for adjustment or when they will be removed and replaced repeatedly. These inserts may be threaded directly into the tool body or cast in place with an epoxy compound (Figure 21–18).

When casting these inserts into the tool body, an oversized hole should be drilled and the insert should be installed in the hole. The space between the hole wall and the insert is then filled with the epoxy compound. When used for drill jigs or other tools where a drill bushing is needed, either a serrated bushing or a knurled bushing may be used. The serrated bushing may be pressed directly into the mounting hole, while the knurled bushings are cast in place in the same fashion as that used for the thread insert, as shown in

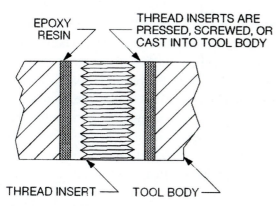

Figure 21–18 Thread insert.

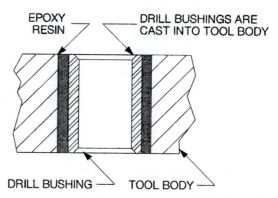

Figure 21–19 Drill bushing cast into tool body.

Figure 21–19. Likewise for applications where dowels are needed, either groove pins or roll pins should be used.

Building workholders and other special-purpose tools from this material is quite simple and often faster than is possible with many standard tool materials. *Richlite* is available in standard 48.00" by 96.00" sheets in thicknesses between 1/8" and 3.000". Sheets less than 1.00" thick are made in 1/8" increments, and sheets between 1.00" and 3.00" are made in 1/4"increments. Cut sheets are available on request, and special, nonstandard thicknesses may also be ordered. This array of sizes permits individual tools to be built from a variety of different pieces, rather than a single thick piece, machined to the necessary tool or workholder configuration. Simply sand the joined surfaces lightly and apply a coating of any

resorcinol or epoxy adhesive to permanently join the various pieces.

Figure 21–20 shows a sample tool body fabricated from individual pieces of *Richlite*. Depending on the complexity of the tool body, these pieces may be rough cut and joined together into a shape that approximates the tool body. Once the adhesive is cured, the tool body can then be machined to the exact size and shape. With simpler tools, the individual pieces may be machined to their exact sizes and then assembled.

Richlite is well suited for any number of different applications. In addition to being used to fabricate complete tool bodies, it can be used for a wide range of tooling elements where the unique attributes of *Richlite* are better suited than other metallic or nonmetallic materials. Its light weight makes it quite

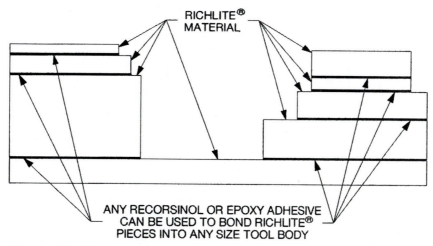

Figure 21–20 Tool body fabricated from *Richlite*.

valuable for building large templates, customizable faces for tombstone tooling, vacuum chuck faces, or similar applications where other materials might be too heavy. Smaller pieces of the material may be used for nonmarring clamping elements, chuck inserts, or vise jaws. It can also be used as other applications where abrasive resistance properties are required.

DESIGNING WITH RELATION TO HEAT TREATMENT

Tool designers normally do not concern themselves with the actual mechanics of heat treating. However, a general understanding of how this process affects design is important. Occasionally failures that seem to be due to improper heat treatment are in fact caused by poor design. The primary factors that control how a part will react during heat treatment are the material itself and the design features of the part.

Tool Material

The material selected for a particular part has a great deal to do with how well the part responds to heat treatment. Such material must be carefully selected and matched to its intended use. New materials should be thoroughly analyzed and tested before use. Manufacturers' representatives or industrial distributors should be questioned about the capabilities and limitations of new materials before the material is used for a tool. Do not be caught up in the belief that since a material is new, it must be better. Weigh each material against its past performance with similar parts and choose the one that best suits the need.

Design Features

The way the part is designed has a major effect on how well it responds to heat treatment. Even if the perfect material were used, a poorly designed part would still fail during either heat treatment or use. Several factors must be considered, but the three most common errors in design are unequal mass, sharp corners, and poor surface conditions.

Unequal mass causes the part to cool at varying rates. This could result in cracking or total failure. Whenever possible, the part should be fairly uniform in cross section. It may be better to make a part in more than one piece to avoid heat-treating problems. Figure 21–21 illustrates how unequal mass can be avoided.

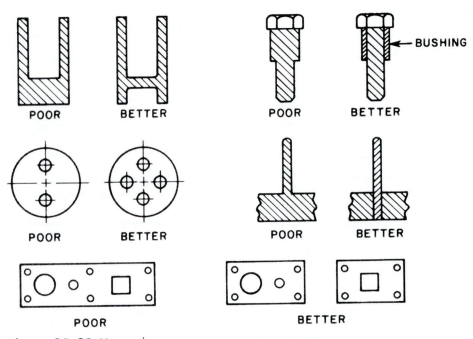

Figure 21–21 Unequal mass.

Sharp corners are prime spots for tool failure. Whenever possible, corners and fillets should be rounded, not sharp. Even countersunk or counterbored holes should not have sharp edges. Sharp edges and corners set up stresses that could cause a part to crack when cooled. Figure 21–22 illustrates ways to avoid sharp corners.

Poor surface condition results from tool marks, burrs, or any other form of surface irregularity (Figure 21–23). Scratches and burrs set up stresses on the surface of a material that could affect the entire part.

In most cases of tool failure, the first reaction is to fault the heat treatment. Before doing this, the tool designer should reevaluate the material and the design.

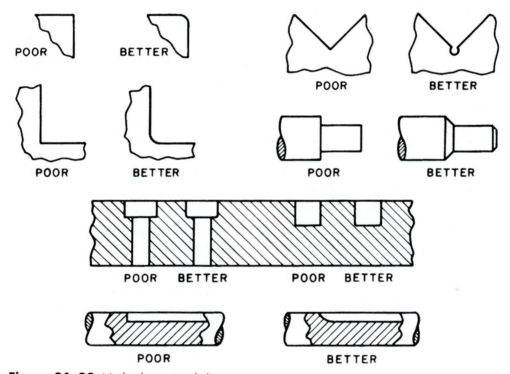

Figure 21-22 Methods to avoid sharp corners.

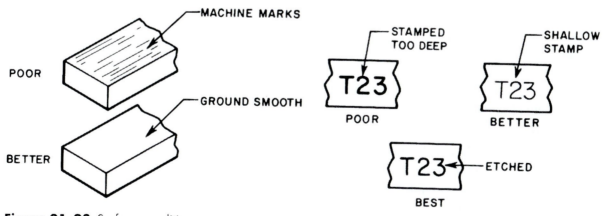

Figure 21-23 Surface conditions.

SUMMARY

The following important concepts were presented in this unit:

- Tool designers must select the proper materials to ensure the suitability of their tool designs. The major considerations here are the adaptability, durability, and economy of the materials.

- The properties of tool materials are the features that influence the behavior of the material. The properties that should be considered are:
  - Hardness
  - Toughness
  - Wear resistance
  - Machinability
  - Brittleness
  - Tensile strength
  - Shear strength

- The materials used for workholders are divided into three general groups: ferrous, nonferrous, and nonmetallic.

- Ferrous materials are those whose major alloying element is iron. These materials include:
  - Cast iron
  - Carbon steels
  - Alloy steels
  - Tool steels

- Nonferrous materials are those whose major alloying element is any metal but iron. These materials include:
  - Aluminum
  - Magnesium
  - Bismuth alloys

- Nonmetallic materials are those that have no metal in their composition. These materials include:
  - Wood
  - Urethane
  - Epoxy and plastic resins

- One primary concern in preparing the tooling materials is the proper design of elements that are to be heat-treated. Both the tool material and the design features must be carefully analyzed to ensure the desired results.

REVIEW

1. Match the property with the definition that best describes it.
   a. The ability to absorb sudden applied loads or shocks.
   b. The measure of tool life, cutting speed, and surface finish.
   c. Resistance to being pulled apart.
   d. The property that is the direct opposite of toughness.
   e. The ability of a material to resist abrasion.
   f. The measure of resistance to penetration.
   g. The resistance of forces applied in opposite and parallel directions.
      1. Hardness
      2. Toughness
      3. Wear resistance
      4. Machinability
      5. Brittleness
      6. Tensile strength
      7. Shear strength

2. What two factors control the properties of metallic tool materials?

3. What are ferrous metals?

4. What are the three types of carbon steel used for jigs and fixtures?

5. Briefly describe the effects each of these elements has on alloy steel:
   a. Aluminum
   b. Boron
   c. Carbon
   d. Chromium
   e. Cobalt
   f. Copper
   g. Lead
   h. Manganese
   i. Molybdenum
   j. Nickel
   k. Phosphorus
   l. Silicon
   m. Silicon and manganese
   n. Sulfur
   o. Tungsten
   p. Vanadium

6. Explain the meaning of the following types of steel:
   a. 1020 steel
   b. 10105 steel
   c. 4340 steel
7. To which class of tool steels do W, O, A, and M belong?
8. What nonferrous tool materials are common to jig and fixture design?
9. Which nonferrous material best fits the listed descriptions?
   a. Lightest
   b. Heaviest
   c. One-third the weight of steel
   d. Low-melt alloy
   e. Highest cutting speed
   f. Most widely used
10. What nonmetallic tool materials are common to jig and fixture work?
11. Which nonmetallic tool material best fits the listed descriptions?
    a. Must be sealed to prevent swelling
    b. Can be drilled and tapped
    c. Normally used for nests
    d. Requires special bushings
    e. Uses ground walnut shells as a filler
    f. Good for secondary action clamping
12. What two elements must be considered in designing a part to be heat-treated?
13. What are the primary design errors?
14. Who should be consulted about a tool material's problem?

APPENDIX

SI to U.S. Customary Conversion Table

mm	Decimal	mm	Decimal	mm	Decimal	mm	Decimal	mm	Decimal
0.01	.00039	0.41	.01614	0.81	.03189	21	.82677	61	2.40157
0.02	.00079	0.42	.01654	0.82	.03228	22	.86614	62	2.44094
0.03	.00118	0.43	.01693	0.83	.03268	23	.90551	63	2.48031
0.04	.00157	0.44	.01732	0.84	.03307	24	.94488	64	2.51969
0.05	.00197	0.45	.01772	0.85	.03346	25	.98425	65	2.55906
0.06	.00236	0.46	.01811	0.86	.03386	26	1.02362	66	2.59843
0.07	.00276	0.47	.01850	0.87	.03425	27	1.06299	67	2.63780
0.08	.00315	0.48	.01890	0.88	.03465	28	1.10236	68	2.67717
0.09	.00354	0.49	.01929	0.89	.03504	29	1.14173	69	2.71654
0.10	.00394	0.50	.01969	0.90	.03543	30	1.18110	70	2.75591
0.11	.00433	0.51	.02008	0.91	.03583	31	1.22047	71	2.79528
0.12	.00472	0.52	.02047	0.92	.03622	32	1.25984	72	2.83465
0.13	.00512	0.53	.02087	0.93	.03661	33	1.29921	73	2.87402
0.14	.00551	0.54	.02126	0.94	.03701	34	1.33858	74	2.91339
0.15	.00591	0.55	.02165	0.95	.03740	35	1.37795	75	2.95276
0.16	.00630	0.56	.02205	0.96	.03780	36	1.41732	76	2.99213
0.17	.00669	0.57	.02244	0.97	.03819	37	1.45669	77	3.03150
0.18	.00709	0.58	.02283	0.98	.03858	38	1.49606	78	3.07087
0.19	.00748	0.59	.02323	0.99	.03898	39	1.53543	79	3.11024
0.20	.00787	0.60	.02362	1.00	.03937	40	1.57480	80	3.14961
0.21	.00827	0.61	.02402	1	.03937	41	1.61417	81	3.18898
0.22	.00866	0.62	.02441	2	.07874	42	1.65354	82	3.22835
0.23	.00906	0.63	.02480	3	.11811	43	1.69291	83	3.26772
0.24	.00945	0.64	.02520	4	.15748	44	1.73228	84	3.30709
0.25	.00984	0.65	.02559	5	.19685	45	1.77165	85	3.34646
0.26	.01024	0.66	.02598	6	.23622	46	1.81102	86	3.38583
0.27	.01063	0.67	.02638	7	.27559	47	1.85039	87	3.42520
0.28	.01102	0.68	.02677	8	.31496	48	1.88976	88	3.46457
0.29	.01142	0.69	.02717	9	.35433	49	1.92913	89	3.50394
0.30	.01181	0.70	.02756	10	.39370	50	1.96850	90	3.54331
0.31	.01220	0.71	.02795	11	.43307	51	2.00787	91	3.58268
0.32	.01260	0.72	.02835	12	.47244	52	2.04724	92	3.62205
0.33	.01299	0.73	.02874	13	.51181	53	2.08661	93	3.66142
0.34	.01339	0.74	.02913	14	.55118	54	2.12598	94	3.70079
0.35	.01378	0.75	.02953	15	.59055	55	2.16535	95	3.74016
0.36	.01417	0.76	.02992	16	.62992	56	2.20472	96	3.77953
0.37	.01457	0.77	.03032	17	.66929	57	2.24409	97	3.81890
0.38	.01496	0.78	.03071	18	.70866	58	2.28346	98	3.85827
0.39	.01535	0.79	.03110	19	.74803	59	2.32283	99	3.89764
0.40	.01575	0.80	.03150	20	.78740	60	2.36220	100	3.93701

Glossary

Accumulator. A device installed between the clamps and the power source in a power-workholding system. It is charged with either fluid or gas and maintains the necessary pressure in the system when the power is disconnected.

Adjustable Locator. A locator used on jigs and fixtures and made to be adjusted after installation on the tool body.

Adjustable Support. A support, or locator, positioned under the part and made to be adjusted to suit individual parts.

Air-Assisted Hydraulic System. A form of power-workholding system where compressed air is used to activate the hydraulic pressure through a booster. This is the most common form of power-workholding system.

Allowance. The intentional difference between the sizes of mating parts in an assembly.

Alloy. A mixture of metals that have been combined to form the alloy in order to gain desired characteristics.

Alloy Steel. A steel made up of several substances in addition to carbon and iron.

American Sequence. A sequence of information in a feature control symbol. The sequence is geometric characteristic symbol, datum reference, and tolerance value.

Angle-Plate Fixture. A modified form of plate fixture used to machine a part at an angle to its locator.

Angle-Plate Jig. A modified form of plate jig used to machine a part at an angle to its locator.

Angularity. The geometric characteristic that specifies a specific angular relationship between two surfaces of a part.

Assembly Drawing. A mechanical drawing that shows all the parts and elements of a tool in their assembled form.

Basic Dimension. A theoretically perfect dimension used to locate part features or datums. This type of dimension is shown enclosed in a box on the part drawing.

Basic Size. The size of a part to which the tolerance values are applied to obtain the limits of size.

Bilateral Tolerance. The tolerance value that may go in both a plus and a minus direction from the basic size.

Bismuth Alloys. A group of metals having a bismuth base which is used primarily for cast tooling parts. These alloys are commonly referred to as low-melt alloys.

Booster. A device that provides the power to operate a power-workholding system. Boosters are classified as either air operated or electrically operated.

Boring Jig. A type of jig specifically designed for boring operations. This jig normally has some type of tool guide, or bushing, on both sides of the hole being machined rather than on just one side, as is the case with a drill jig.

Box Jig. A form of jig that completely encloses the part being machined. This jig is also called a tumble jig.

Bracket Method. A dual dimensioning method in which the converted dimensions are shown in brackets ([ ]).

Break-Even Point. The minimum number of parts required for a tool to pay for itself. Any number of parts less than this break-even point will result in a loss, while any number over this point will result in a profit.

Brinell Hardness. A specific measure and rating of a material's resistance to penetration or indentation. A Brinell hardness test is performed by forcing a steel ball 10 millimeters in diameter into the material with a force, or load, of 3000 kilograms.

Brittleness. The characteristic of a metal that causes it to fail when sudden loads are applied. This is normally considered to be a negative quality, but in some cases, as with shear pins, it may be a positive characteristic of the material.

Built-Up Tool Body. A form of tool body used for jigs and fixtures in which the entire construction consists of a series of parts screwed and doweled together. This is the most common and least expensive tool body.

Burr. The rough ribbon, or ridge, of metal formed on both sides of a machined hole. The larger ridge, normally on the exit side of the hole, is referred to as the primary burr, while the smaller ridge, on the side of the hole where the drill enters, is called the secondary burr.

CAD. The acronym for *computer aided design.*

CAM. The acronym for *computer aided manufacturing.*

Cam. A mechanical device used to clamp parts in a jig or fixture. The two fundamental types are flat and conical.

Cap Screw. A type of mechanical fastener used to assemble the parts of a jig or fixture. Typically these fasteners have a cylindrical head with an internal hexagon drive.

Carbon Steel. A specific type of steel composed mainly of iron and carbon. The three standard types or designations of carbon steel are low, medium, and high carbon.

Cast Iron. A cast ferrous material with a higher carbon content than carbon steel; it is used for both manufactured parts and tooling. Composed mainly of iron, carbon, and other elements.

Cast Tool Bodies. Tool bodies made from cast iron, cast aluminum, or other castable materials.

Channel Jig. A specific style of jig that has a *U*-shaped channel as its major structural element.

Chuck. A commercially available workholder, used mainly on a lathe or grinder, which holds parts with a series of jaws.

Circular Runout. The geometric characteristic that specifies a relationship between two or more concentric diameters of a part measured in single line elements around the part.

Circularity. The geometric characteristic that specifies a condition of roundness in a part feature.

Clamp. A mechanical device used to hold a part securely in a jig or fixture.

Clamping Forces. The forces exerted by the clamping device while holding the part in the workholder.

Closed Jig. A jig that encompasses a part on more than two sides.

CNC. *Computer numerical control.* The abbreviation is used to describe a system where a computer controls the movements of a machine tool through a direct interface, floppy disk, or a magnetic tape used as an input medium.

Collet. A mechanical device used to hold parts, normally of a specific diameter or size, that are to be turned or rotated around a central axis.

Component Libraries. The groups of individual types of components, such as clamps, locators, etc., used for fixturing applications with CAD.

Concentricity. The geometric characteristic that specifies a concentric relationship between two or more diameters of a part.

Conical Wedge. A wedge that has a tapered form, similar to a solid mandrel.

Continuous Path. A programming method used for numerically controlled machine tools in which the position of the tool is constantly controlled throughout the complete machining cycle.

Conversion Chart Method. A dual dimensioning method in which the converted dimensions are shown in a chart along with the design units they reference.

Critical Dimension. A part dimension that controls the size or location of a part feature that is very important to the overall function of the part.

Cutting Forces. The forces exerted against the part by the cutter during the machining cycle.

Cylindrical Cam. A cam design that uses a cylindrically shaped element to activate the clamping device. This cam may use the outside surface of the cylinder or a groove cut into the surface of the cylinder.

Datum. A point, line, surface, or feature considered to be theoretically exact that is used to locate the part or the geometric characteristic features of the part.

Datum Feature Symbol. The datum reference letter contained in a rectangular box and used to note the datum surfaces on the part.

Datum Reference. The letter values contained in the feature control symbol which reference the control to the appropriate datum feature symbols.

Degrees of Freedom. The 12 degrees, or directions, of movement an unrestricted part is free to make. These directions include six radial and six axial directions of movement.

Detailed Drawing. A mechanical drawing that shows each part of a larger assembly drawn in a series of individual views. This is the primary type of drawing used to manufacture each part.

Diamond Pin. A specific type of relieved locating pin used for jigs and fixtures. The pin, when viewed from the end, has a diamond shape.

Dowel Pin. A hardened steel pin used to locate assembled parts of a workholder or as a locating device to accurately position the parts in a jig or fixture.

Dowel-Pin System. A modular tooling system that uses a series of precisely located holes and dowel pins to accurately position and locate the various elements used to build the modular tool.

Drill Bushing. A hardened steel or carbide bushing used to guide and support a cutting tool, such as a drill, throughout its machining cycle.

Drill Jig. A special purpose tool used to accurately position, support, and hold a workpiece so that a variety of operations may be performed. Drill jigs also use some type of cutter guiding device, such as a drill bushing, to accurately position and support the cutting tool.

Dual Dimensioning. A dimensioning system used to show both inch and millimeter dimensions on the same print.

Duplex Tools. Jigs and fixtures designed and constructed so that two parts may be machined. As one part is machined, the other station is loaded or unloaded. When the first part is complete, the second part is moved to the cutter and the first is unloaded and a new part loaded. This process is continued until all the parts are completed.

Eccentricity. The condition of a part in which two diameters do not share the same center-line positions.

Ejector. A device, contained within a workholder, that is used to aid in the removal of the part from the tool.

Epoxy Resins. A family of plastic tooling compounds used to prepare cast nests or other types of workholding or locating devices. They normally consist of two elements, a resin and a hardener, that are mixed and cast to prepare the required element.

Equalizing Support. A supporting device used to spread the support over a larger area of the part. It normally has two contact points; as one is depressed, the other rises.

Expanding Mandrel. A mechanical holding device that grips the workpiece in an internal diameter or other shaped hole and exerts outward pressure to hold the part.

Feature Control Symbol. The symbol used in geometric dimensioning and tolerancing which contains the geometric characteristic symbol, tolerance value, and datum reference.

Feeler Gauge. A flat metal strip of precise thickness that is used to reference the cutters to a fixture by establishing their relationship to the set block.

Ferrous Metals. Metals and alloys that have a base primarily of iron.

F.I.M. The abbreviation for *full indicator movement*.

Fixed-Limit Gauge. A gauge that has fixed limits of size. Many times these gauges are go–no-go and check the upper and lower limits of the part size.

Fixed-Renewable Bushings. Easily replaceable drill bushings designed to be used in a liner bushing and which are held in place with a lock screw or clamp.

Fixed-Stop Locators. External locators that are not adjustable. All types of nonadjustable external locators are fixed-stop locators.

Fixture. A workholding device that holds, supports, and locates the workpiece while providing a referencing surface or device for the cutting tool.

Fixture Key. A square or rectangular block attached to the base of a jig or fixture which locates and aligns the workholder in the "T"-slots of the machine tool on which it is used.

Flat Cam. A cam design that uses the outside profile of a flat plate to generate the desired movement. The two principal types of flat cam are the spiral and the eccentric.

Flatness. The geometric characteristic that specifies an amount of permitted variation from a perfectly flat surface.

Flush-Pin Gauge. A type of inspection fixture that uses a pin to indicate the part's compliance to specific sizes.

Foolproofing. The process of installing a device on a jig or fixture to prevent improper loading of parts in the tool.

Form Characteristics. The geometric characteristics that control the form of a part. These include straightness, flatness, circularity, and cylindricity.

Gang Milling. A milling operation using more than two milling cutters to obtain a specific part size, shape, or form.

Gauge. A mechanical device of a known and precise size, used to check or compare part sizes.

Gauging Fixture. A fixture that uses some type of gauge to check the sizes of a part.

Geometric Characteristic Symbols. The symbols used to show the specific type of feature control required for the part.

Geometric Dimensioning and Tolerancing. An exact dimensioning method used to show the exact specifications of part size and form.

Go–No-Go Gauges. Gauging devices used to check the upper and lower limits of a part size.

Groove Pin. A pin made with a series of grooves or ridges around its profile. Its use is similar to that of a dowel pin, but it is not as accurate because it is positioned in a drilled rather than a reamed hole.

Hardness. The property of a material that permits it to resist penetration or indentation.

Heat Treatment. The process of using heat to modify the properties of a metal.

Hook Clamp. A type of commercially available clamp used for holding parts in jigs and fixtures.

Inch. The base unit used for linear measurements in the U.S. Customary System of measurement.

Indexing. The process of accurately spacing holes or other details around a central axis.

Inspection. The process of checking a workpiece to verify its conformity to the specifications in the part print.

Inspection Tooling. Fixtures and other tools used to aid in the process of inspecting parts.

International Sequence. The sequence of information in a feature control symbol. The sequence is geometric characteristic symbol, tolerance value, and datum reference.

Jaws. The mechanical devices or elements of a workholder, such as a vise or chuck, that actually contact the part being held.

Jig. A workholding device that holds, supports, and locates the workpiece while providing a guiding device for the cutting tool.

Jig Pin. A pin used to hold approximate alignment between two or more parts of a tool. It can act as a hinge pin or locator pin and can also be used to attach removable parts to a jig or fixture.

Jig Plate. The part of a jig that contains the drill bushings.

Knurled Bushings. Drill bushings made with a knurled outside diameter to provide a positive gripping surface in soft materials or to hold the bushings when they are cast into the jig plate.

Lead Time. The time between the design and the construction of a jig or fixture.

Leaf Jig. A specific jig design that uses a hinged leaf as a jig plate.

Least Material Condition. The condition of a part feature in which it has its least amount of material allowed by the tolerance. It is the largest size of an internal feature and the smallest size of an external feature. It is also referred to as the LMC.

Limit Dimensions. Dimensions shown on a part drawing that specify the maximum and minimum sizes of a part feature.

Limits of Size. The maximum and minimum sizes of a part as specified by the tolerances and the basic size.

Liner Bushing. A drill bushing specifically used to provide a hardened, wear-resistant hole in a jig plate in which either slip- or fixed-renewable drill bushings are installed.

Location Characteristics. The geometric characteristics that control the location of part features. These characteristics include position and concentricity.

Locator. A device used to establish and maintain the position of a part in a jig or fixture to insure the repeatability of the workholder.

Low-Melt Alloys. A group of metals having a bismuth base that are used primarily for cast tooling parts.

Machinability. The measure of a material's ability to be machined. Some materials are soft and have good machinability characteristics, while others are harder and have poor machinability.

Machine Vise. A vise designed and intended to be used on a machine rather than on a bench. The most common types of machine vise are the milling machine vise and the drill press vise.

Magnetic Chuck. A mechanical workholder that uses magnetic force to hold a ferrous workpiece. These chucks may use either an electromagnetic or a permanent-type magnet.

Mandrel. A solid bar used to mount and drive parts from a central hole. This bar appears to be cylindrical, but is actually conical since it has a .006-inch taper per foot.

Manifold. A unit used to split or redirect the hydraulic fluid in a power-workholding system. The manifold has one inlet and many outlets for several power clamps.

Maximum Material Condition. The condition of a part feature in which it has the maximum amount of material permitted by the tolerance. It is the smallest size of a hole or the largest size of an external part feature. This is also referred to as the MMC.

Metric System. The common term used to identify the Systeme Internationale d'Unites, or the International System of Units.

Millimeter. The base unit of linear measurement in the Systeme Internationale d'Unites for manufacturing purposes.

Modifiers. The letter values used to modify tolerance values or datum references in a feature control symbol. These include the circled letters *M*, *L*, and *S*.

Modular Tooling. A system of individual parts assembled for a variety of jigs and fixtures. Another term frequently used to describe these tools is *erector set tooling.*

Multistation Tooling. Workholders that are capable of holding and positioning more than one part at a time.

N/C. *Numerical control.* N/C systems typically use either punched paper or mylar tape to convey the instructions to the machine tool.

Nest Locator. A type of locating device that encloses all or part of the outer profile of a part.

Numerical Control. An automated manufacturing system that uses numerical data, usually on a punched tape or in a computer, to operate a variety of machine tools.

Nuts. Mechanical fastening devices used to secure bolts in assembled units.

Oil-Groove Bushing. A special-purpose drill bushing that has grooves cut into the inner surface to provide complete lubrication to the cutting tool or boring bar used in the drill bushing.

Open Jig. A variation of a drill jig, used to machine parts on less than two sides.

Orientation Characteristics. The geometric characteristics that control the orientation of part features. These include angularity, parallelism, and perpendicularity.

Parallelism. The geometric characteristic that specifies a relationship between two parallel surfaces of a part.

Part. The object, or workpiece, machined in a jig or fixture.

Part Drawing. The drawing, or print, of the object to be made or machined.

Perpendicularity. The geometric characteristic that specifies a perpendicular relationship between two surfaces of a part.

Pin. A cylindrical fastener that is normally pressed into a premachined hole for the purpose of aligning or holding parts of an assembled workholder.

Planes of Movement. The three principal planes, or axes, or a part used to reference the twelve degrees of freedom. These are generally identified as *X*, *Y*, and *Z*.

Plate Fixture. A fixture that has a plate as its main structural component.

Plate Jig. A jig with a plate as its main structural component.

Pneumatic Workholding System. A form of power-workholding system that uses air pressure to operate all workholding devices.

Point-to-Point. A programming method used for numerically controlled machine tools, in which the position of the tool is controlled between a series of specific locations, or points. Only the position of these points is important; the path the tool takes between these points is not important.

Position. The geometric characteristic that specifies a relationship between a part feature and a central axis, or center

Position Method. A dual dimensioning method in which the converted dimensions are shown either below or to the right of the design units. The units are separated with a slash (/) for horizontal dimensions and a straight line (–) for vertically placed dimensions.

Power Clamping. A clamping system, either hydraulic, pneumatic, or a combination of both, used to activate and hold the clamping pressure.

Precast Tool Bodies. Commercially available tool bodies made from cast aluminum, cast iron, or cast steel; they need only minimal machining to accommodate the various clamps and locators required for the part being held.

Precision Ground Materials. Commercially available tooling components used in the construction of built-up tool bodies. The two principal types are cylindrical, called drill rod, and flat.

Preformed Materials. Commercially available materials made in a variety of precast forms which are used to construct built-up tool bodies.

Press-Fit Bushings. Drill bushings specifically designed and intended to be pressed into a hole in the jig plate. The friction and pressure of the press fit are the forces that hold the bushing in place.

Primary Locators. The locators used to locate, or reference, the primary locating surface of the part.

Process Planning. The act, or process, of planning the step-by-step procedure to be used to manufacture or machine a workpiece.

Production Plan. The document specifying the exact means and methods to be used to manufacture a part or product.

Production Run. The specific group, or number, of parts to be made at one time.

Profile. The external edges, or surfaces, of a part.

Profile Characteristics. The geometric characteristics that control the form, or shape, of a part profile. These include the profile of a line and the profile of a surface.

Profiling Fixture. A fixture specifically designed to aid in machining the external edges of a part.

Projected Tolerance Zone. The imaginary zone projected above a part feature, such as a hole, to insure assembly with a mechanical fastener passing through both the tolerance feature and the projected thickness of the mating part.

Projection. A means of establishing the relative position of the views in a mechanical drawing. The two main types of projection in use today are first angle (Europe) and third angle (United States and Canada).

Pump Jig. A commercially available tooling device used as the major structural element in some types of jigs. This device consists of a base and a top plate connected and aligned with two or more posts. The top plate is moved up and down with an integral handle.

Quarter-Turn Screw. A commercially available fastening device used for clamping on some types of workholders. The head of this screw is flat and is usually used to clamp tooling elements, such as leaves, where the screw is positioned in a slot and turned 90 degrees (one quarter turn) to grip the top plate.

Reference Dimension. A dimension shown on a part print for information purposes and used for reference rather than manufacture. Such a dimension is shown in parentheses or with the abbreviation "REF."

Reference Surface. A part surface used to reference the part. It may also be referred to as a datum surface.

Referencing. The proper positioning of the workpiece with reference to the tool.

Regardless of Feature Size. A modifier used to indicate that the tolerance value shown applies regardless of the size of the feature or datum.

Releasing Agent. A compound used in conjunction with epoxy resins to prevent the resins from adhering to the part used as a pattern.

Renewable Bushings. Easily replaceable drill bushings designed to be used in liner bushings and held in place with a screw or clamp. The two principal variations are the fixed-renewable and the slip-renewable bushing.

Repeatability. The feature of a workholder that permits the parts being machined to be duplicated within the stated limits of size, part after part.

Resins. The compounds, usually epoxy, mixed together to make resin tools.

Retaining Rings. Spring steel rings that are used in place of nuts or screws to hold the part on which they are installed in some relative position.

Rise. The term used to describe the movement of a cam per degree of arc.

Rockwell Hardness. A specific measure and rating of a material's resistance to penetration or indentation. A Rockwell hardness test is performed by forcing either a diamond cone or a steel ball into the surface of a part under a known and controlled load.

Rotary Tooling. A term used to describe indexing jigs and fixtures.

Runout Characteristics. The geometric characteristics that control the runout of a part. These include circular runout and total runout.

Sandwich Jig. A type of drill jig that holds the workpiece on two opposite sides. It may be used to support the part or to machine the part from two sides.

Screw Clamp. A clamping device that uses a screw thread to clamp the part.

Secondary Locator. A locator used to reference the secondary locating surface of the part.

Serrated Bushing. A drill bushing made with a serrated outside diameter to provide a positive gripping surface in soft materials or to hold the bushing when it is cast into the jig plate.

Set Block. A gauging device included on many fixtures to aid in setting the cutters to suit the workpiece.

Set Screw. A mechanical fastening device that is generally made without a head and used for adjustable locators or as a locking device for other applications.

Setup Gauge. A special-purpose gauge used in conjunction with a set block to establish the proper cutter position. A feeler gauge is a form of setup gauge.

Shear Strength. The measure of a material's resistance to forces acting in opposite directions, causing a shearing effect on the part.

Shop-Air System. The compressed air system found in machine shops. The shop-air system provides power to a variety of shop equipment. Air tools, coolant systems, and power-workholding systems rely on this power source. The most common shop-air systems use compressed air at a pressure of 90 to 100 pounds per square inch.

Sight Locators. Locators used on jigs and fixtures that are intended to be used for approximate location. The part may be aligned against milled slots or scribed lines.

Slip-Renewable Bushings. Easily replaceable drill bushings designed to be used in a liner bushing and held in place with a screw.

Slot Nuts. A term used to describe T-nuts.

Socket-Head Cap Screws. A type of mechanical fastener used to assemble the parts of a jig or fixture. Typically these fasteners have a cylindrical head with an internal hexagon drive.

Software. The medium used to convey specific instructions to a computer.

Solid Locators. Locators used on jigs and fixtures which are solid and nonadjustable.

Spiral Cams. A type of flat cam that has a shape similar to an involute curve.

Split Pin. A pin made from flat material formed into a cylindrical shape with a split on one side. Its use is similar to that of a dowel pin but it isn't as accurate because it is used in a drilled rather than a reamed hole. Also called *roll pin*.

Straddle Milling. A milling operation using two milling cutters to mill two sides of a part at the same time.

Straightness. The geometric characteristic that specifies a relationship between a part feature and a true straight plane or line.

Strap Clamp. A specific type of clamping device that uses a metal bar, similar to a strap, to hold the part securely to the workholder. The basic elements of this clamp are the strap, the hold-down bolt, and the heel pin or block.

Structural Sections. Commercially available materials made in a variety of preformed shapes which are used to construct built-up tool bodies.

Subplate. A plate containing many drilled and tapped holes to which a variety of different components and clamps may be attached to form a crude type of modular tooling.

Support. A locator placed under a workpiece.

Swing Clamp. A variation of a screw clamp consisting of an arm that swings about a stud at one end and contains the screw clamp at the other end.

Systeme Internationale d'Unites. The system of measurement frequently called the *metric system*.

T-Bolt. A mechanical fastener that has a head designed to fit into the "T"-slots of machine tools. Usually used to mount workholders to a machine tool.

T-Nut. A nut made specifically for use in the "T"-slots of machine tools. Using these nuts permits standard bolts to be used to attach the workholder or other elements.

"T"-Slot System. A modular tooling system that uses a series of precisely located "T"-slots to accurately position and locate the various elements used to build the modular tool.

Table Jig. A variation of the basic plate jig that uses legs to elevate the jig off the machine table.

Tacking. A preliminary welding process in which the joined edges are tacked, or welded, with small widely spaced weld deposits to hold the relative position prior to finish welding.

Template. A plate that duplicates the desired part configuration and is used as a guide for layout work or for light machining.

Template Jig. A jig that has a template as its major structural element.

Tensile Strength. The measure of a material's resistance to being pulled apart. The tensile strength rating is the principal means used to classify the strengths of various metals.

Tertiary Locator. A locator used to reference the tertiary locating surface of the part.

Thread Inserts. Steel inserts used to provide a more wear-resistant threaded hole in soft tooling elements, such as aluminum tool bases.

Throw. The radial movement of a cam about its mounting point or hole.

TIR. The acronym for *total indicator reading*.

Toggle Clamp. A commercially available style of clamping device that uses a series of levers to operate and hold the position of the clamp.

Tolerance. The amount of variation, or deviation, permitted in the size of a part.

Tombstone. A commercially available tooling block that is mounted on machining centers and similar machine tools to provide multiple mounting surfaces for workholders.

Tool Body. The major structural element of a jig or fixture.

Tool Design. The science, or process, of designing jigs, fixtures, or any other special-purpose tools required for manufacturing purposes.

Tool Drawings. The mechanical drawings made and used to build special-purpose tools.

Tool Force. The force exerted against a workpiece by the cutting tool.

Tool Steel. A type of alloy steel made to very close specifications and primarily used for tools or similar types of parts.

Tooling Estimate. A management guide to the overall cost and benefit of a special-purpose tool. This estimate shows the cost of the tool and the relative cost of the operation with the tool and with alternate methods of manufacture.

Toolmaker. A skilled craftsman who specializes in making the special-purpose tools required for manufacturing.

Total Runout. The geometric characteristic that specifies a relationship between two or more concentric diameters of a part measured totally around and across the part feature.

Toughness. The property of a material that allows it to resist fracture from sudden loads.

Trunnion Jig. A type of jig used to accommodate large parts that must be machined on several sides.

Tumble Jig. A form of jig that completely encloses the part being machined. It is also called a *box jig*.

Unilateral Tolerances. Tolerance values that may go only in a single direction, either plus or minus, from the basic size.

U.S. Customary System. The inch-based system of linear measurement.

Urethane. A nonmetallic tooling material used for applications in which the tool must be nonmarring.

Vacuum Chuck. A mechanical workholder that uses the force of a vacuum to hold a nonporous workpiece. This chuck may be used on any ferrous, nonferrous, or nonmetallic material that will hold a vacuum.

Vee Locators. Locating devices that have a *V* form. They are very useful for cylindrically shaped parts.

Vise-Jaw Tooling. Special-purpose jigs and fixtures made from blanks that are the same dimensions as those of the replaceable jaws of a machine vise.

Washers. Mechanical fastening devices used under nuts, screws, and bolts either to prevent damage when the fastener is tightened or to provide a mechanical lock to prevent loosening.

Wear Block. Any device used to prevent premature wear to any surface of a workholder.

Wear Resistance. The ability of a material to resist wear or abrasion.

Wedge Clamp. A type of clamping device that uses a wedge to activate and hold the clamped part.

Wedge. A mechanical clamping device that uses the principle of the inclined plane. The two primary variations of the wedge are flat and conical.

Workholder. A term used to describe jigs and fixtures.

Professional Organizations

Aluminum Association
750 Third Avenue
New York, NY 10017

American Iron and Steel Institute
1000 16th Street NW
Washington, DC 20036

American National Standards Institute
1430 Broadway
New York, NY 10018

American Society of Mechanical Engineers
United Engineering Center
345 East 47th Street
New York, NY 10017

American Society for Metals
9513 Kinsman Road
Metals Park, OH 44073

National Machine Tool Builders Association
7901 Westpark Drive
McLean, VA 22101

National Tool, Die & Precision Machining
Association
9300 Livingston Road
Washington, DC 20022

Society of Manufacturing Engineers
One SME Drive
P.O. Box 930
Dearborn, MI 48128

CPSIA information can be obtained
at www.ICGtesting.com
Printed in the USA
FFOW04n2239170914
7458FF